Snail-Transmitted Parasitic Diseases

Volume I

Author

Emile A. Malek, Ph.D.
Professor of Parasitology
Department of Tropical Medicine
School of Public Health and Tropical Medicine
Tulane University Medical Center
New Orleans, Louisiana

CRC Press, Inc.
Boca Raton, Florida

Library of Congress Cataloging in Publication Data

Malek, Emile Abdel, 1922-
Snail-transmitted parasitic diseases.

Bibliography: p.
Includes index.
1. Parasitic diseases—Transmission.
2. Gasteropoda as carriers of disease. I. Title.
RC119.M25   616.9'6   79- 15250
ISBN 0-8493-5269-X (v. 1)
ISBN 0-8493-5270-3 (v. 2)

This book represents information obtained from authentic and highly regarded sources. Reprinted material is quoted with permission, and sources are indicated. A wide variety of references are listed. Every reasonable effort has been made to give reliable data and information, but the author and the publisher cannot assume responsibility for the validity of all materials or for the consequences of their use.

Direct all inquiries to CRC Press, 2000 N.W. 24th Street, Boca Raton, Florida, 33431.

International Standard Book Number 0-8493-5269-X (v. 1)
Internationl Standard Book Number 0-8493-5270-3 (v. 2)

Library of Congress Card Number 79-15250
Printed in the United States

PREFACE

Research on snail-transmitted diseases has continued at a considerable pace during the last twenty years, and several investigators have made valuable contributions to our knowledge of these diseases. I have, therefore, felt for some time that a text is needed which contains only information about the major infections or diseases of the world which are transmitted by snails, and in which each disease is considered in much greater detail than is usually provided by textbooks on general, medical or veterinary parasitology.

The objective of this book is to present an up-to-date account of these infections or diseases. Its scope of interest encompasses research workers, graduate students in medical and veterinary parasitology and malacology, practicing physicians and veterinarians, those interested in general parasitology and zoonosis, and government health planners, especially those interested in the epidemiology and control or prevention of these diseases.

The majority of the snail-transmitted infections are caused by digenetic trematodes; some are due to certain metastrongyle nematodes. The very few tapeworms which utilize snails as intermediate hosts are not considered because they do not cause significant infections. Although it is realized that parasitic diseases comprise those caused by viruses, bacteria, fungi, protozoa and helminths, rarely are any except helminths transmitted to humans and domestic animals. Therefore I feel that *Snail-Transmitted Parasitic Diseases* is an appropriate and correct title for the book.

The infection caused by each of the trematodes or nematodes considered in this book is called by a name derived from that of the parasite which causes it. In each chapter, particular emphasis is placed on the geographical distribution of the parasites, their developmental cycle, the morphology of the adult parasites and their larval stages, the snail species involved, the snail-helminth relationships, the epidemiology or epizootiology of the infection, its zoonotic significance, the pathogenesis and pathology, immunity, diagnosis, treatment and control. Throughout the chapters a distinction has been made between infection and disease. An infection with a helminth does not always produce a disease.

I have attempted to provide information in each chapter based on my own experiences and those of colleagues, together with that found in the vast world literature on the subject. In addition to valuable consultations with many of my colleagues in the United States, visits abroad during the last three years have enabled me to exchange information with investigators in other countries. Actually, some sections of the book were partly written during three assignments to Colombia and to Senegal, and during brief stays in Iran, Egypt, England, the Netherlands, and Brazil.

In the preparation of the book, some limitations have had to be imposed: first, on the number of parasites to be included, and secondly, on the literature to be cited. Of the very large trematode group, only those forms which infect humans, subhuman primates and domestic animals are considered. Only some of those infecting wild mammals and birds are included. Also excluded are those which infect lower vertebrates. The literature on the parasites which are considered is so voluminous that I had to exclude some reports. However, this does not reflect on the significance of the information contained in the publications excluded. This is especially the case with Volume I, Chapter 1, where the trematodes are treated as a group, and a discussion is included of their biology and other aspects. In the same chapter, the snail hosts are also treated as a group, and various aspects are discussed, but only a few literature citations are made. In coverage of such a vast subject, errors and omissions are to be expected. I should be thankful if these would be brought to my attention for possible consideration should there be a revision of the text.

Acknowledgments are due to many. My gratitude goes first to the late Dr. George R. La Rue and the late Dr. Asa C. Chandler, for their instruction and for their stimulating me to acquire a combined knowledge of trematodology and malacology. My deep appreciation is due to my wife Carolyn and to our children Rebecca, Christopher, and Steven, for their constant encouragement, assistance, understanding and patience. Moreover, my wife and Rebecca have given invaluable editorial help. I should thank all my colleagues and past and present students in the Tulane Department of Parasitology for consultation and help, especially Dr. Paul C. Beaver. During my teaching in the Sudan and my twenty years at Tulane, advanced parasitology students and public health and medical students have brought up interesting and stimulating points which were taken into consideration in the information provided in this book. Valuable literature, especially old publications, was obtained from the Faust Parasitology Library in our department, and many thanks are due to Miss Helen Day, the librarian. The invaluable assistance of Mrs. Andrée M. Cucullu Smith, in the preparation of the line drawings, is gratefully acknowledged. I would like to thank my colleague Dr. Wojciech A. Krotoski for many stimulating discussions and for sparing the time to assist with the photographic work. My students Frank Cogswell, Jean-Paul Chaine, and Dr. G. H. Sahba have also assisted in photography. Many associates have contributed material or photographs, and grateful acknowledgments are made in the captions accompanying these items. Special thanks are due to Mrs. Myra B. Hamm for aid and patience in the preparation of the typescript. Finally, grateful appreciation is due to the publishers, CRC Press, particularly the coordinating editors Ms. Marsha Baker and Mrs. Benita Segraves, for their interest and constructive criticism.

GENERAL INTRODUCTION

This book covers the infections or diseases that are caused by certain helminths which are transmitted by snails. These conditions fall under two categories viz., the trematodiases and the nematodiases. The trematodiases are dealt with first, in Volume I and in Chapters 1 through 15 of Volume II, and then the nematodiases will follow in Chapter 16, Volume II. Trematodiases which are transmitted by snails are caused by flukes of the digenetic trematodes of flatworms. Nematodiases are infections, usually causing diseases, of which the etiologic agents are some species of parasitic nematodes (round worms) of the superfamily Metastrongyloidea, members of which are commonly known as metastrongyles. These are parasitic nematodes of vertebrates, and most of them require gastropods (terrestrial and aquatic snails and slugs) for the development of their early larval stages. The trematodiases are by far more numerous and of wider geographic range than the nematodiases which are transmitted by gastropods.

It is realized that many parasitic helminths can cause serious diseases and consequent economic loss. However, although the title of the book refers to diseases, a distinction is made, in the various chapters, between infection and disease. This applies to the helminths of humans and those of animals. Infection rates in a human or in an animal population represent those individuals who are positive by recognized diagnostic methods, e.g., examination of excreta or by immunodiagnosis. The disease, however, is represented by the morbidity, especially in the advanced stages. Thus an infection with a fluke or a nematode need not necessarily be a cause of disease. It should be noted that some parasitologists have used, for a parallel distinction, the terms "subclinical disease" and "clinical disease".

Examples of the parasitic nematodes which are treated in Chapter 16, Volume II, are the metastrongyles *Muellerius capillaris,* a parasite in the lungs of sheep; *Aelurostrongylus abstrusus* and *Anafilaroides rostratus,* both parasites of the lungs of cats, and *Angiostrongylus cantonensis,* which is a parasite of the lungs of rats and can infect humans, causing parasitic meningoencephalitis. Terrestrial snails of the genera *Helix, Hygromia, Monacha* and *Cepaea,* and slugs of the genera *Limax, Deroceras* and *Arion* transmit *Muellerius capillaris. Aelurostrongylus abstrusus* and *Anafilaroides rostratus* are transmitted by their terrestrial snail intermediate hosts, *Bradybaena similaris* and *Subulina octona. Angiostrongylus cantonensis* causes human disease in a number of islands in the Pacific, including Hawaii, Tahiti, the New Hebrides, the Loyalty Islands and New Caledonia. Cases have also been reported in Taiwan, Thailand, the Philippines, Sumatra and Vietnam. The parasite utilizes a number of aquatic and terrestrial snails and slugs. Among the most important hosts are the terrestrial giant African snail, *Achatina fulica* and the slugs *Vaginulus plebeius* and *Veronicella alte* which are all eaten by natives in the Pacific Islands, Malaysia, and Thailand, and cause the disease. Especially in Thailand, the large aquatic snails *Pila* spp., harboring the third stage larva, are the source of the disease when consumed by humans. Other species of *Angiostrongylus* which are dealt with are *A. costaricensis* and *A. vasorum.*

Digenetic trematodes are of world-wide distribution, and infect a variety of hosts, which are members of all classes of vertebrates. Usually there are well-marked lines of compatibility and incompatibility between some classes of vertebrate hosts and whole families of trematodes. Their distribution, however, is dependent on the occurrence of certain specific molluscan hosts for the completion of their life cycle. In addition, certain human infections with trematodes are determined by the consumption of particular plants, molluscs, crayfish, crabs, and fish. Thus food habits and peculiarities of diet are important factors in the maintenance of some digenetic trematodes.

The trematodes which are considered in this book are parasites of humans, domestic animals, subhuman primates and other wild mammals, poultry and other birds. They include the schistosomes (blood flukes) of mammals and birds, the liver flukes, *Fasciola, Fascioloides, Clonorchis, Opisthorchis, Dicrocoelium,* and *Platynosomum;* the intestinal flukes, *Fasciolopsis, Heterophyes, Metagonimus, Nanophyetus, Echinostoma, Phaneropsolus, Gastrodiscus, Gastrodiscoides,* and *Watsonius;* the rumen flukes *Paramphistomum* and *Cotylophoron;* and the lung flukes *Paragonimus* species and other flukes. In Chapter 1, the trematodes are treated together, as a group. Information is provided as to their systematics, gross morphology, microstructure, ultrastructure, biology and physiology of the adult worms as well as the successive stages in their life cycle. In the same chapter, and in a simila way, the molluscan intermediate hosts, in particular the gastropods, are treated as a group, and information is provided about their systematics, gross morphology, ecology, histology, ultrastructure; their physiology and the effect of larval trematodes on them.

Trematodes not only differ in their morphology and life cycles, but also in the syndrome which each causes, and in the epidemiology and control of infections. Accordingly, in this book the infections or the diseases are treated separately in a number of chapters. Every infection is called by a name derived from that of the trematode which causes it. This is especially the case with the major etiologic agents of the trematodiases, which are considered in most of the chapters on the digenetic trematodes. Those of lesser significance are treated in groups in Chapters 12 through 15.

THE AUTHOR

Emile A. Malek, Ph.D., has been on the Faculty of Tulane University since 1959. He is Professor of Parasitology, and Director of the Laboratory of Schistosomiasis and Medical Malacology, Department of Tropical Medicine, Tulane Medical Center, New Orleans, Louisiana.

Dr. Malek graduated in 1943 from Cairo University, Egypt and obtained the M.S. degree from the same university. In 1952 he received the Ph.D. degree from the University of Michigan, Ann Arbor, after an additional two summers at the University of Minnesota.

Dr. Malek became a member of the American Society of Parasitologists in 1949; the American Society of Tropical Medicine and Hygiene, and the Royal Society of Tropical Medicine and Hygiene (London) in 1960. He has been a consultant to the U.S. Agency for International Development, the State Department, and consultant on several occasions to the Pan American Health Organization and the World Health Organization. He was on the Faculty of the University of Khartoum, Sudan; a former Visting Professor, the Federal University of Minas Gerais, Belo Horizonte, and the Federal University of Ceará, Fortaleza, Brazil, and former Staff member of the World Health Organization in Geneva, Switzerland.

He is the recipient of the U.S. Public Health Service, the National Institutes of Health, Research Career Award since 1962, and a member of the Expert Advisory Panel on Parasitic Diseases of the World Health Organization since 1964.

Dr. Malek has presented invited papers at international meetings, and scheduled papers at several national meetings. He has published more than 50 research papers and has contributed to scientific publications by the World Health Organization and the Pan American Health Organization. He has also written chapters in four books, and has a book, Malek and Cheng, *Medical and Economic Malacology,* Academic Press, 1974.

His current major research interests include the epidemiology and control of schistosomiasis, and other snail-transmitted parasitic diseases, especially fascioliasis and paragonimiasis. His research interests and consultations have taken him to Michigan, Minnesota, New Jersey, Brazil, Venezuela, Colombia, Saint Lucia, Central America, the Sudan, Egypt, Kenya, Uganda, Tanzania, Ethiopia, Senegal, Mali, and Iran.

TABLE OF CONTENTS

VOLUME I

SNAIL-TRANSMITTED PARASITIC DISEASES

VOLUME I

Trematodes and their Molluscan Hosts

Schistosomiasis

VOLUME II

Paragonimiasis

Nanophyetiasis

Clonorchiasis

Opisthorchiasis

Heterophyidiasis

Fascioliasis

Fascioloidiasis

Fasciolopsiasis

Paramphistomatidiasis

Dicrocoeliasis

Echinostomatidiasis

Notocotylidiasis, Plagiorchiidiasis, and Prosthogonimiasis

Microphallidiasis and Lecithodendriidiasis

Strigeidiasis and Clinostomatidiasis

Brachylaemidiasis and Cyclocoelidiasis

Nematodiases Transmitted by Snails: Angiostrongyliasis

Chapter 1

TREMATODES AND THE MOLLUSCAN HOSTS

I. GENERAL CONSIDERATIONS

There is evidence that there was a knowledge of some of the snail-transmitted trematode diseases by ancient peoples. The Ebers Papyrus in Egypt (16th century B.C.) is the oldest record in which a helminth was regarded as a pathogenic organism; the disease "AAA" is believed to have been schistosomiasis. Egyptian mummies of the 13th century B.C. contained eggs of *Schistosoma haematobium,* which is responsible for the urinary form of the disease.

The first trematode or fluke to be reported was *Fasciola hepatica,* by Jehan de Brie in 1379,[34] and this fluke is the causative parasite of sheep liver rot, more accurately described by Gabucinus in 1547.[51] This was followed by descriptions of other similar parasites, and the name "Trematoda" was given to them by Rudolphi in 1808. The complete life history of *Fasciola hepatica* was described by Leuckart (1881, 1882)[83,84] and by Thomas (1881, 1882).[164,165] It was shown by the latter authors that the life cycle of this sheep liver fluke involves an alternation of generations and requires a snail as an intermediate host. Meanwhile Bilharz (1851), had discovered the human blood fluke, *Schistosoma haematobium,* and the small intestinal fluke, *Heterophyes heterophyes.* There followed the finding of *Clonorchis sinensis* by McConnell in 1874,[116] of *Paragonimus* by Kerbert in 1878[72] and Ringer in 1879, of *Schistosoma japonicum* by Katsurada in 1904,[70] and the differentiation of *Schistosoma mansoni* from *S. haematobium* by Sambon in 1907.[147] The elucidation of the life cycles of these human and animal infections with snail parasites came during the first three decades of this century and interest in determining their exact geographical distribution and epidemiology has also increased. The life cycle of other trematodes became known later, while details of the life cycle of many other species are still unknown.

Owing to the high prevalence of human and domestic animal infections with some trematodes in certain areas of the world and the easily recognizable detrimental effects on the health and economy of large populations, early efforts were made to control them. Results in some limited areas have been successful, but generally not, and the control of these diseases remains an important and almost unsolved problem facing public health authorities in areas where the diseases are endemic. The difficulties in shaping control policies are that the factors responsible for the endemicity of these infections are tied up with people's food and other habits which are culturally conditioned and very hard to change. Another factor is the low economic level in these countries, which necessitates the contact with and use of infested waters and the use of nightsoil as fertilizer.

The helminths discussed in the major part of this book are digenetic trematodes, or flukes. In addition to their medical and veterinary importance they are an interesting group to the zoologist and the parasitologist because their physiology and morphology have been altered to meet their new parasitic existence. The estimate of known species of digenetic trematodes is well over 40,000. Most of them are parasitic in lower vertebrates, and even those which are parasites of mammals other than humans and domestic animals, and those of birds, are so numerous that it is naturally impossible to consider all of them in this book. The digenetic trematodes or flukes require development in one or more intermediate hosts during the period between leaving the mammalian host as eggs and returning to the mammalian host as *cercariae* or encysted *metacercar-*

iae. The snail is an essential first intermediate host, or the only intermediate host. Since these snails have special habitat requirements and geographical distribution, it follows that the flukes also have a spotty distribution even within a country. The presence of the susceptible snail host is the primary requirement for the establishment of a focus of infection. Absence of the disease in other areas where the same species of snail is present is sometimes difficult to explain. The strain of the parasite in adjacent areas may not be infective to the strain of the snail present; environmental conditions and habits of the people may not be suitable for transmission of the infection; or the topography of the country may be such that movements of reservoir hosts from neighboring endemic foci are restricted.

Climate is an important factor in the distribution of plants and animals. Climate is determined by and varies with latitude, longitude, and altitude. It affects parasites in general, not only through the host, but also directly by rainfall and temperature. For example, in the case of fascioliasis, some investigators were able to show a close correlation between the prevalence of this disease in sheep and the climate, especially rainfall. They suggested means of forecasting disease in any particular year. The climate zones of the world are divided into tropical, subtropical, midlatitude or temperate, polar, and mountain climate. Although the tropics and subtropics are perhaps the most favorable regions for the propagation of the snail-transmitted trematodiases, the nosogeographic range of some of them extends into temperate regions, and some even extend into frigid zones, such as schistosomiasis japonica, opisthorchiasis, clonorchiasis, paragonimiasis, fascioliasis, metagonimiasis, nanophyetiasis, echinostomiasis, and dicrocoeliasis. On the other hand there are extensive areas in the tropics which are too dry for these helminths to establish themselves, while in other tropical areas the susceptible snail hosts are absent. Unlike many of the vertebrates, arthropods, and molluscs, the nosogeographic range of snail-borne diseases is rarely coincident with faunistic regions. While fascioliasis is cosmopolitan in its distribution, schistosomiasis haematobia is African and western Asiatic, schistosomiasis mansoni is African and neotropical, and schistosomiasis japonica is confined to the Orient as is clonorchiasis. Stoll[160] gave estimates of the prevalence of the major trematodiases and their geographical locations (Table 1-1).

As has been stated above, the presence of a susceptible snail host is a primary requirement for the establishment of a focus of infection. However, the perpetuation of a snail-borne infection is dependent on other things, such as environmental factors and food and sanitary habits of the human population. Water is essential for the establishment of snail populations and fish, crabs, crayfish, and aquatic vegetation, as well as for the survival of the eggs and larval stages of the parasite in the water and in the snails. Expansion of the areas occupied by water bodies, for example through introduction of irrigation, increases the habitats of the intermediate hosts, and thus aggravates a preexisting moderate prevalence of the infections. An example is the increase in prevalence of schistosomiasis in many parts of Africa, the Near East and the Americas. On the other hand, filling of swampy and overflooded areas reduces prevalence of the infection, as happened in habitats of the amphibious oncomelanid snails, the major hosts of schistosomiasis japonica in the Orient.

Among the factors, other than environment and occurrence of the snail hosts, that govern the dissemination of snail-transmitted trematodiases and their prevalence in man are food habits of the human populations, nightsoil, drinking water, and migration. In China, Taiwan, Thailand, Borneo, Sumatra, Assam, and other parts of the Orient man becomes infected with *Fasciolopsis buski* by peeling off, between the teeth, the hull or the skin of infected fruits of aquatic plants. Eating raw watercress results in epidemics of human fascioliasis in several European and Central and South Ameri-

Table 1-1
THE CALCULATED NUMBER OF THE MAJOR HUMAN TREMATODE INFECTIONS IN MILLIONS

	Middle and South America	Africa	Europe without USSR	USSR in Europe	USSR in Asia	Asia without USSR	Oceania	Total
Clonorchis sinensis						19.0		19.0
Opisthorchis felineus			0.1	0.4	0.6	b		1.1
Opisthorchis viverrini						3.5		3.5
Fasciolopsis buski						10.0		10.0
Paragonimus spp.	b	b				3.2	b	3.5
Fasciola spp.	b	b	b					
Schistosoma japonicum						46.0		46.0c
Schistosoma mansoni	6.2	23.0						29.2c
Schistosoma haematobium		39.0	b			b		39.0c

Note: Heterophyes heterophyes, Metagonimus yokogawi, Echinostoma spp., and Gastrodiscoides hominis to be added to the table with a few hundred thousands each.

Modified from Stoll (1947), with additions.

a   Represents less than 100,000

b   Wright (1973) gives 124,905,800 as total of infections with the three species of human schistosomes, instead of 114.2 million in this table.

can countries. The liver fluke *Clonorchis sinensis* is prevalent among populations in the Orient who eat their fish uncooked. Reservoir hosts maintain the infection in areas where it is rare among the human population. Similarly, populations in the Orient, West Africa, and South America who eat uncooked crabs and crayfish are infected with the lung fluke *Paragonimus* spp. Under unfavorable situations, such as wars which are accompanied by food shortages, the human population resorts to unaccustomed kinds of food and thus can increase the prevalence of snail-transmitted diseases. An example is the civil war in Nigeria in the late 1960s where fresh-water crabs were consumed by humans resulting in a considerable increase in prevalence of paragonimiasis in some parts of that country.

In countries where nightsoil is used as fertilizer, the viability of parasite eggs depends on how long nightsoil is stored before being placed in the cultivated field and ditches. In China, Japan, Korea, India, and elsewhere, nightsoil is used as a fertilizer, and in other countries where it is not made use of as manure a great amount of infection nevertheless is spread by unclean habits of defecation, promiscuous defecation, and improper methods of disposal of human excreta. Through these means, human excreta containing parasite eggs reach water bodies directly or indirectly and thus may be brought into the snail's habitats.

As to drinking water as a factor in maintaining snail-borne infections, it has been reported that metacercarial cysts of *Fasciola* can be present in the water and swallowed when the water from contaminated waterbodies is used for drinking. Moreover, although the usual transmission of schistosomiasis is by the cercariae penetrating directly into the skin of the mammalian host, cercariae in drinking water may also be infective if they penetrate through the mucous membrane of the mouth.

Importation of slaves from some parts of Africa to the Western Hemisphere introduced schistosomiasis mansoni in certain countries and islands of the Americas. The parasite has apparently become adapted to certain species of the snail genus *Biomphalaria* which are found in these American endemic foci. No doubt *Schistosoma haematobium* was also brought with the slaves to the Western Hemisphere, but because of the absence of snails of the genus *Bulinus*, the parasite did not become established. It is also believed that nomadic tribes in Africa introduced schistosomiasis, and in some cases the snail intermediate hosts, from one body of water to the other through their migratory habits. Snails can be transported with drinking water kept in goat skin or other containers.

II. THE BIOLOGY OF DIGENETIC TREMATODES AND THEIR MOLLUSCAN HOSTS

A. The Digenetic Trematodes
1. Introduction

The Class Trematoda of the Phylum Platyhelminthes, or flat worms, is divided into three orders: Monogenea, Aspidobothrea (= Aspidogastrea), and Digenea. These can be separated on the basis of clear differences in morphology, development, and life cycle.

The monogenetic trematodes, or monogeneids, are mainly ectoparasites of fishes, amphibians, and turtles. Some have been found attached to crustaceans, cephalopods, and aquatic mammals. A few species have affinities to the bladder, cloaca, ureter, bucal cavity, and other body spaces which open directly to the exterior. Whether this is true endoparasitism or not has been disputed. As ectoparasites, the monogeneids live attached to the gills and fins of their fish hosts. They are usually of medium or fairly large size. The posterior part of the worm is modified to form an adhesive organ, the "opisthaptor", which is generally provided with hooks and hooklets and sometimes suckers. A "prohaptor" near, or at, the anterior extremity, is a less conspicuous adhesive organ of a glandular or muscular nature. The monogeneids possess paired excretory pores that are located anteriorly on the dorsal surface of the body, and comparatively short uteri containing few eggs. Monogeneids have direct life cycles which do not involve any intermediate hosts.

The aspidogastrid trematodes are endoparasites of clams, snails, and fishes; some are in turtles. They do not have an oral sucker, but they possess a very large ventral adhesive disc, in the form of a strong hold-fast organ, which occupies almost the entire ventral body surface, and is divided into compartments or alveoli. There is a single excretory pore located posteriorly. The Aspidobothria are intermediate between the Monogenea and the Digenea. They resemble the Monogenea in that none of them has asexual reproduction, and resemble the Digenea in that some of them have an alternation of hosts. In fact, some aspidogastrids have been considered by various parasitologists as members of the Digenea or the Monogenea. Although the aspidogastrids resemble members of each of the other orders, they are sufficiently different as to be recognized as a distinct group.

The adult Digenea are endoparasites in all classes of vertebrates. They occur free in certain organs of the body, or they may be encysted. They have been found in the intestine, stomach esophagus, mouth, pancreas, bile ducts, gall bladder, urinary bladder, lung, blood, under the skin, on the peritoneal wall, and in other locations. Some Digenea are more or less flattened dorsoventrally, leaf-like, but others are circular or oval in cross section. Some flukes, such as the strigeids, have a body which is divided into an anterior and a posterior portion as a result of a transverse constriction. The surface of the body is covered with a tegument which may be spined, serrated, or corrugated   esence or absence, and the number and distribution, of spines or hooks is   ecific.

Like   tyhelminthes, the Digenea are usually bilaterally symmetrical and posse   cell or protonephritic type of excretory (osmoregulatory) system, which   und in some other invertebrate groups. The digestive system is incomplete,   w species with a complete alimentary tract are known. The adult digenean, as   e cestodes or tapeworms, differs from the related free-living turbellarians in   ng a readily visible cellular or syncytial epithelial covering on the body surface, but they have a tegument consisting of a thin layer of syncytially arranged cytoplasm which is joined with underlying cells. This was previously termed "cuticle" before the advent of the electron microscope. By histochemical tests, the tegument of the Digenea has been found to consist of proteins, an acid mucopolysaccharide, lipids, lipoproteins, mucoids, alkaline and acid phosphatases, and other enzymes. Adult Digenea differ from the aspidogastrids and monogenetic trematodes in usually possessing two prominent suckers as organs of attachment, an anterior sucker (oral sucker, when it surrounds the mouth) and a ventral sucker or acetabulum. The suckers may be supplemented with hooks, proboscids, tentacular appendages, pseudosuckers, adhesive glandular pits, and other structures. Some Digenea lack the ventral or the oral sucker, or both. The mouth is usually near the anterior tip of the body, but in some it is near the middle. Accordingly, the Digenea are divided into two main suborders, the Gastero stomata, with a centrally located mouth, and the Prosostomata, comprising the majority of the Digenea, with a terminally or subterminally located mouth opening. The Gasterostomata comprises one family, the Bucephalidae, which are parasites of fishes and will not be dealt with in this book. The Prosostomata is divided into several families, members of which parasitize a variety of vertebrate hosts. These flukes fall under descriptive, but distinctive, morphological types which are commonly known by such names as distome (the most common), amphistome, monostome, echinostome, holostome, and schistosome.

The Digenea have complex developmental cycles in which the fluke utilizes two, three, four, or more hosts; one is the definitive or final host, the others are intermediate hosts, and in some cases paratenic hosts are also involved. In the life cycle there are usually free-living stages in the water, the miracidium and the cercaria, but larval forms are present and develop inside the intermediate hosts. The development inside the snail first intermediate host (in some cases the only intermediate host) is asexual, one larval form developing from the preceding one, but in the definitive host the worm undergoes sexual reproduction. The majority of the digenetic trematodes are hermaphroditic. They have a distinct ovary and paired testes (usually, but also one, or many, testes in some species) in which haploid female and male gametes are produced by normal meiotic processes. The diploid zygote formed by union of the gametes develops into a miracidium either while the egg is still in the host or in the external environment. The sexes in the schistosomes (family Schistosomatidae, the blood flukes of mammals and birds) are separate, and usually the male possesses a gynaecophoric canal to hold the female. One digenean family, the Didymozoidae, comprises some hermaphroditic flukes, species of the genera *Didymozoan*, *Didymocystis*, *Nematobothrium*, while in others, there are male and female flukes, such as species of the genera *Wedlia* and *Köllikeria*.

It has been stated above that there are morphological types of the Digenea and although names designating these types were used by early parasitologists, some of them are still in use. They are only useful in describing adult flukes, and they include the following:

The distome type — This is the most common type. The body is usually elongate oval; there is an anterior oral sucker surrounding the mouth opening; and a ventral sucker or acetabulum is present in the anterior third of the body.

The echinostome type — This is similar to the distome type but the   llar of comparatively large spines around the oral sucker and the mouth opening.

The monostome type — This type has only one sucker, the oral at the a   tremity of the body.

The amphistome type — This type is characterized by having the acetabulum very near to, the posterior border of the body; it might be called the posterior suc

The gasterostome type — This type has the mouth located in the center of the ventr surface of the body instead of being at the oral sucker.

The holostome type — This type is characterized by the division of the body into a forebody and a hindbody. In the forebody there are the oral sucker and acetabulum arranged as in the distome type; in addition there is a large glandular adhesive organ, the tribocytic organ, located posterior to the acetabulum.

The schistosome type — This type is characteristic of the blood fluke family Schistosomatidae, and in this type the body of the male worm is split longitudinally along the ventral surface to form a gynaecophoric canal.

In addition to the above differences there are also slight variations in the morphology of the Digenea, but the anatomy of the various forms or groups is basically similar. Therefore it is preferable to describe the morphology of a generalized digenetic trematode as illustrated in Figure 1-1.

2. Generalized Adult Morphology and Life Cycle

The adult worm varies in length from a few millimeters (*Heterophyes heterophyes*, *Metagonimus yokogawi*, and other heterophyid species) to about seven centimeters (*Fasciolopsis buski*). The worm is leaf-shaped (i.e., dorso-ventrally compressed),

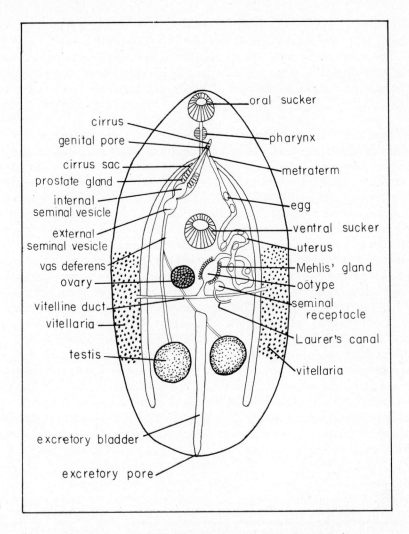

FIGURE 1-1.   Generalized morphology of a digenetic trematode.

ovoid, or at times nearly cylindrical in cross section. There is one muscular organ of attachment around the oral opening, the anterior or oral sucker, and another sucker on the ventral surface of the worm known as the acetabulum or ventral sucker. The worm is covered by a protective tegument which may be provided with scales or spines. The simple digestive system begins with the oral opening at the oral sucker and leads to a muscular pharynx that may be lacking, as in *Schistosoma*. The pharynx in turn leads into an esophagus, which bifurcates into a pair of ceca. The ceca proceed to the posterior end of the worm and end blindly. The ceca may be simple (*Clonorchis, Fasciolopsis, Echinostoma, Dicrocoelium, Heterophyes*), or branched (*Fasciola, Fascioloides*). In *Schistosoma* the two ceca unite and form a single tube; the place of the union varies from one species to the other. The unit of the excretory system is a flame cell or solenocyte. The number and distribution of these cells is characteristic for each species, and is the same in related species. The cells collect excretory material from various parts of the body of the fluke, pour them into collecting capillaries which in turn lead to a number of excretory tubules, and these terminate in a pair of excretory ducts. The latter pour the material into an excretory bladder, which is situated along the mediad in the posterior part of the body and opens to the outside through an excretory pore. The fundamental flame cell pattern of a certain trematode species is also present in the cercarial stage, and is easier to study in this stage. The miracidial stage also possesses flame cells. The miracidium of most trematodes possesses one pair of flame cells while that of the schistosomes, and related flukes, possesses two pairs.

The reproductive organs of the adult fluke are well developed and occupy the major part of the trematode body. All species of flukes of medical and veterinary importance, with the exception of the schistosomes or blood flukes of birds and mammals, are hermaphroditic. In general, there are two testes, but in some flukes there may be one only, while in others there may be several (multiple testes in the schistosomes.) The testes vary in position in relation to each other and in relation to the ovary. The testes may be rounded, lobed, or dendritic in contour. A vas efferens arises from each testis and the two vasa efferencia unite to form a vas deferens, which proceeds towards a genital atrium. Before it reaches the genital atrium it enlarges to form a seminal vesicle. Anterior to the seminal vesicle there is, in some flukes, a cluster of prostate glands and a cirrus which together with the seminal vesicle are enclosed in a cirrus sac.

The female organs consist of an ovary which is usually spherical, but in some cases it may be lobed or dendritic. An oviduct arises from the ovary, receives a duct from vitelline glands on both sides of the worm, then enlarges to form an ootype. The ova are produced in the ovary and pass through the oviduct where they become surrounded by yolk secreted by the vitelline glands. The ootype is surrounded by a group of glandular cells referred to as "Mehli's gland". The uterus arises from the ootype and usually exhibits some coiling before it reaches the genital atrium, which opens to the outside via a genital pore. The pattern of coiling of the uterus is characteristic of groups or families of flukes. A seminal receptacle opens in the oviduct near the ovary. A Laurer's canal extends from the dorsal surface of some species to the seminal receptacle, and is believed to serve during cross insemination as a vestigial vagina. In many species, however, Laurer's canal is rudimentary and does not open to the outside, and the metraterm (the distal portion of the uterus at the genital atrium) serves as a vagina.

a. Life Cycle

In general, a large number of eggs are produced by digenetic trematodes, and these eggs leave the host with feces, but in the case of *Schistosoma haematobium* the eggs are in the urine, and in the case of *Paragonimus* the eggs leave the body of the host with the sputum; when the sputum is swallowed, they leave with the feces. In some species, (*Schistosoma* spp., *Clonorchis, Opisthorchis*, heterophyid forms), the eggs already contain a miracidium when they are discharged from the body. *Schistosoma* eggs hatch soon after the excreta are diluted with water; opisthorchoid eggs only after ingestion by the aquatic snail host; dicrocoelid and brachylaemid eggs only after ingestion by terrestrial pulmonate snails; and *Fasciola, Fasciolopsis, Echinostoma*, and *Paragonimus* eggs require a period for embryonation after deposition in water. The latter eggs mature in water in about 2 weeks, but this period depends on environmental factors, mainly temperature, and then they hatch. The miracidium inside the embryonated egg is covered with a ciliated epithelium. It has a characteristic basic excretory system, a nerve center, penetration glands, an apical gland, and a number of reproductive or germ cells.

The miracidium of *Schistosoma, Fasciola, Fasciolopsis*, and *Paragonimus* spends a free-living period in the water after hatching. It swims freely in the water to locate and penetrate the snail host. Its span of life is about 24 hr, but it is vigorous only during the first 8 hr. It has been reported that the miracidium penetrates susceptible and non-susceptible snails, as well as other organisms in the water. Although at first the swimming movement of the miracidium may appear to be random, it is now the general agreement that once in the vicinity of a susceptible snail the miracidium is attracted by a substance(s) emitted by the snail.

In a nonsusceptible snail the miracidium is enveloped by infiltrating host cells and soon dies. However, once inside a susceptible snail it loses its ciliated epithelium while entering and within a day or two it takes the form of a simple sac, without mouth or other distinct external or internal structures, but retains the group of germ cells and is known as the "first generation sporocyst" or "mother sporocyst". The germ cells divide steadily and group together to form "germ balls", and these develop into the next generation, which is a "second generation sporocyst" or "daughter sporocyst" (*Schistosoma* spp., strigeoids), or a redia (*Fasciola, Fasciolopsis, Paragonimus, Clonorchis, Echinostoma*). The daughter sporocyst is similar in morphology to the mother sporocyst, being in the form of a simple elongated sac, but longer. However, the redia is provided with a mouth, a muscular pharynx, and an intestine varying in length from one species to the other. Whether it is a daughter sporocyst or a redia, the germ cells inside divide and subdivide through the same asexual process followed by the germ cells in the mother sporocyst and known as "polyembryony" or germ cell lineage. The cells form germ balls which develop into the next stage, known as "cercaria". In some species (*Fasciola, Paramphistomum, Gastrodiscus*) the redia might form other redial generation before forming cercariae. When mature, the cercariae escape from the redia or the daughter sporocyst and then break out of the snail host to assume a free-living existence in the water. The process of formation of cercariae and their release continues in such a way that 100,000 and more of these cercarial progeny are formed in the case of *Schistosoma mansoni*, for example. Thus, the intramolluscan phase of the life cycle is a multiplication process, from one single miracidium to the formation of thousands of cercariae. It takes one month at about 25°C from the time the miracidium of *S. mansoni* penetrates the snail host to the escape of the first cercariae from the snail, but the higher the temperature, the shorter this period is. The incubation period in the snail is also about a month at 25°C in the case of *Fasciola hepatica*. Cercariae exhibit periodicity in their emergence from the snail. Those of *Fasciola* emerge from the snail at night, those of *Paragonimus* emerge in the late afternoon and evening, and so do those of *Schistosoma japonicum*, while those of *S. mansoni* and *S. haematobium* emerge in the morning, reaching maximum numbers at noon and early afternoon.

The cercariae of many of the digenetic trematodes possess a tail, which may be small and ovoidal (*Paragonimus*), have a simple trunk (*Fasciola, Fasciolopsis*), may be provided with a fluted keel (*Clonorchis, Opisthorchis, Heterophyes*), or may be bifurcated at its end (schistosomes). The tegument of the cercariae is nonciliated but often carries spines. The cercariae possess penetration and cystogenous glands, and in some species mucoid glands, suckers which may resemble those of the adult, and a pattern of excretory system which also resembles that of the adult, but simpler.

The cercaria swims freely in the water and its fate depends on the species of trematode. The cercariae of the schistosomes penetrate into the skin of the definitive or final host, reach a venous capillary or the lymphatics, and migrate through the bloodstream until they reach the portal or caval system where they mature. If they do not find the definitive host in a day or two they die. Although they survive in the water for that period they are, however, powerful penetrants only during the first 12 hr or thereabouts.

The cercariae of *Fasciola, Fasciolopsis, Paramphistomum*, and *Gastrodiscus* encyst in the open on aquatic vegetation in the snail's habitat, forming metacercarial cysts on this vegetation. Cercariae of notocotylids and some philophthalmids encyst in the open on hard surfaces, on the snail's shell, or on vegetation. Those of the cyclocoelids do not emerge but encyst inside the redia in which they had developed. The majority of digenetic trematode species, however, require a second intermediate host in which they shed the tail and encyst. The second intermediate host may be fresh-water fishes

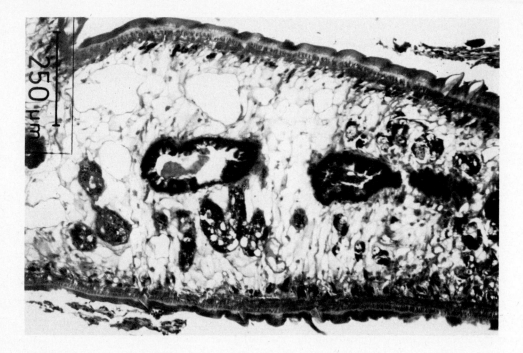

FIGURE 1-2.   Cross section in a digenetic trematode, showing tegument, spines, and some internal organs embedded in the parenchyma.

(*Clonorchis, Opisthorchis, Heterophyes, Nanophyetus*, and heterophyid species). Certain heterophyid species require brackish or salt-water fishes. *Paragonimus* cercariae encyst in crabs or crayfishes, and *Echinostoma ilocanum* cercariae and those of some other echinostomes encyst in a second molluscan host. The definitive host acquires the infection with these trematodes by consumption of the infected second intermediate hosts, raw or inadequately cooked. A metacercarial cyst when eaten excysts in the upper levels of the small intestine and migrates to the tissues or organs where they develop into mature adult trematode worms. The intestinal forms (*Fasciolopsis*, heterophyid, and echinostome species) do not need to migrate, but attach themselves to the wall of the intestine and mature; *Clonorchis* and *Opisthorchis* migrate through the ampulla of Vater to the distal bile passages; *Fasciola* penetrates the intestinal wall, then Glisson's capsule and the liver parenchyma to the bile ducts; while *Paragonimus* metacercariae penetrate the intestinal wall, move through the peritoneal cavity, penetrate the diaphragm, move through the pleural cavity, and then penetrate the lungs. The excysted metacercariae of *Fasciola* and *Paragonimus* do not always reach their typical habitats and thus settle in ectopic sites.

3. Morphology and Biology of Life Cycle Stages
a. The Adult Fluke
1. Morphology
a. The Tegument

By the use of the light microscope the surface of adult trematodes has been reported to be covered by a "cuticle" in which no nuclei or cellular structures could be discerned (Figure 1-2). As to the origin of this layer, some helminthologists believed that the cuticle is the degenerate remnant of an epidermis, or the basement membrane of an epidermis subsequently lost, while others were of the opinion that the cells producing the cuticle were retained and were in fact ordinary mesenchymal cells or, alternatively, epidermal cells which have "sunk" into the mesenchyme. Because this surface layer represents the area of contact and metabolic exchanges between host and parasite, special attention was given to elucidate its nature and structure. Even before the advent of the electron microscope, histological and histochemical techniques were utilized for this purpose.

With the use of the electron microscope it was revealed that the surface of the fluke's body is a continuous layer of cytoplasm, 15 to 21μ thick, and not the amorphous, nonliving "cuticle" described by light microscopy.[166] This cytoplasmic layer is covered on the outside by a plasma membrane about 10μ thick; there are many invaginations and many underlying pinocytotic vesicles. The outer surface is usually thrown into folds to form microvilli, apparently to increase the absorptive surface. Underneath the dense outer zone there are numerous mitochondria which have a dense matrix. Because of the presence of mitochondria and endoplasmic reticulum, the surface layer is no inert layer nor a secretory product of cells, as had been believed. The outer zone is bounded by a plasma membrane and rests on a thick basal lamina. Several processes connect the outer zone of the tegument with flask-shaped cells in the parenchyma, underneath the tegument, and known as cytons. These cells contain a nucleus, several mitochondria, some Golgi vesicles, and small vacuoles; thus the cells are in cytoplasmic continuity with the outer cytoplasmic layer and form part of the syncytial epithelium (Figures 1-3 and 1-4). Thus the superficial zone of the epithelium and the cells correspond, respectively, to the "cuticle" and "subcuticular cells", "glands" or "myoblasts" of traditional helminthology. Electron microscopy has thus shown that the superficial zone is not partitioned by intact membranes, and so the definitive tegument of trematodes can sometimes be referred to as a symplasm, i.e., a large multinucleated body of cytoplasm.

Light microscope studies, for some time, showed the presence of tubercles, papillae, and other projections from the surface of adult digenean flukes. The use of the electron microscope has provided more information about these structures. For example, in the case of some schistosome species, the surface of the male is covered with tubercles and these, seen by the scanning electron microscope (SEM), give a rough surface for the male of *Schistosoma haematobium* (Figure 1-5) and *S. mansoni*. The integumental surface of the female is relatively smooth. SEM reveals certain basic features such as spines in the oral sucker and the acetabulum of both sexes, which may facilitate rasping and/or attachment of the parasite for residence in the circulatory system of the definitive host. SEM studies conducted on *S. haematobium* by Kuntz et al.[81] showed marked differences between various parts of the same worm, male or female. They also showed the presence of a gynecophoral fold on the gynecophoral canal of the male which may enhance anchorage of the female in the grasp of the male.

b. Muscles

Underneath the basal lamina of the tegument there is an outer transverse, and an inner longitudinal, muscle layer. In addition there are muscle fibers, crossed in all directions, which surround the gut, especially the intestinal ceca. Dorsoventral muscle fibers are also present. The position and structure of the transverse, longitudinal, and dorsoventral muscles show very clearly in electron micrographs (Figure 1-6).

c. The Parenchyma

The bulk of the fluke between the tegument and the internal organs is filled with an area of mesenchymatous cells known as the parenchyma. Electron microscope studies of the parenchyma of some digenetic trematodes show the parenchymal cells to have large areas without cell organoids, but containing considerable quantities of α and β glycogen. These cell organoids tend to be concentrated in the region of the nucleus, and include a few narrow cisternae of granular endoplasmic reticulum and mitochondria. Some lipid inclusions and smooth membranes are also present.

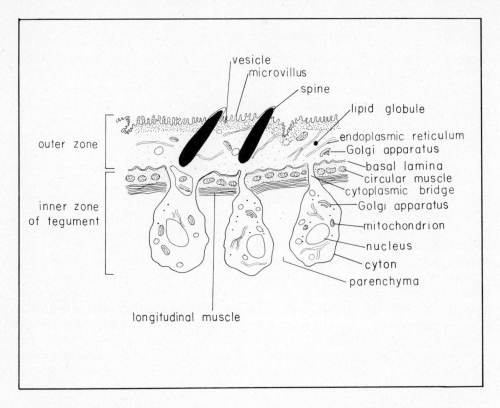

FIGURE 1-3.   Schematic representation of the tegument of an adult trematode as revealed by electron microscopy.

On the basis of several studies of Threadgold and his co-workers[166a] on *F. hepatica*, three possible functions were suggested for the parenchymal cells, namely: (1) the synthesis and secretion of the matrix and fibers of the interstitial material; (2) the transport of substances throughout the fluke in the absence of any recognizable circulatory system; and (3) the storage of food reserves, essentially glycogen, and their mobilization and resynthesis when required.

d. The Alimentary Tract

Relative to the alimentary tract, the pharynx serves primarily as a masticatory organ, from which the ingested food particles pass into the esophagus. The lumen of the latter is lined with a layer of cells, as is that of the intestinal ceca. The cells lining the lumen of the ceca not only form a lining epithelium but they go through a secretory and an absorptive phase according to which their shape and size vary. The material secreted is believed to be rich in enzymes, and the cells themselves are rich in enzymes and in ribonucleic acid.

e. The Reproductive System

The general plan of organization of the reproductive system was outlined above.

f. Egg Formation

Aside from the ovary and the beginning of the uterus, the ovary-ootype area has the following structures: oocapt, vitelline duct, vitelline reservoir (in some forms), ootype, Mehli's gland, and Laurer's canal. The "oocapt" or "ovicapt" is a ciliated dilation of the proximal end of the oviduct at the ovary-oviduct junction. The oocapt

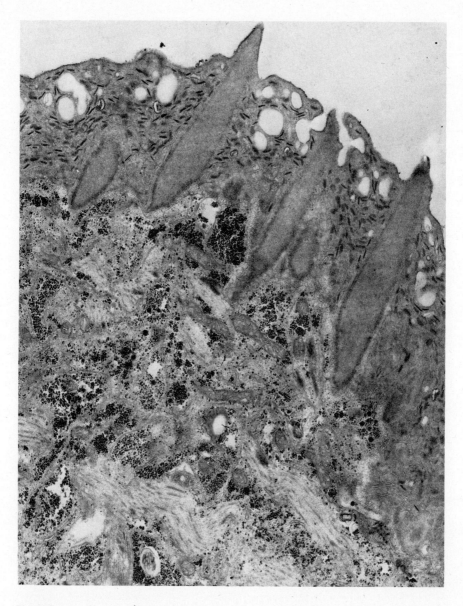

FIGURE 1-4. Electron photomicrograph of adult *Schistosoma mansoni* showing ultrastructure of the tegument. For details see text. (Courtesy of Dr. James Byram.)

has a muscular wall and valve-like sphincters, which seem to control the passage of the oocytes from the ovary into the oviduct. Often the oocytes leave the ovary one by one. In a few species, however, they are in groups surrounded by a membrane, which is ruptured by the oocapt, and thus even in such cases they will pass one by one in the oviduct. The ootype is a muscular walled dilation of the proximal end of the uterus or the distal end of the oviduct, in which the egg capsule is formed. The Mehli's gland, known previously as the "shell gland", consists of glandular cells which surround the ootype. The oocytes enter the ootype, encounter spermatozoa and become surrounded by vitelline cells from the common duct of the vitellaria. Only part of the function of the vitelline cells is to provide nutritive yolk material to the oocyte, but the globules constituting a major part of the vitelline cell are of raw shell material. Secretions from

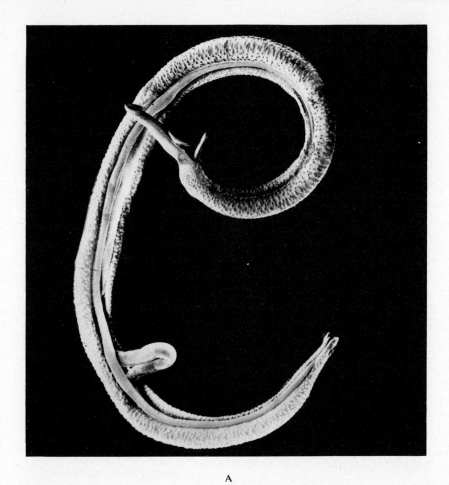

A

FIGURE 1-5.   Scanning electron micrographs of *Schistosoma haematobium* (Iran). (A) Male and female schistosome in copula. (Magnification × 34.) (B) Dorsal surface of male near midbody showing arrangement of large and small bosses. (Magnification × 266.) (C) Dorsal surface of female near midbody showing texture and scattered surface elevations. (Magnification × 646.) (D) Lateral view of male and female worms at midbody. (Magnification × 120.) (E) Inner surface of gynecophoral canal near midbody showing rough texture and striae. (Magnification × 1280.) (F) Portion of (E) at greater magnification showing rough surface. (Magnification × 2000.) (Courtesy of Drs. Kuntz, Tulloh, Davidson, and Huang.)

the Mehli's gland are believed to harden the newly formed egg shell, and probably also initiate the breakdown of the vitelline cell thus releasing the globules. The globules, also known as shell globules, envelop the developing egg and eventually coalesce and become hardened to form the shell. This hardening process involves the tanning of the protein or sclerotin present within the coalesced globules by quinone. As the eggs leave the ootype they are usually lighter in color than the eggs reaching the terminal portion of the uterus, due to the tanning of the eggshell.

Other functions attributed to the secretions of the Mehli's gland are the lubrication of the uterus and the activation of the spermatozoa, which are passed down to the ootype. Thus it is now definitely known that the Mehli's gland does not form the shell of the egg, as was thought earlier. The cells of the Mehli's gland lack various proteins, phenols, and phenolases which are present in large amounts in the shell of trematode

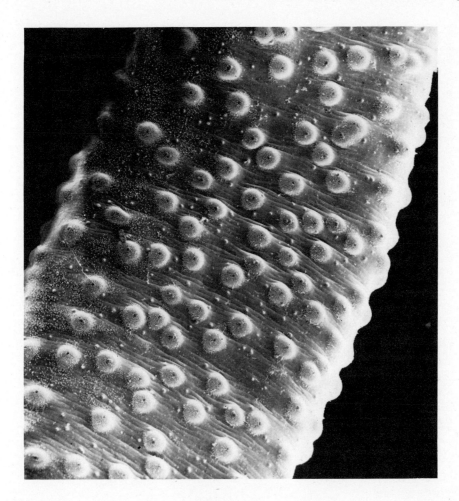

FIGURE 1-5B

eggs. On the other hand, it is the vitelline cells which are responsible for the eggshell formation, in addition to their contribution of yolk material to be incorporated within the egg. Although this is the accepted method of eggshell formation in many trematodes, it was reported that in some species the uterine wall may produce a substance necessary for shell formation.

The eggs are transported to the exterior via the uterus. In some trematodes, the terminal segment of the uterus is muscular, known as the "metraterm". By peristaltic movement the eggs are expelled into the genital atrium, then to the exterior via the genital pore. During their development in the exterior, some trematode eggs increase in size. Eggs of the schistosomes which normally mature in the tissues of the host after they are deposited by the female worm increase gradually in size as the miracidium develops inside (Figure 1-7).

g. The Excretory System

The excretory or osmoregulatory system of the Digenea is of the protonephritic type. The basic unit is the flame cell, and these cells have a characteristic arrangement pattern known as the flame cell pattern. The pattern can be expressed in a formula, the flame cell formula, which in many cases is important for species identification. Electron microscopy has shown that the flame cells bear true cilia with the typical 9—2 arrangement of microfilaments. The shape of the excretory vesicle (excretory bladder) is also diagnostic. The vesicle may be Y-, V-, or I-shaped.

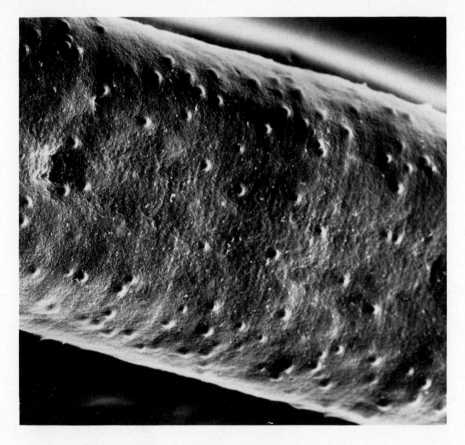

FIGURE 1-5C

h. The Nervous System

In general the nervous system is similar in the various groups of Digenea studied (Figure 1-8). It consists of two prominent cerebral ganglia situated on either side of the pharynx, connected by a dorsal cerebral commissure. From the cerebral ganglia extend three pairs of anterior and posterior peripheral nerves: the dorsal nerves, the ventral nerves, and the lateral nerves.

The dorsal anterior and posterior nerves are usually not as well developed as the ventral nerves. Posterior ventral trunks, by far the best developed of all the nerves, extend the full length of the body and, at least in *Postharmostomum helicis*[168] anastomose just anterior to the bladder. Each lateral anterior nerve branches shortly after its origin near the junction of the ventral nerves and the cerebral ganglion, and sends forward two rami. There are well-formed commissures, termed "lateral commissures", and dorsal commissures connecting the anterior and posterior lateral nerves and posterior dorsal nerves (Figure 1-8).

i. Physiology

Nutrition of adult digenetic trematodes has received the attention of various investigators, because of their varied habitats within the host. Gut-dwelling trematodes feed on the superficial epithelial tissues and associated mucoid secretions of the host, and those living within the respiratory and circulatory system feed exclusively on blood. The mode of feeding is suctorial, brought about by the muscular pharynx and normal attachment process of the oral sucker. In some trematodes Halton[56] found evidence that this purely mechanical process is supplemented with enzymic secretions produced

17

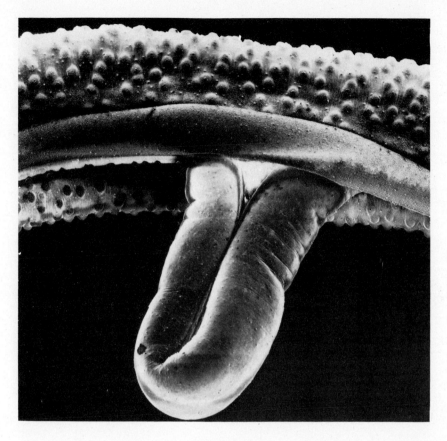

FIGURE 1-5D

by the trematode which have an histolic effect upon host tissues. Digestion in the Digenea is predominantly an extracellular process, but the exact sequence varies according to the nature of the food ingested. A number of enzymes have been found in several species studied. Alkaline phosphatase, acid phosphatase, and protease have been found in the majority of flukes. Amino peptidase was found only in *Cyathocotyle bushiensis. Schistosoma mansoni* lacks lipase and esterase but these enzymes are present in *Fasciola hepatica. Haematoloechus medioplexus* lacks lipase and esterase. The schistosomes possess a protease of high specificity which releases tyrosine from globin and hemoglobin; this protease has an optimum pH of 3.9.

Microvilli (Figure 1-9) of one form or the other occur in the gut of most species and provide a greatly increased internal surface area to the otherwise simple ceca.

On the basis of his histological and histochemical studies and from the work of other authors, Halton[56] found varying degrees of adaptation exhibited by the Digenea to a diet of blood, where numerous attempts have been made to eliminate the unwanted iron component of the hemoglobin molecule. In *Fasciola hepatica* extracellular digestion renders the blood meal soluble, the products are then absorbed, and digestion is completed in the gastrodermis. *F. hepatica* is primarily a tissue feeder and is not adapted to a diet of blood. A more successful adaptation to a diet of blood is present in the schistosomes. In these flukes digestion is predominantly extracellular and results in the formation of insoluble hematin. Since the formation of the pigment occurs in the lumen and its elimination from the body is readily achieved by simple regurgitation, this avoids the unnecessary storage of iron, or its elimination from the body through the cuticle and excretory system.

FIGURE 1-5E

FIGURE 1-5F

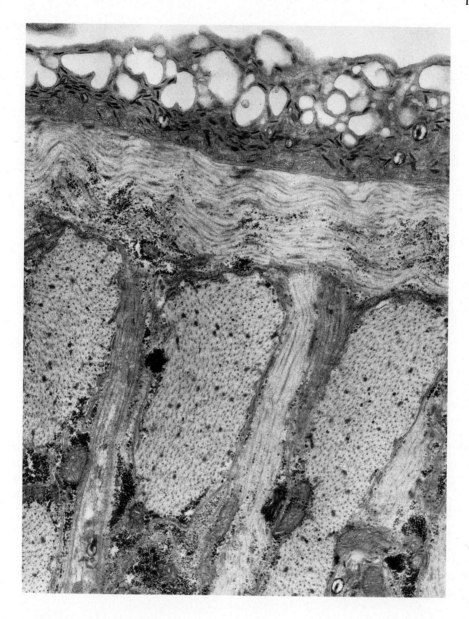

FIGURE 1-6.   Electron photomicrograph of adult *Schistosoma mansoni*, showing tegument, circular muscle fibers, longitudinal muscle fibers, and dorsoventral muscle fibers. (Courtesy of Dr. James Byram.)

It should be noted that it is not necessary that flukes should reside in the vascular system to be exclusively blood ingestors. *Diplodiscus subclavatus*, a rectum parasite of amphibians, and *Haematoloechus medioplexus*, a lung parasite of frogs, also feed on blood. The acquisition of host hemoglobin by a trematode is not unlikely, since ferritin, a larger molecule than vertebrate hemoglobin, can readily enter *Fasciola hepatica* through the tegument. Trematodes may therefore be capable of absorbing host hemoglobin and altering its physico-chemical characteristics. Sections of *Philophthalmus megalurus* and *Fasciolopsis buski* show hemoglobin generally distributed in the parenchyma and concentrated near the excretory channels and proximal uterine coils. Hemoglobin has also been found to be abundant near the vitellaria and distal uterine

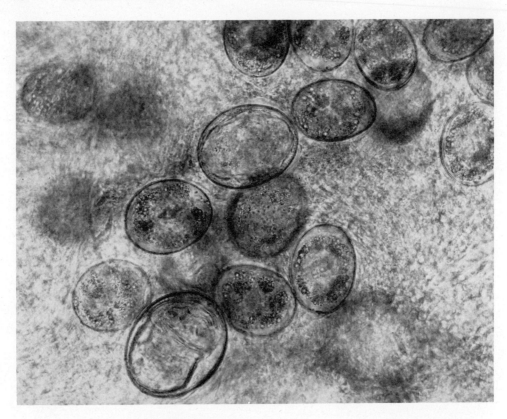

FIGURE 1-7.   Eggs of *Heterobilharzia americana* deposited in the tissues of the host. Note various stages in the development of the miracidium inside, and that eggs increase in size as they progress in development.

coils in *Fasciola hepatica*, and near both suckers in *F. gigantica*. Although there are few studies on trematode hemoglobins, investigations have shown them to be electrophoretically and spectrophotometrically different from host hemoglobins.[20]

The majority of flukes contain large amounts of glycogen. Glycogen is primarily stored in the parenchymal cells and muscular tissues but there is also glycogen in the cells of the vitellaria and in the uterine eggs. It has been estimated, for example, that *Fasciola hepatica* may contain over 5% (25% dry weight) of glycogen; male schistosomes contain 13.9 to 29% dry weight, but the females have comparatively little glycogen (2.7 to 5%). The main end products of carbohydrate metabolism of *F. hepatica* are propionic and acetic acids but small amounts of isobutyric acid and isovaleric and 2-methyl butyric acids are also excreted. It is natural that liver flukes and blood flukes differ in their carbohydrate metabolism because of differences in their environments in the mammalian host. Coles,[28] who reviewed the available information on the metabolism of *Fasciola hepatica* and the schistosomes, noted that liver flukes live in a primarily anaerobic environment, whereas the schistosomes live in a glucose-rich aerobic habitat. Schistosomes obtain the major part of their energy by glycolysis and excretion of lactic acid rather than by oxidative phosphorylation.

2. Culture of Flukes

Various attempts have been made to culture digenetic trematodes, and various advances have been made in this field. Several investigators were successful in maintaining adult schistosomes for extended periods, with production of viable eggs. Several media were used, and the addition of glucose, small amounts of mammalian red and

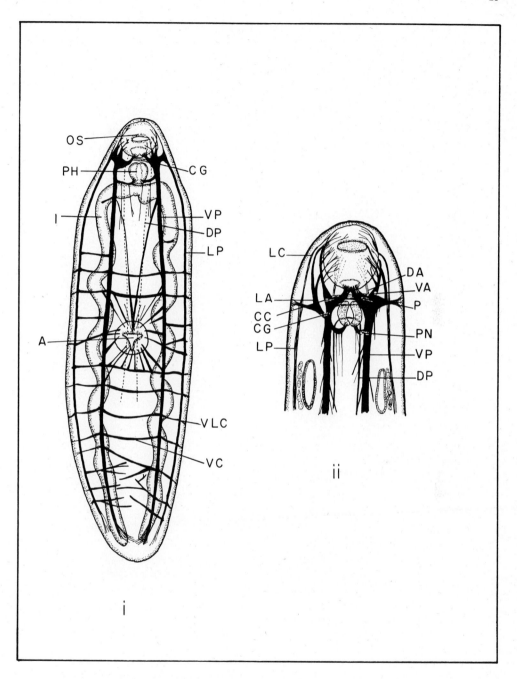

FIGURE 1-8. Nervous system of the brachylaemid *Postharmostomum helicis*. (i) Reconstruction from frontal sections of nervous system of an extended specimen showing major nerve trunks; dorsal commissures not shown. (ii) Anterior end of fluke, showing the major nerves associated with the cerebral ganglia. A, acetabulum; ATD, anterior transverse dorsal commissure; CC, cerebral commissure; CG, cerebral ganglion; DA, dorsal anterior nerve; DC, dorsal commissure; DLC, dorso-lateral commissure; DP, dorsal posterior nerve; DPC, dorsal pharyngeal commissure; EV, excretory vessel; I, intestine; LA, lateral anterior nerve; LC, lateral commissure; LP, lateral posterior nerve; OS, oral sucker; P, palatine nerve; PH, pharynx; PN, pharyngeal nerve; VA ventral anterior nerve; VC, ventral commissure; VLC, ventro-lateral commissure; VP, ventral posterior nerve; and VPC, ventral pharyngeal commissure. (Redrawn after Ulmer, 1953.)

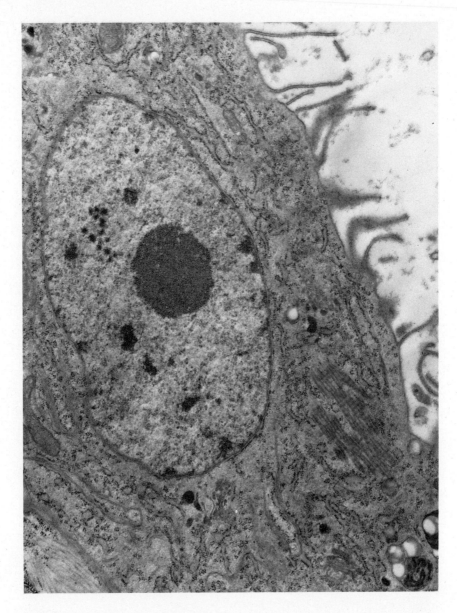

FIGURE 1-9.   Electron photomicrograph of adult *Schistosoma mansoni*, showing epithelial
lining of the intestinal ceca, with microvilli projecting into the lumen. (Courtesy of Dr. James
Byram.)

white blood cells, and amino acids of the types and proportion found in globulin enhanced survival and egg production. Information on the culture of other flukes is also
available. Excysted metacercariae of *Sphaeridiotrema globulus* were successfully cultivated in vitro to adults which produced eggs capable of embryonation, miracidium
formation, and hatching.[14] Tetracotyles of the strigeid *Cotylurus lutzi* were cultured
in vitro to adult flukes when incubated in a basic medium of 40% NCTC-135 and 40%
chicken serum, supplemented with 20% extract of chicken heart, liver, muscle, intestinal wall, or mucosa.[9] Fully developed miracidia were obtained when upper small
intestinal mucosal extract was used. The miracidia were able to infect the planorbid
snail *Biomphalaria glabrata*, and cercariae were produced which encysted readily as
tetracotyles, and later developed into normal adults in zebra finches.

b. The Egg
1. Egg Passage

Not all the eggs produced by some Digenea leave the host with the excreta. A certain number of eggs of the schistosomes as well as the lung flukes, the heterophyids, and some liver flukes are retained in the host tissues. After the female schistosome worm deposits the eggs in the fine mesenteric capillaries or those of the vesico-prostatic plexus, some eggs, instead of gaining access to the wall of the intestine or the urinary bladder, are swept away by the circulating blood and are later filtered into various tissues, especially the liver. Schistosome eggs have been found in almost every organ of the body and in subcutaneous lesions. Heterophyid worms, because of their burrowing habits in the intestinal mucosa of their hosts, deposit eggs which are carried by the circulation to various organs of the body. Lung fluke eggs have also been found in various organs in addition to the lung.

The number of those eggs passing in the excreta of the host varies considerably from one digenean to the other, but the majority produce a large number of eggs. *Heronimus chelydrae* is known to produce only a few eggs, whereas among the fasciolids, for example, the figures for the daily egg output are very high: 20,000 for *Fasciola hepatica* and 20,000 to 25,000 for *Fasciolopsis buski*. In spite of the number of heterophyid eggs retained in the host's body, the number passing in the feces is high. For *Metagonimus yokogawai* it was found in Korea that the number of eggs per day per worm is about 1505 during one year of infection in humans. The estimates are 1400 to 3500 for the daily egg production by a single *Schistosoma japonicum* female, 250 to 350 for *S. mansoni*, and 50 to 300 by *Schistosoma haematobium*. These estimates among the schistosomes vary depending on the strain of the parasite and the age of the female worms. Old female worms (2 to 5 years) may produce fewer eggs than younger ones.

Usually there is a regularity in the passage of trematode eggs in the excreta of their hosts; however among some species daily variations have been observed. With the schistosomes, it was found that the egg output of *S. mansoni* is no more variable than that of hookworms if measured over an extended period, for example 2 to 3 weeks. However, the pattern of egg passage in the feces, in the case of humans and other animals infected with schistosomes, has been found to be characteristic, inasmuch as a peak number of eggs is reached shortly after potency and remains as such for a few months, after which the number shows a marked decrease. When grivet monkeys (*Cercopithecus aethiops aethiops*) were infected with *S. mansoni*, by only a single dose of cercariae, the number of eggs per day passed in the feces decreased sharply after the 3rd month, from 30,000 to 15,000 per day, as shown in Figure 1-10. Thereafter no significant change in the passage of eggs in the feces was noted.[73]

The egg output by *Schistosoma haematobium* has another characteristic pattern, in addition to the above, and that is that the output depends on the time of day at which the urine sample is collected. The morning sample has few eggs, while the early afternoon sample has more eggs, and the variability in the number of eggs at that time is much less pronounced than in the morning sample. Thus for epidemiological surveys urine samples should be taken in the early afternoon. *Clonorchis sinensis*, the Chinese liver fluke, shows another characteristic pattern, and that is periodicity in egg production and passage in the feces of experimentally infected animals. When seven rabbits were inoculated with metacercariae of *C. sinensis* egg production increased evenly from the beginning of patency at the 22nd day to the 13th week, after which there was cyclic fluctuation at roughly 10-week intervals. The first peak of egg production occurred between weeks 16 and 17 for five of the seven animals; the others at weeks 13 to 14 and at week 19.[192] At these times the egg production ranged from 8,000 to 30,000 eggs per gram feces (EPGF), with a mean of 18,000.

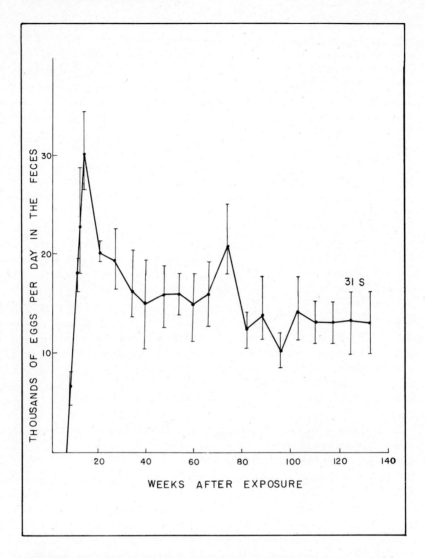

FIGURE 1-10.   The mean number of *Schistosoma mansoni* eggs per day passed in the feces of grivet monkeys infected only once is shown for the course of the infection. The vertical bars indicate ±1 standard deviation of the mean. Each point on the graph is the mean of eggs in seven stool collections from each of six monkeys. (Redrawn from Cheever and Duvall, 1974. With permission.)

2. Morphology

The egg of digenetic trematodes is usually oval in shape and is surrounded by an operculate capsule which is light to dark brown in color. The eggs of the schistosomes are not operculated. The egg capsule of most trematode eggs is simple, though it may have one or two filaments or a spine. Eggs of several trematodes are included in Figures 1-11 and 1-12 to show their relative size and the stage of their development when they leave the host with the excreta. Interspecific variations have been reported for the eggs of *Paragonimus*, the schistosomes, and other flukes. Interspecific variations in eggs of the schistosomes are in size, shape, length to breadth ratio in the nonspined eggs, and position and length of spine. For the schistosome eggs with a terminal spine, the width of the egg at a distance of 40μ from the tip of the spine is diagnostic. Figures showing these differences are in Chapter 2.

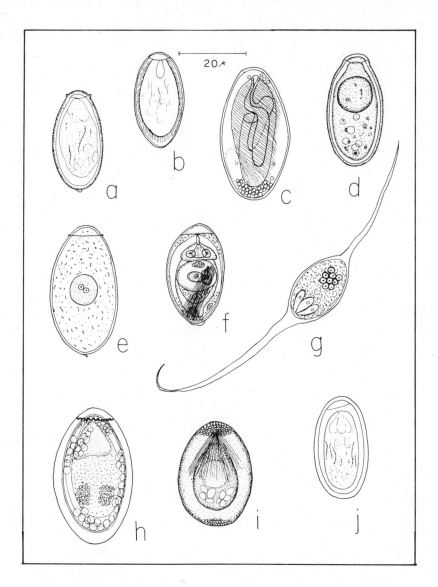

FIGURE 1-11.   Eggs of some Digenea; the "small" eggs, drawn to the same scale. Some original, others adapted after various authors. The measurements given after each egg are in microns. (a) The opisthorchid, *Clonorchis sinensis*, 27 to 35 × 12 to 20 (av. 29 × 16). (b) The heterophyid, *Heterophyes heterophyes*, 28 to 30 × 15 to 17. (c) The heterophyid *Cryptocotyle lingua*, 42 to 56 × 21 to 30 (av. 46 × 24). (d) The microphalid, *Microphallus opacus*, 35 × 17. (e) The plagiorchid, *Plagiorchis goodmani*, 47 × 25. (f) The brachylaemid, *Postharmostomum helicis*, 30 to 36 × 17. (g) The notocotylid, *Notocotylus* sp., 20 × 14. (h) The dicrocoelid, *Brachylecithum orfi*, 36 to 47 × 25 to 32. (i) The dicrocoelid, *Platynosomum fastosum*, 30 × 25. (j) The dicrocoelid, *Dicrocoelium dendriticum*, 37 to 44 × 21 to 30.

The egg capsule of some trematodes is smooth, at least by light microscopy, but in others it might have minute structures which are revealed by the electron microscope. The egg capsule of *Fasciola hepatica* has a smooth surface. Stereoscan studies also show the surface to be smooth, but the surface has a slightly elevated circle marking the fracture of the operculum, and the operculum and the aperture have serrated edges.[76] As visualized with scanning electron microscopy, the shape of the egg shell of the heterophyid *Cryptocotyle lingua* is that of a concavo-convex ellipsoid. Drawings

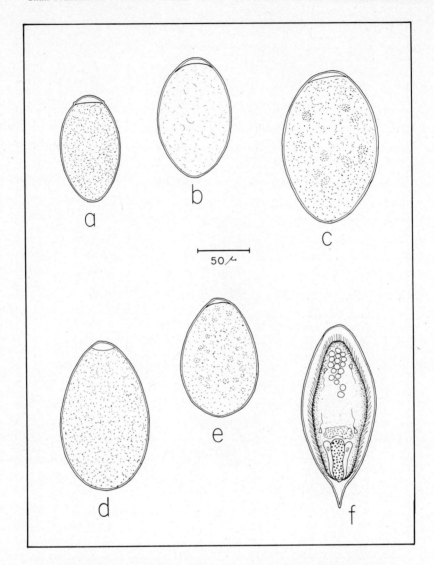

FIGURE 1-12.   Eggs of some Digenea; the ''large'' eggs, drawn to the
same scale. Some original, others adapted after various authors. The
measurements given after each egg are in microns. (a) The paragonimid
Paragonimus kellicotti, 83 to 111 × 55 to 65 (av. 90 to 58). (b) The echinostome, *Echinoparyphium flexum*, 85 to 116 × 60 to 70. (c) The fasciolid, *Fasciola hepatica*, 130 to 150 × 62 to 90. (d) The amphistome,
Gastrodiscus aegyptiacus, 150 × 100. (e) strigeid, 100 × 70. (f) The schistosome, *Schistosoma haematobium*, 120 to 170 × 40 to 70 (av. 140 × 60).

based on early light microscope observations often depicted the shape of *C. lingua*
eggs in this fashion. Scanning electron micrographs also show a longitudinal ridge-like
demarcation extending along the convex, but not the concave, surface.[79]

Scanning electron micrography of the egg capsule of *Schistosoma mansoni*, *S. hae-*
matobium and *S. japonicum* have shown that the surface is covered with microspines
about 3 to 5μ in length (Figures 1-13 and 1-14). The microspine on the shell of *S.*
haematobium and *S. mansoni* can be flexed by contact with the surrounding host cell
membranes. It has been suggested that the microspines probably allow the eggs to cling
to the vascular endothelial lining and facilitate their penetration into adjacent tissues.

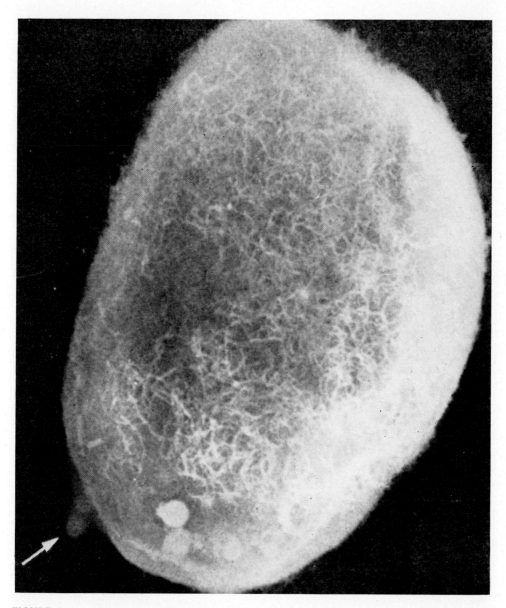

FIGURE 1-13. Scanning electron micrograph of the egg of *S. japonicum,* showing cobweb-like surface and minute spine (arrow). (Magnification × 1600.) (From Sakamoto, K. and Ishii, Y., *Am. J. Trop. Med. Hyg.,* 25, 842, 1976. With permisson.)

It has also been suggested that the microspines form a mode of increasing the shell surface area, and might serve as conduits to localize egg or worm secretion on the shell surface. The microspines of *S. japonicum* (also termed microvilli-like chitinous projections) are different in length, thickness, and density from the microspines of *S. mansoni* and *S. haematobium* eggs; consequently, their functions may also be different.[146] The egg capsule of schistosome eggs also has minute pores located principally in the anterior portion of the egg (Figure 1-14). Each pore is lined by a trilaminar unit membrane. The presence of these pores offers a possible explanation for the route of exchange of metabolites and antigenic materials, and might also function as the route of exit of the enzymatic secretions that are thought to be responsible for the eruption of the eggs from veins into the tissues.[136]

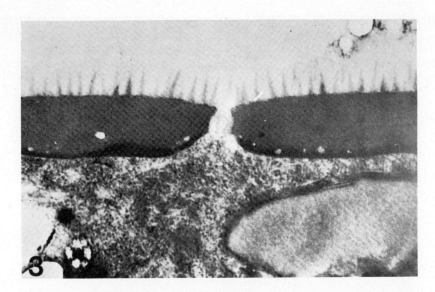

FIGURE 1-14. Eggshell of *Schistosoma mansoni.* Transmission electronmicrograph
showing shell pore lined by biological membrane. Epicuticular microbarbs are present on
the surface. (Magnification × 40,000.) (After Race, G., Martin, J., Moore, D., and Larsch,
J., *Am. J. Trop. Med. Hyg.,* 20, 916, 1971. Micrograph by courtesy of *Am. J. Trop. Med.
Hyg.,* and Dr. George Race.)

3. Chemical Composition

Several studies have shown that the trematode eggshell is composed of sclerotin type
proteins, and a tanning process takes place by which the color turns light brown. The
precursors of the quinon tanning system, protein, phenols, and phenolase, have been
demonstrated by histochemical tests to occur in the globules of the vitelline cells of
the vitellaria.

4. Development and Hatching

The majority of trematode eggs leave the host in an undeveloped stage, while, in
some, a fully developed miracidium is contained in the egg found in the host's excreta.
Eggs of *Fasciola, Fascioloides, Fasciolopsis, Echinostoma, Paramphistomum,* and
Paragonimus, among others which leave the host to the outside environment in an
undifferentiated condition, require a developmental period before hatching which varies from about 12 days to 3 weeks, depending on the trematode species and on environmental factors, mainly temperature, moisture, and dissolved oxygen. The higher the
temperature the shorter the developmental period. At low temperature, usually below
10°C development is arrested and in nature eggs of several trematodes, in temperate
climates, over-winter (where the winter is mild) and hatch in the spring when the temperature rises. In the laboratory, eggs of fasciolids can be stored for several months
in the refrigerator, and when brought out later at room temperature they resume development and can be induced to hatch. In our laboratory similar observations were
made on eggs of *Fasciola hepatica* from St. Lucia; they developed, after storage for 6
months at 5°C, when they were later maintained at a temperature of 23 to 25°C. Moreover, eggs of *Neodiplostomum intermedium* were maintained by Pearson[128] for 3
months at 5°C. It was shown in the case of eggs of *Fasciola hepatica* that they have
to be freed from the feces before they commence development and, later, hatching;
they also have to be surrounded by a film of moisture, in addition to a temperature
higher than 10°C. The developed miracidium inside fasciolid eggs is enclosed in a vitelline membrane as well as other membranes, and there is a viscous cushion at the

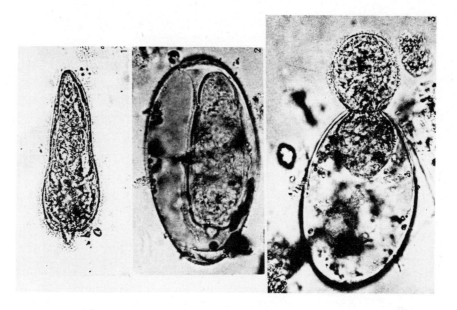

FIGURE 1-15. Hatching of miracidium of *Fasciolopsis buski.* (1) Miracidium, (2) Fully developed miracidium inside egg, (3) Hatching miracidium half way outside the egg. (After Wu, 1937. Photographs by courtesy of *Ann. Parasitol.,* and Masson, Paris.)

opercular end of the egg. This cushion is a colloid containing some protein; it changes from the contracted gel condition to an expanded sol as the ionic constitution of the fluid surrounding it changes. The process of hatching was reviewed by Erasmus.[43] Rowan[141,142] was of the opinion that the expansion of the cushion, and accordingly hatching, is produced by the action of a hatching enzyme. Wilson[181] opposed the idea of action of an enzyme in the hatching process, and was of the opinion that it is light that first stimulates the miracidium inside the egg; by its resulting activity the permeability of the membrane on the inner concave surface of the viscous cushion is altered, which in turn causes the hydration of the cushion (which is made of a fibrillar mucoprotein complex) and an increase in its volume. The resulting internal pressure ruptures the operculum and releases the miracidium. Hatching of the miracidium of *Fasciolopsis buski* is shown in Figure 1-15.

Several operculate trematode eggs do not contain a viscous cushion, for example, strigeids and some amphistomes, and the miracidium in these cases emerges after the rupture of the vitelline membrane or the miracidium is extruded while still enclosed in the vitelline membrane. Apparently light is not essential for the mechanism of hatching of some operculate trematode eggs, because those of *Fibricola cratera* hatch in both dark and light surroundings.

Operculate eggs of some trematodes are fully embryonated when they reach the outside environment but they do not hatch until they are ingested by the molluscan intermediate host. Example of these are those of the opisthorchids *Clonorchis sinensis* and *Opisthorchis felineus,* those of several heterophyid species which are ingested by freshwater prosobranch snails, and those of dicrocoelids and brachylaemids, which are ingested by terrestrial snails. Inside the snail host's stomach the miracidium hatches about 1 hr after ingestion, penetrates the wall of the stomach and commences its metamorphosis, outside the digestive tube.

Schistosome eggs are nonoperculate and are fully mature before leaving the host. Maturation of the eggs takes place in the tissues of the host, and the maturation period, after being deposited by the female worm, varies according to the species. This period is 6 days for the eggs of *Schistosoma mansoni*, 12 days for the eggs of *S. japonicum* and 16 days for those of *Heterobilharzia americana*. However, it has been my experience with *Schistosome mansoni, S. haematobium, S. bovis, Schistosomatium douthitti*, and *Heterobilharzia americana*, that not all the eggs which pass in the host's excreta are fully developed, and the percentage of the undeveloped eggs varies from one species to the other. More *S. mansoni* and *S. bovis* eggs in the feces and *S. haematobium* in human urine or in feces of experimental animals pass partly developed than in the case of *H. americana* and *Schistosomatium douthitti*. Among the two latter species more partly developed eggs of *S. douthitti* are excreted than *H. americana*. Being nonoperculate, the emergence of the miracidium to the outside environment necessitates the splitting of the shell. Rupture of the egg capsule of *S. mansoni* may be vertical, oblique, or horizontal, and in the latter case the shell sometimes splits completely into two halves.

Factors believed to be stimulatory or inhibitory for the hatching of schistosome eggs have been studied by several investigators, but there is little information as to the interaction of these factors in triggering hatching or preventing hatching while the eggs are still in the host's tissues. Light is not regarded as a significant factor by some,[44,6,69] while others believe that it plays an important role.[93,163] It was shown that eggs of *S. japonicum* and *S. mansoni* are inhibited from hatching at 37°C,[154] whereas Bair and Etges[6] found that eggs of *S. mansoni* maintained at 39°C hatch as well as eggs at 26°C, and that thermal shock does not increase hatching. The reducing agent ascorbic acid and cysteine inhibited egg hatching of *S. mansoni*; however, the oxidized forms of these compounds inhibited hatching as well, indicating that the reducing conditions they provided were not responsible for the inhibition.[6] The ionic composition of the water was also considered. Salinity of the surrounding medium has always been regarded as an important factor in the hatching process. Hypotonicity of the medium is found by several investigators to cause schistosome eggs to hatch. Thus dilution of the urine or feces with water will bring about hatching of some of the eggs; further dilution causes more eggs to hatch.[96] On the other hand, a hypertonic medium will stop hatching. The majority of the eggs of *S. haematobium* remain unhatched and viable for several days if 0.75 to 0.9% NaCl solution is added to sterile urine. For *S. mansoni*, salinity as low as 0.05% has some inhibitory effect and hatching normally ceases in saline above 0.6%.[154] Eggs of *Orientobilharzia turkestanicum* in cattle feces, which were washed and sedimented in 0.75% NaCl solution, remained viable for 6 days when hand carried from the enzootic area in Iran.[111]

Like the operculate eggs of other trematodes, schistosome eggs are also exposed in some cases to unfavorable conditions outside the host. If the human or animal feces are not deposited directly in the water, many of the eggs die when the exposed feces dry up in the heat of the sun before the excreta is washed down into the water by rain. Results of experimental work on eggs of *S. mansoni* support these presumptions. When the viability of the eggs of *Heterobilharzia americana* in raccoon and dog feces was investigated under conditions simulating natural ones (on moist mud in a wooden box) all eggs were dead after 55 hr in direct sunlight at 19 to 27°C, and after 4 days in the shade at 19 to 23.5°C with a relative humidity of 73 to 90%. It was found that alternating rain and fair weather was favorable for the survival of the unhatched miracidia, as a few were recovered after 5 days in the sun and 9 days in the shade at 10 to 25.5°C. Placing infected dog or raccoon feces at the edge of water of a partly shaded lagoon resulted in a 9-day survival.[111] It is believed that the longevity of the eggs is enhanced in nature when droppings from raccoons fall on the moist mud and ooze at the edge of water habitats. Apparently the feces are leached in these situations and this produces favorable conditions for the survival of the eggs of this, or other, schistosomes.

Freezing or near-freezing temperatures, in contrast to merely cold temperatures, seem to be injurious to eggs of some mammalian and bird schistosomes. Of the schistosomes which infect man, *S. japonicum* is exposed to such adverse conditions, in some areas only of its geographical distribution. The writer found that freezing temperatures are damaging to the eggs of *H. americana* in raccoon or dog feces, as they die in 1 or 1½ days; at 8 to 10°C they survive for 4 days. Freezing temperatures are also detrimental to the eggs of *Schistosomatium douthitti* in mouse feces, and they die in 1 day, whereas at 5 and 9°C, about 60% survive for up to 10 days, 25% for 17 days, and they all die after 22 days.[109] Effects of some environmental factors on the eggs of *Schistosoma mansoni* are summarized in Table 1-2.

c. The Miracidium

The actively swimming miracidium is almost cylindrical, with a rounded anterior end and a tapering posterior end. Anteriorly there is a conical apical papilla, which can be everted or introverted. There are a number of epidermal plates to carry the cilia, which cover the entire body except the apical papilla, sensory papillae, and the excretory pore (Figure 1-16). The scanning electron microscope shows smaller cilia on the apical papilla. Most miracidia show this morphology; however, the cilia in the brachylaemid *Leucochloridiomorpha constantae* are carried on plumose rod-like appendages;[2] in the brachylaemid *Postharmostomum helicis* the cilia are in three groups, one of which is caudally located and is smaller (Figure 1-16), and in the miracidia of the bucephalid (gasterostome) trematodes the cilia are in groups at the end of stalks. The cilia of most miracidia are placed on a definite number of plates, and these plates are arranged in four or five transverse tiers around the body. There are usually six in the anterior tier. The arrangement and number of plates in each row are often characteristic for groups of trematodes. For the paramphistomes, for example, the pattern is 6,9,4,3; the echinostomes, 6,6,4,2; the fasciolids, 6,6,3,4,3; and the strigeids, 6,8,4,3. In all miracidia the edges of the plates do not meet, and there are spaces between the plates representing extensions from the underlying cytoplasmic layer. It should be noted that some Digenea whose eggs do not hatch until they are ingested by aquatic or by terrestrial snails also have miracidia covered with ciliated plates. In these, the cilia may be evenly distributed over the body in the same way as in the majority of trematodes or they may be in groups, as the example given above for *P. helicis*. There are two large lateral papillae situated on either side between the first and second tiers of epidermal plates; they are assumed to be sense organs since they are connected to the neural mass.

The fine structure of the tegument of the miracidium of *F. hepatica* has been described.[182] The ciliated epithelial plates overlie a thin discontinuous layer of cytoplasm, which is thickened between the plates to form prominent intercellular ridges. Beneath the thin layer are circular and longitudinal muscle fibers; these are followed by large vesiculated cells. The latter communicate with the thin cytoplasmic layer by means of narrow cytoplasmic connections running between adjacent muscle fibers. The apical papilla has an array of circular and longitudinal muscle fibers. (See Chapter 6.)

Additional information about the external morphology of the miracidium has been gathered from stereoscan studies. Such studies of the miracidium of *F. hepatica* (Figure 1-17) showed the apical papilla to be provided with a dorso-ventral furrow, multiciliated pits, and isolated sensory cilia. The narrow intercellular ridge is smooth, whereas the epithelial cells have small cytoplasmic knobs between the cilia.

The apical papilla of the miracidium of *S. mansoni* is covered with cytoplasmic folds which form an anastomosing network. Short cilia protrude above the folded surface. In the transmission electron microscope these cilia are revealed to be simple ciliated

Table 1-2

ENVIRONMENTAL FACTORS INFLUENCING SURVIVAL OF *SCHISTOSOMA MANSONI* LARVAL STAGES AND *BIOMPHALARIA* SNAILS AND COMPLETION OF THE LIFE CYCLE

Schistosoma mansoni

	Egg	Miracidium	Cercaria	*Biomphalaria* spp.
Temperature°C	Optimum for hatching, 23-28; continue to hatch for 2 to 3 days by further dilution of the medium. (Malek, 1950).[96] No hatching below 7; survive at 5.	At 4 lives for 24 hr; at 20—25 lives for about 9 hr; at 30—37 lives for 1½ hr; at 45 lives for ½ hr. Bennex & Deschiens (1963)[13] When *B. sudanica* was exposed at 9 to 39 percentage infection of snails increased linearly with temperature up to thermal death point of snail (Purnell, 1966).[134] No infection with miracidia at 10 and 13. Infection rates in snails rise with increasing temperature. Optimum 25—34 (Upatham 1973).[170]	Cercarial penetration optimum, 30 to 35; no penetration over 40 or below 10, (De Witt, 1965).[37] Optimum for penetration 27—28; good penetration over 16—35, and still occurs between 7 and 45 (Stirewalt and Frejeau, 1965).[158]	Can tolerate temperatures from little above freezing to 36. Optimum 22—26; 45 to 50 kill them. Low temperatures inhibit growth and breeding. Where warm season is long, breeding goes on all year. Strain and age differences influence optimum and maximum temperature (Malek, 1961)[103] High mortality of *B. glabrata* on St. Lucia at 37 and 40. (Upatham, 1973)[170] Optimum temperature for a rapid population expansion of *B. pfeifferi* in Tanzania appears to be close to 25; whereas at 19 survival is good, at 30 it is poor. Maximum temperature tolerated by *B. pfeifferi* appears to be 32 (Sturrock, 1966).[162]
Gravity		Negatively geotropic	The majority are near surface of water, but some are halfway to bottom.	
Rainfall	Favors survival of eggs in the stools, and increases the chance of the eggs reaching the water and miracidia reaching and infecting snails (Malek, 1961.)[103] Alternating fair and rainy weather is most favorable to survival in stools, but prolonged dry weather is as unfavorable as exposure to direct sunlight (Maldonado et al., 1949).[95]			Favors survival and breeding of snails, and in some situations disperses them over a wide area.

	Positively phototropic	Positively phototropic	Positively phototropic
Light	Although hatching of eggs and emergence of cercariae from snail can take place in total darkness at a very slow rate, the rate increases considerably in natural or artificial light (Malek, 1961).[154] Light stimulates egg hatching (Standen, 1951).[154] Hatching of eggs is unaffected by light or dark conditions (Bair and Etges. 1973).[6] Under strong illumination, the majority of eggs of mouse and human origin hatch within 1 and 2 hr, respectively; under room light, the majority hatch within 2 and 24 hr, respectively. Although light stimulates hatching; it is not essential to hatching, 39% of eggs of human origin hatch when kept in darkness for a week, but at a slow rate (Maldonado et al. 1950a).[93]	The usual pattern of cercarial emergence, i.e., in morning and afternoon can be reversed by reversing the light cycle (Luttermoser, 1955).[92]	Can be raised in total darkness, but population remains small. However, light affects the snail directly (reproductive organs) and indirectly (microflora, the snail's food; and aquatic weeds); increases decomposition rate of animal and plant remains and thus a healthier habitat for the snails.
Current velocity	Water velocity may enhance the effective scanning capacity of miracidia. High infection rates are obtained at water velocities of between 0.5 and 3.5 ft/sec (Webbe, 1966).[175]	Velocity is important in mixing cercariae in the water and in effecting contact with host, for at velocities below an optimum value fewer cercariae appear to make contact, and at higher velocities some of those that do are swept off (Webbe, 1966).[176]	A flow of 33 cm/sec causes immobilization, and 65 cm/sec causes dislodgement of *B. glabrata* attached to a smooth surface (Jobin and Ippen, 1964).[67]
Turbidity	0—100 ppm turbidity result in sharp decrease in snail infection rate, but decrease slowly from 100—500 ppm (Upatham, 1972).[169]		No adverse effect of turbidity on snails (*B. glabrata*, St. Lucia) within 1 hr (Upatham, 1972).[169] Silt turbidity affects snails through inhibitions of growth of aquatic weeds and microflora by obstructing light penetration (Malek, 1961).[103]
Gamma-radiation	Eggs irradiated with 10-200 Krad or with 1.0—50 miracidia hatch, penetrate snails, but no development occurs. Eggs irradiated with only 0.5 Krad. miracidia hatch and a few develop in snails and produce cercariae (Antunes et al., 1971).[4]		

Table 1-2 (Continued)
ENVIRONMENTAL FACTORS INFLUENCING SURVIVAL OF *SCHISTOSOMA MANSONI* LARVAL STAGES AND
BIOMPHALARIA SNAILS AND COMPLETION OF THE LIFE CYCLE

	Egg	*Schistosoma mansoni* Miracidium	Cercaria	*Biomphalaria* spp.
Centrifugation		A maximum speed of 18,000 rpm for 5 min has no effect on the vitality or penetration capacity of miracidia (Holliman et al., 1972).[65]		
Substratum	Probably influences the schistosome stages through influencing water quality.			Composition affects snails through affecting water quality. Firm mud substratum rich in organic matter, and also limestone favor the snails existence. Granite rocks and sandstone unfavorable. Gradient of substratum is also important. In Puerto Rico populations of *B. glabrata* are not maintained in reaches of streams steeper than 20 m fall per 1000 m of length (Harry and Cumbie, 1956).[60]
pH	Eggs hatch uniformly within the range of 5—9. Hatchability not affected through a rather wide range of pH (5.16—8.35). (Maldonado et al., 1950).[94]	5 and 10.5 lethal; 5.5 or 10, half die; 7.5—8.5 optimum (Bennex and Deschiens, 1963)[13] pH 4; miracidia active for 1—2 min; pH 5; active for 5—10 min; pH 6, active for 3 hr; pH 7—9: active for 5—6 hr (Upatham, 1972).[169] Highly sensitive to changes in pH. A medium at 5—6 curtails the life of the miracidium; 7—8 is fairly satisfactory; 8—9 allow for maximum longevity (Maldonado et al., 1950).[94]	The range of 5 to 10 is tolerated by the cercariae and their emergence takes place uniformly throughout this range.	*B. glabrata* of St. Lucia tolerate pH 5—10 (Upatham, 1972).[169] The range in the majority of world habitats is 6—8, but varies from 4.8 to 9.8 (Malek, 1961).[103]

Dissolved oxygen	Important for survival of egg, miracidium, and cercariae. Temperatures over 40°C may kill cercariae or reduce their span of life indirectly through reduction of O_2 content of the water, rather than direct effect of heat.	Emergence of cercariae from infected snails is greatly reduced under anaerobic conditions and most freshly shed cercariae die rapidly when deprived of O_2. Cercariae inside the snail, however, can survive without O_2 (Olivier, von Brand, and Mehlman, 1953).[124]	Habitats of *B. rüppellii* in the Sudan contain 4.7 to 7 ppm O_2 in certain particular foci may reach 10 ppm (Malek, 1958).[102] There is a wide range of O_2 tension through which *B. glabrata* maintains a steady rate of O_2 consumption (von Brand, Nolan, and Mann, 1948).[174] Although infected snails withstand 6 hr under anaerobic conditions most of them die when kept anaerobically for 16 hr (Olivier, von Brand, and Mehlman, 1953).[124]
Chlorine	No unhatched eggs in sewage effluent with a BOD of 20—48 mg/l survive an initial dosage of chlorine of 2.4 mg/l with a residual level after 15 min of 1.5 mg/l or higher (Rowan, 1964).[143]	Chlorination for 30 min contact with a residual of 0.5 ppm of chlorine kill all cercariae (Coles and Mann, 1971).[29] With Na hypochlorite solution, the cercaricidal activity of chlorine is affected by pH than by temperature (Frick and Hilleyer, 1966).[48]	
Cations		Na and K stimulate cercarial proteolytic activity with maximum stimulation at 20—40 mM concentrations; Mg and Ca stimulate activity at low concentrations (between 0 and 10 mM) but inhibited activity at higher concentrations; Zn, Cu, Fe are inhibitory even at very low concentrations (Dresden and Edin, 1974).[40] Calcium and/or Mg ions essential for successful invasion of mammalian hosts (Lewert et al., 1966).[86]	*B. pfeifferi* egg-laying rates much lower in Ca/Mg ratio of 7:19 than in 4:12. Moreover, snails absent near Salisbury, Rhodesia, in streams high in dissolved Mg but low in dissolved Ca (Harrison et al., 1966).[58]

Table 1-2 (Continued)
ENVIRONMENTAL FACTORS INFLUENCING SURVIVAL OF *SCHISTOSOMA MANSONI* LARVAL STAGES AND
BIOMPHALARIA SNAILS AND COMPLETION OF THE LIFE CYCLE

| | *Schistosoma mansoni* | | | *Biomphalaria* spp. |
	Egg	Miracidium	Cercaria	
Cations			Cercarial penetration in dechlorinated tap water (92%), creek water (86%), single-distilled water 76%); triple distilled water (68%). Mineral reconstitution of distilled water increased penetration (Stirewalt and Fregeau, 1965).[158]	*B. rüppellii* present in waters with Ca/Mg ratio of 30/10; 25/15, but still absent in waters with 25/10 in the Sudan (Malek, 1958).[102] Growth and low mortality of *B. pfeifferi* obtained with Na/Ca ratio of 1:0 (Frank, 1963).[45]
Carbonate hardness (indicates normal carbonates and bicarbonates)				Distribution of *B. pfeifferi* in Rhodesia is determined by calcium bicarbonate content of the water. The snail is found in medium and hard waters, and the snail is less tolerant of soft water. "Soft": less than 5 mg/l Ca and less than 20 mg/1 bicarbonate as CaCO₃ "Medium": 5 to 40 mg/1 Ca and 20 to 200 mg/1 bicarbonate as CaCO₃, "Hard": above 40 mg/1 Ca and above 200mg/1 bicarbonate as CaCO₃ (Williams, 1970).[179] The intrinsic rate of natural increase for *B. pfeifferi* showed higher increase rates of experimental populations at medium concentrations of bicarbonates and calcium ions than at the upper and especially the lower extremes (Williams, 1970; Harrison et al. 1970).[59,180] *B. pfeifferi* in Kenya is found in waters with a carbonate hardness range of 8—200 ppm (Van Someren, 1946).[171]

	B. rüppellii in the Sudan found in waters with a temporary hardness (bicarbonates) of 100—180 ppm (Malek, 1958).[102] Significant differences in O_2 uptake rates due to different calcium bicarbonate content of water. Highest uptake rates by *B. pfeifferi* in Rhodesia at bicarbonate content of 35 mg/ℓ (Harrison, 1968).[57]	Maximum tolerated concentration of chlorides as NaCl 3641 ppm; lethal concentration 6000 ppm (Deschiens, 1954).[36] The maximum chloride content in planorbid habitats in Brazil is 2562 ppm (De Andrade, 1954).[3] 4800 ppm snails become nonmotile; above 4800 ppm snails die 16 hr (*B. glabrata* St. Lucia) (Upatham, 1972).[169]	
Salinity (NaCl)	Salinity as low as 0.05% has some inhibitory effect on hatching of eggs and hatching normally ceases in saline above 0.6% (Standen, 1951).[154]	In 3% saline similar to control; in 5% saline well tolerated; 7% saline kill 3/4 in ½ hr. (Bennex and Deschiens, 1963).[13] 0.5 ppm Na Cl, 78.7 infection of *B. glabrata*; above 4200 ppm, 2.1%; above 4200 ppm no infection; above 1200 ppm affect activity of miracidium (Upatham, 1972).[169]	Remain alive in a 1% salt solution but its span of life is reduced.
Salinity (total salts)	In the laboratory, transmission of *S. mansoni* to and from *B. glabrata* can occur in artificial sea water at or below a concentration of 12.5%. Miracidia can also emerge and infect snails in 25% sea water, but this concentration is inimical to the survival of the snails or their eggs. The parasite thus appears to be better adapted to brackish water than is the snail intermediate host. Artificial sea water contains Na Cl, KCl, Ca Cl2, Mg Cl2, Mg SO4, NaHCO3 (Chernin and Bower, 1971).[26]		

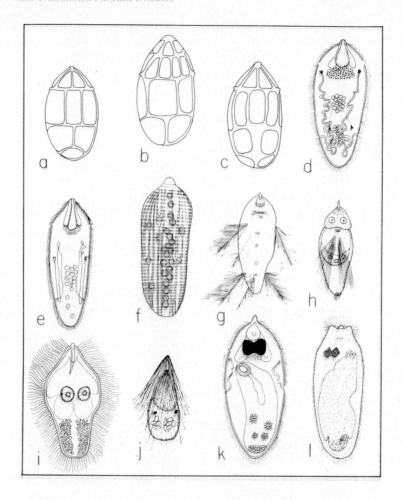

FIGURE 1-16.   Miracidia of some Digenea. Some original, others adapted after various authors. (a) Epidermal plates of *Paragonimus kellicotti* miracidium, (b) Epidermal plates of the echinostome *Paryphostomum segregatum* miracidium, (c) Epidermal plates of the schistosome *Heterobilharzia americana* miracidium, (d) *Schistosoma mansoni*, (e) The amphistome, *Gastrodiscus aegyptiacus*, (f) A diplostomatid miracidium showing the musculature, (g) The bracylaemid, *Leucochloridiomorpha constantiae*, (h) The brachylaemid, *Posthrarmostomum helicis*, (i) The dicrocoelid, *Brachylecithum orfi*, (j) The dicrocoelid, *Platynosomum fastosum*, (k) The philophthalmid, *Parorchis avitus*, (l) The cyclocoelid, *Cyclocoelum* sp.

nerve endings which penetrate the cytoplasmic surface; they are found in the posterior part of the papilla. Shorter cilia are found about the middle, but none are present at the apex. The lateral papillae of the miracidium of *S. mansoni* appear by stereoscan studies as bulbous projections on either side between the first and second tiers of epidermal cells. There is a ciliated pit nerve ending close to each lateral papilla, a few ciliated pits occur between the cells in the first tier, and up to 12 ciliated pits with long cilia are found between the second and third tiers. The miracidium usually remains for about 10 min with the anterior end embedded as far as the lateral papillae. During this period, the secretory products from one or both gland types may help to loosen and digest the host tissue. The subsequent rapid penetration of the remainder of the body suggests that a loosening or digesting of the host tissue must have taken place. Figure 1-18 shows features of *S. haematobium* and *S. japonicum* miracidia by SEM.

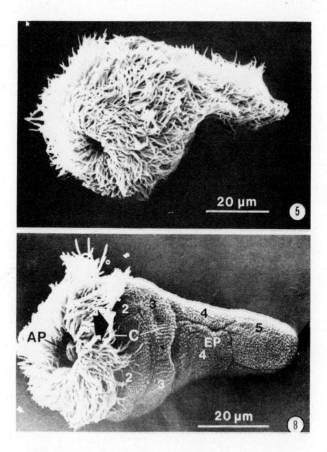

FIGURE 1-17. Stereoscan electron microscope photographs of miracidium of *Fasciola hepatica*. (5) Miracidium with retracted apical papilla, (8) Miracidium that had been vibrated in an ultrasonic cleaner. Only the cilia of the anterior tier are retained. The dorsal median line follows the lower profile of the miracidium and arrows show the right lateral line. (Photograph after Koie, M., Christensen, N., and Nansen, P., *Z. Parasitenkd.*, 59, 82, 1976. By courtesy of Springer-Verlag, New York.)

Inside the anterior part of the miracidium the following structures are observed by light microscopy:

1. A granular mass, the apical gland, referred to also in the literature as a gut, or a primitive gut: it is not used for feeding, however. Its secretion is poured out at the apical papilla, and probably has to do with adhesion to, or penetration into, the snail host. Thus it functions in the lyses of the snail tissues.

2. A pair of pear-shaped penetration glands, one on each side of the apical gland; these are also found in the literature as salivary or pharyngeal glands.

3. A pair of pigmented eyespots in most miracidia, each partially enclosing a spherical lens and having sensory cells as revealed by electron microscopy. Like several other miracidia, the miracidium of *S. mansoni* does not have eyespots but possesses certain cells believed to be photoreceptors.

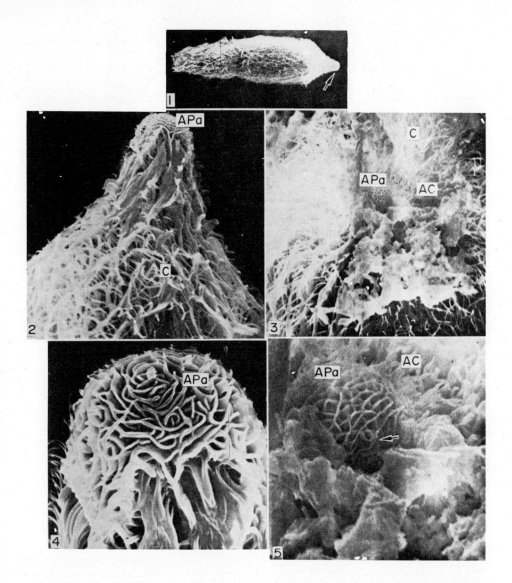

FIGURE 1-18.   Scanning electron micrographs of the miracidia of *Schistosoma haematobium* and *S. japonicum*. (1) Miracidium of *S. haematobium*; arrow denotes location of apical papilla. (2) Apical papilla (A pa) of *S. haematobium* miracidium surrounded by locomotor cilia (c). (3) Apical papilla (A Pa) of *S. japonicum* miracidium. (4) Apical papilla (A Pa) of *S. haematobium* miracidium showing a mini-sucker pad composed of tiny sucker-like cups. (5) Apical papilla (A Pa) of *S. japonicum* miracidium showing a mini-sucker pad composed of tiny sucker-like cups, with a secretory pore (arrow) surrounded by tactile apical cilia (AC). (Photograph after Lo Verde, P. T., *Int. J. Parasitol.*, 5, 96, 1975. By courtesy of Pergamon Press, New York.)

4.   A quadrangular cerebral mass, the "brain", lying beneath the eyespots, and giving out processes for innervation of the eyespots, the apical papilla, the lateral papillae, and the posterior part of the miracidium: electron microscope studies have confirmed the nervous nature of this mass, and that it is similar to nervous elements of the cercaria of *F. hepatica*.

In the remaining part of the miracidium there are a number of large hyaline cells and a number of germ cells, some of which are in groups to form germ balls. There is also one pair of flame cells in lower forms such as the fasciolids, paramphistomes, and echinostomes, and two pairs of flame cells in higher trematodes such as strigeids and schistosomes. The flame cells are connected to excretory tubules which open at two excretory pores.

The miracidia of the philophthalmids *Parorchis* and *Philophthalmus*, the paramphistome *Stichorchis*, and some cyclocoelids show precocious development of the germ cells, inasmuch as each contains a well-developed redia (Figure 1-16), a feature which would naturally shorten the intramolluscan incubation period.

1. Histochemistry and Physiology

Histochemical tests show that the germ cells of the free miracidium have higher nucleic acid and acid-basic protein content than other embryonic cells, and thus resemble the adult reproductive cells. In addition the germ cells do not contain storage substances such as fat or glycogen.

Glycogen is stored in some cells of the miracidium (such as the hyaline cells, the lateral penetration glands, and the apical gland of some species), and, together with fat, has been demonstrated histochemically. No doubt glycogen is necessary as an energy source for the actively swimming miracidium. Small molecules such as glucose can be taken up by the miracidium inside intact eggs. Autoradiography of miracidia inside eggs of *Schistosoma mansoni* incubated in a solution containing isotopically labeled glucose has demonstrated the passage of glucose through the sclerotin shell and its uptake by the miracidium within.[88] Developing schistosome eggs inside the female worm have also been labeled. It appears then that miracidia inside eggs in host tissues utilize compounds such as glucose and amino acids from the host for their maturation and extended survival. A high level of glucose utilization is characteristic of other stages of the schistosomes and other trematodes. The process of uptake of glucose, amino acids, calcium, tetracycline, and other substances takes place through the submicroscopic pores in the shell of the schistosome egg. Also, the circumoval precipitates which occur in vitro in the presence of immune serum, and the in vivo occurrence of the radiating precipitates of the Hoeppli phenomenon are presumed to result from miracidial antigens (secretions and excretions) passing outward through the pores in the shell.

In their free-living existence in the water the miracidia utilize endogenous sources and the rate at which this is done is governed by environmental conditions, mainly temperature.

2. Longevity and Behavior

Miracidia of various species are usually very active and vigorous during the first 6 or 8 hr after hatching; then they slow down and in general succumb sometime during the first day. In their free-swimming existence in the water the miracidium is exposed to certain physical and chemical factors in the environment which affect its longevity, behavior, and host-finding ability. Some of these factors are summarized in Table 1-2 for the miracidium of *Schistosoma mansoni* and, naturally with some differences, apply to other miracidial species.

In addition to its endogenous sources, the miracidium probably needs supplement from the environment. Addition of glucose to the water was found to lengthen the life span of the miracidia of *Fasciola hepatica* and the paramphistome *Cotylophoron cotylophorum*. Longevity of some schistosome miracidia has been studied, especially those of some species infecting humans. About half the miracidia of *S. mansoni* survive for 12 hr at a temperature of 22 to 24°C after which time mortality increases until they all succumb after 20 or 22 hr. The writer[109] used different types of water in determining the longevity of the miracidia of *S. douthitti*. In dechlorinated tap water (pH about 8.0) 7 out of 100 miracidia died within the first 5 hr, about 50% after 12 hr, and all were dead after 25 hr at a temperature of 22 to 24°C. Many of the miracidia were sluggish after about 16 to 18 hr. The 50% mortality of the miracidia in dechlorinated tap water and in pond water was slightly lower than in distilled water and a well water (commercially known as ozone water), where in the latter it was 11 hr.

The ionic composition of the water not only affects the survival of the miracidia but also their movement in certain directions. A recent study indicated that the miracidia of *S. mansoni* are sensitive to the Ca/Mg ratio in the water and swim toward, or aggregate in, areas with lower Ca/Mg ratios; in effect they are attracted by the fall in Ca ions as compared to Mg ions.

Relative to the effect of light on miracidia it has been shown that miracidia of *S. mansoni* exhibit a peaked response in the red (650 mm) region of the spectrum.[190] Thus the miracidia of this schistosome seem to be able to respond to wavelengths of monochromatic light which make up the greatest percentage in composition of spectral intensity in clear and muddy fresh water.

Various behavioral patterns have been described for the miracidia of the Digenea. The first phase of the host-finding pattern is the location of the host environment by reaction to a physical stimulus; in the case of some trematodes, including some *Schistosoma* species, there is a positive phototropism and a negative geotropism, which bring the miracidia into the surface layers of the water, where their snail hosts are most concentrated. Certain strains of *S. haematobium*, however, have been found not to exhibit negative geotropism, and this behavior is probably in the miracidium's favor, since the bulinid snail hosts are often found closer to the bottom of several habitats. Moreover, the miracidia of *S. japonicum* follow the responses of their oncomelanid snail hosts to stimuli such as temperature and light. When the snails are sometimes close to the surface because temperature and light conditions are such as to bring them there, the miracidia will rise; otherwise the miracidia will remain near the bottom if the snails are there. When the host environment is reached, a period of random movement follows, which continues either until the death of the miracidium or until it comes within the orbit of chemical attraction of a snail. Snails of any species are known to emit a water-soluble substance which alters the swimming behavior of the miracidium. Once under the influence of this chemical stimulus, the movements of the miracidium do not become definitely directional; the miracidium still continues to exhibit random movements, but it will tend to be held within the orbit of the chemical stimulus. Entry into the snail does not usually occur on first contact; the miracidium usually moves away but returns for further attempts at contact.

The excited swimming behavior of the miracidium and its attempts at penetration are not only evoked when the snail is present, but this behavior is also observed when the miracidia are placed in water where the snails had been maintained for a few hours. This is what has been termed "Snail Conditioned Water" (SCW). This water can be prepared when the snails are present, and can then be stored at 5°C, for further use and experimentation. The stimulant in the SCW is not known with certainty; it is referred to as "miraxone", a term coined by Chernin to refer to this chemical substance(s) emitted by the snail which alter the swimming behavior of the miracidium. Indications as to the chemical composition of the stimulant were made in a recent study.[115] The stimulant in the water was unaffected by trypsin but was protease-sensitive, suggesting its possible identity as a peptide. Whatever the stimulus is, it has been found that it is definitely nonspecific: it has definitely been demonstrated that the miracidium penetrates susceptible and nonsusceptible snails, as well as other invertebrate organisms in the water. Moreover the behavioral responses of free-swimming miracidia to SCW are also nonspecific as to the source of the stimulant and the species of miracidium responding.

d. The Sporocyst (Mother and Daughter)
1. Miracidial Metamorphosis

The stage that penetrates the snail is considered by Dawes,[31] and a few others, to be the sporocyst rather than the miracidium. Dawes' opinion was based on his work with *Fasciola hepatica* where he expressed the opinion that the shedding of the ciliated epidermis by the miracidium, thus transforming it to a sporocyst, is a prerequisite for successful penetration of the snail. Earlier, the shedding of the ciliated epidermal cells had been reported by other workers for *Fasciolopsis buski* in 1925, *Fascioloides magna* in 1955, and later in 1968 for *Fasciola gigantica*. On the other hand, workers using the light microscope in their studies on the miracidia of *Schistosoma bovis*, *Schistosoma mansoni*, *Echinostoma audyi*, and *E. malayanum* observed that the miracidia do not shed their plates until after penetration. However, by the use of the electron microscope and by stereoscan studies it was shown at least for *S. mansoni*, that the miracidia when placed in a culture medium or in snail hemolymph readily shed the apical ciliated part of the epithelial cells.

The penetration of *F. hepatica* miracidium into its snail host *Lymnaea truncatula* and its transformation into a sporocyst have been reported upon, based on both electron microscope and stereoscan studies (Figure 1-19).[183,76] The miracidial body performs extensive contraction and relaxation following attachment to the snail. This coincides with the start of secretion by the apical gland and accessory gland cells. The snail's columnar epithelium is rapidly cytolyzed, and after about 5 min most cilia are broken off near the cell surface. The miracidium remains for about 10 to 15 min embedded as far as the intercellular ridge receptors (lateral papillae and sheathed ciliated nerve endings). Simultaneously with further penetration into the snail tissues, the epithelial cells of the four posterior tiers loosen, form globules, and fall off. This shedding apparently removed a protective barrier against osmosis, which is probably the acid mucopolysaccharide present in the epithelial cells. The miracidium metamorphoses into the sporocyst as it penetrates the snail, by forming a new body surface in not more than 2½ hr after attachment; the new body surface is contributed by the vesiculated cells which lie beneath the musculature of the body wall. The new surface is smooth and in cytoplasmic continuity with the intercellular ridge and the apical papilla, but these connections, and also those with underlying cells, become broken and the cytoplasm is then underlain by a thin fibrous basal lamina. In the first 24 hr after penetration the surface of this syncytium becomes thrown into folds, and metamorphosis into the sporocyst is completed after a few days when the tegument of the sporocyst becomes provided with microvilli, and the apical papilla and sensory structures are lost.

The miracidium-sporocyst transition has been elaborated for *Schistosoma mansoni* under aseptic culture conditions.[10] The transformation of the miracidium of *S. mansoni* to mother sporocyst has also been investigated by the use of the stereoscan electron microscope.[77] Miracidia of *S. mansoni* placed in hemolymph from the snail *Planorbarius corneus* (a nonintermediate host for this schistosome) cast off the apical ciliated part of the epithelial cells, and large scars appeared where the ciliated plates had been. Later, the syncytial intercellular ridge dispersed throughout the surface of the mother sporocyst, and small cytoplasmic knobs appeared on the surface.

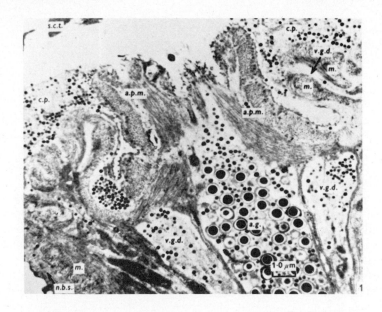

FIGURE 1-19.   Penetration of the miracidium of *Fasciola hepatica* into the snail *Lymnaea truncatula*. High magnification of anterior of miracidium embedded in snail. The apical papilla can be clearly seen with large vesicles in the lumen of the apical gland. A duct, carrying vesiculated cell material, can be seen passing through the muscle layers. (After Wilson R. A., Pullin, R., and Denison, J., An investigation of the mechanism of infection by digenetic trematodes: the penetration of the miracidium of *Fasciola hepatica* into its snail host *Lymnaea truncatula, Parasitology,* 63, 491, 1971. With permission of Cambridge University Press.)

2. Morphology

   (Figure 1-20) After penetration, the metamorphosing miracidium settles near the site of entry, and the developing mother sporocyst is found in the head-foot region, in the tentacles, near the buccal mass, and esophagus (*Schistosoma*), the mantle collar or edge, the wall of the anterior part of the pulmonary sac, or near the buccal mass (*Fasciola*). These and other trematode species may have their sporocyst (or mother sporocyst) located in various hemolymph sinuses adjacent to the digestive tract and sometimes in the kidney. The sporocyst varies slightly in shape, size, and structure among various trematodes, but in general is an oval or elongated structure with one end rounded and the other conical. The conical end might contain a birth pore. The sporocyst is enclosed in a thin wall, and the body cavity usually has a well-developed excretory system with a larger number of flame cells (Figure 1-21) than the miracidium. The basic pattern of development which involves separate somatic and germinal lineages continues in the mother sporocyst and later in successive larval stages. The germ cells in the body cavity divide and subdivide and form germ balls which become differentiated later into the successive stage, sporocysts, or rediae depending on the trematode species. Each germ ball is enclosed in a thin membrane containing a few flattened nuclei, and from this outer membrane the wall of the successive stage is formed. The wall consists of an outer tegument which is cytoplasmic in nature, without spines, and then there is a thin layer of circular and longitudinal muscles, underneath which is a cellular epithelium lining the lumen. In some trematodes, for example *Plagiorchis muris*, the mother sporocyst becomes divided into compartments by ingrowths from the body wall. These compartments contain germ cells and, later, embryos, which also become enclosed by an ingrowth of the septal wall.

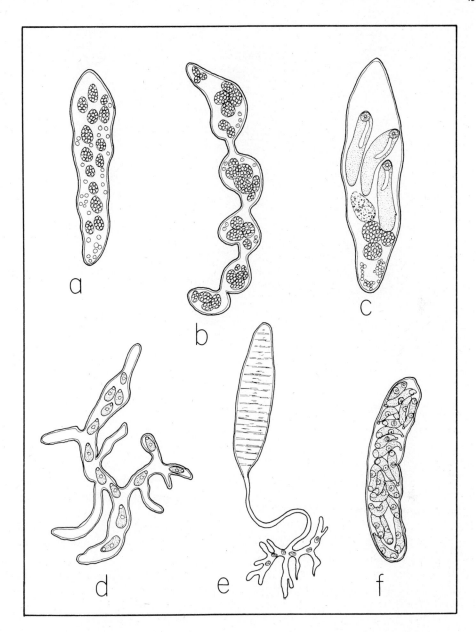

FIGURE 1-20.   Sporocysts of some Digenea. (a) Mother sporocyst of *Schistosoma mansoni*, about 7 days old containing germ cells and germ balls, (b) Mother sporocyst of *Heterobilharzia americana*, about 7 days old, (c) Sporocyst of *Fasciola hepatica* about 14 days old, containing germ cells, embryos and rediae, (d) Branched brachylaemid (*Postharmostomum helicis*) sporocyst containing cercariae, (e) Characteristic sausage-shaped sporocyst of *Leucochloridium* sp., (f) Sporocyst of a dicrocoelid (*Platynosomum fastosum*), containing numerous cercariae.

The intramolluscan larval development has been studied in various groups of trematodes, such as echinostomes, psilostomes, fasciolids, plagiorchiids, strigeids, and schistosomes. There were several publications on this subject in the 1940s and 1950s by Cort and co-workers, with a review article in 1954,[30] in which polyembryony or germ-cell lineage has been described for species representing these groups. Other information on the early development and the systematics of echinostomes, psilostomes,

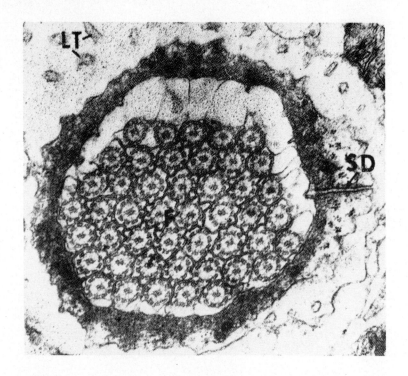

FIGURE 1-21.   Transverse section through the distal part of the barrel of a pro-
tonephridium of a cerceria inside a daughter sporocyst. SD, septate desmosome; F,
cilia of the flame cell; and LT, leptotriches. (After Meuleman, E. A., *Neth. J. Zool.*,
20, 355, 1972.)

and amphistomes has been contributed by Beaver, on the plagiorchiids by McMullen,
on the strigeids and heterophyids by Pearson, and on the heterophyids by Kuntz and
by Martin (the references will be included in other chapters of this book).

The daughter sporocysts which develop inside the mother sporocyst in some members of the xiphidiocercaria group (microphalids, gorgoderids, plagiorchiids, lecithodendriids, dicrocoelids), and in some members of the forked-tail cercaria group (strigeids and schistosomes), are more differentiated than the mother sporocyst. They are
elongate, hollow sacs with a wall, and a cavity containing germ cells and excretory
structures, and many species possess a birth pore. The sporocysts of various species
are developed from about the 12th to the 20th day after penetration of the miracidium.
The daughter sporocysts, once they are outside the mother sporocyst, do not remain
in that site (the proximal or anterior part of the snail) but migrate distad by their
wiggling movements, but mainly by the hemolymph of the snail in its normal flow to
distal (posterior or upper) organs of the snail, especially the digestive gland and ovotestis (or the ovary or testis in prosobranch snails).[101] Some, however, fail to make the
trip to the digestive gland and settle in connective tissue areas of various organs. This
is also the case in heavy infections, in which the daughter sporocysts are found in the
kidney, the rectum, albumen gland, and in tissues associated with the genital tracts.

In the digestive gland and other organs the daughter sporocysts increase in length,
become convoluted, and are intimately associated with the snail's tissues. The germ
cells inside continue to divide and subdivide to form germ balls, and later cercariae.
After their development, and maturation and exit from the sporocyst (through a birth
pore when present), they again, like the daughter sporocysts, depend on the circulating
hemolymph but in the opposite direction, thus moving anteriad. There is a very thin
membranous area near the mantle collar which they pierce and exit from the snail.

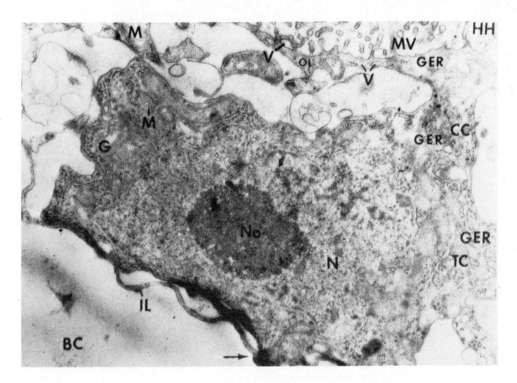

FIGURE 1-22. Electron micrograph of the daughter sporocyst of *Schistosoma mansoni*, 18 days after the infection. The thin outer layer (OL) is joined by a cytoplasmic connection (CC) to a tegumental cell (TC). The outer layer bears irregular microvilli (MV) on the side facing the host hemolymph (HH). The projection (arrow) at the basal cell surface of the tegumental cell forms part of the inner layer (IL) between the tegument and the brood chamber (BC). G, Golgi apparatus; GER, granular endoplasmic reticulum; M, mitochondrion; N, nucleus; No, nucleolus; and V, vesicle. (After Meuleman, E. A., *Neth. J. Zool.*, 20, 355, 1972.)

The length of the incubation period in the snail, i.e., to the cercarial release, for any given species, varies depending on the trematode species, on environmental conditions, on the nutritional state and other conditions in the snail host, and on the life span of the snail. The ultimate number and size of sporocysts produced, and the number of resulting cercariae, is more a function of the size of the molluscan host rather than of the reproductive capacity of the germinal elements.

The ultrastructure of the daughter sporocyst of certain trematode species has been described and the observations of the authors seem to be consistent. The tegument of the daughter sporocysts of *S. mansoni* (Figure 1-22) has the same basic architecture as that of the cercariae, schistosomula, and adult schistosomes. It consists of an outer syncytial anucleate cytoplasmic layer which is joined by cytoplasmic connections to nucleated cell bodies (tegumental cells), situated a little deeper in the body. The cytoplasm of the outer layer is directly continuous with that of the tegumental cells by means of the cytoplasmic connections.[119] The tegument rests on a thin basal lamina. The exposed surface of the outer layer is greatly increased by the presence of numerous closely packed, irregularly shaped, and sometimes branched microvilli. In addition to the microvilli, the outer layer has in places very long processes, which penetrate deeply into the connective tissue sheath surrounding the digestive gland. These processes are possibly sites of attachment. At the time they leave the mother sporocyst, the daughter sporocysts have some spines, about 1μ in length, on the anterior half of the body, and they are numerous on the highly muscular anterior tip of the sporocyst. The musculature of the daughter sporocyst is relatively poorly developed; the musculature is in the form of thin smooth muscle fibers which are loosely arranged in two layers, an outer circular and an inner longitudinal.

3. Physiology and Histochemistry

Glycogen has been reported by histochemical tests in the sporocyst wall and cercarial bodies of several trematode species. Increase in the glycogen content of the sporocyst and the cercariae is usually associated with a decrease in the glycogen content in the digestive gland and other tissues of the snail. Moreover, glucose was demonstrated in the sporocyst wall of *Glypthelmins pennsylvaniensis* as well as in host tissues.[25] This suggests that glucose, obtained as such or derived from host glycogen, passes through the sporocyst wall and becomes stored as glycogen in the cercarial wall. This transport of nutrients is supported by Erasmus[42] findings of alkaline phosphatase activity in the sporocyst wall of a strigeid, and by Dusanic[41] in a schistosome (*S. mansoni*). Fatty acids have been observed to be adhering to and on the interior of the sporocyst wall of *G. pennsylvaniensis*. As to amino acid and protein metabolism, bound and free amino acids detected in the sporocyst of *Glypthelmins quieta* corresponded with those in the digestive gland of the snail host. Moreover, it has been demonstrated in the case of several species of trematodes, including the schistosomes, that the snail (hemolymph) protein level of parasitized snails is less than that in nonparasitized snails, pointing to the utilization of amino acids in the hemolymph by the larval trematodes.

e. The Redia

In a large number of digenetic trematodes a redial stage develops inside the sporocyst; these include among others, the fasciolids, the echinostomes, the psilostomes, the paramphistomatids, the opisthorchids, the heterophyids, the allcreadiids, the troglotrematids, and the clinostomes. In the philophthalmids, cyclocoelids and the paramphistome *Stichorchis* there is also a redial stage but it develops inside the miracidium. There may be one, or two, generations of rediae, depending on the trematode species. Moreover, for the same species, such as fasciolids, and some paramphistomatids, a third generation may be added under certain environmental influences, such as temperature. Accordingly, some redial generations may have inside them a new generation of rediae, in addition to cercariae, as is the case with *Fasciola hepatica, F. gigantica, Carmyerius exporus*, and certain other paramphistomatids.

The body of the redia is elongate, and unlike the sporocyst, has an alimentary system, starting with a mouth which leads to a prominent muscular pharynx and an intestinal sac which varies in length, in some species reaching the posterior extremity of the redia. There is a ridge-like collar at the level of the pharynx in some species; there is usually a birth pore near the collar, and in some trematodes there exist also a pair of lappets or appendages, the procrusculi, about one third of the way from the posterior end (Figure 1-23). The presence of the collar and appendages on the redia e characteristic of some groups and is uually of taxonomic value. Thus, collared rediae with stumpy appendages are present in the life cycle of members of the Superfamily Echinostomatoidea (Families Echinostomatidae, Fasciolidae, Psilostomatidae, Philopthalmidae, Cathaemasiidae, and others), whereas the collar and usually the appendages are lacking in the redia of members of the Superfamily Paramphistomatoidea (comprising the paramphistomes and related forms).

There is a well-developed excretory system (Figure 1-24) terminating in a pair of excretory pores. The wall of the redia is more or less similar to that of the sporocyst. The body surface, such as that of the redia of *Cryptocotyle lingua* is characterized by the presence of numerous broad, cytoplasmic projections, the integumental folds, or flaps.

The redia is more active than the sporocyst, and there is evidence that it feeds through its mouth and simple digestive system as well as through its tegument. Parts of the digestive system show various types of enzymatic activity, including alkaline phosphatase, acid phosphatase and nonspecific esterase activity. Reactivity for acid

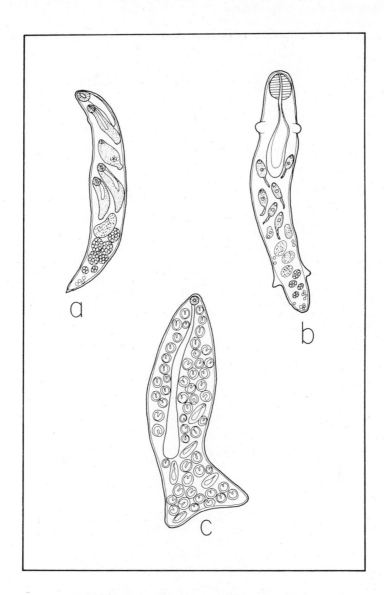

FIGURE 1-23.   Rediae of some Digenea. (a) First generation redia of an amphistome, *Paramphistomum microbothrium.* Note second generation rediae and cercaria in the same redia. (b) Mature echinostome redia containing germ balls, cercarial embryos and cercariae. Note collar in anterior part, and appendages in posterior part. (c) Mature cyclocoelid redia, containing cercariae and encysted metacercariae.

phosphatase has been found in the redial cecal contents of *Echinoparyphium* sp., and in the cecal epithelium, cecal contents, tegument, and pharynx of *Echinoparyphium recurvatum.* Morphological observations on the rediae of *Cryptocotyle lingua* have shown the presence of cytoplasmic folds and pinocytotic-like vesicles, in both the cecal cells and tegumental surface, which may have a function in nutritive phagocytosis. In the rediae of both *Cryptocotyle lingua* and *Sphaeridiotrema globulus* acid phosphatase activity was demonstrated in lysosomal-type cytoplasmic organelles of both surface and perinuclear tegument. Naphthylamidase activity was detected in the tegument of

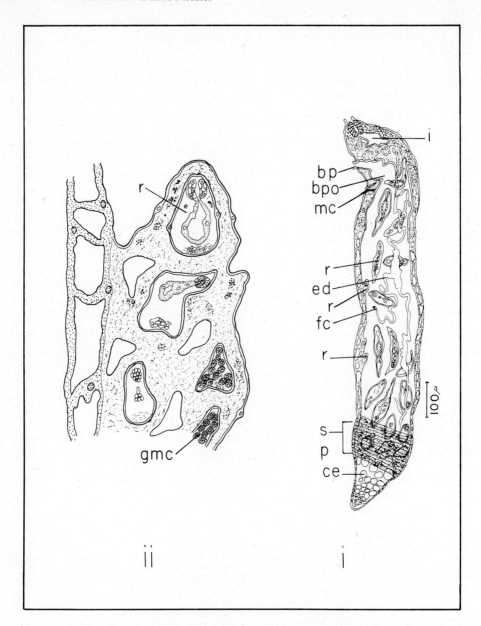

FIGURE 1-24.   Redia of the heterophyid *Heterophyes aequalis* in Egypt. (i) Mature
mother redia showing excretory system, the birth pouch and sites of development for
daughter rediae and cercariae, (ii) Part of lateral wall of mother redia showing germinal
cells and developing redial embryos. bp, birth pore; bpo, birth pouch; ce, cercarial
embryos; ed, excretory duct; fc, flame cells; gmc, germinal cells; i, intestine; mc, muscle
cells; p, pigment; r, redia; and s, surface structure. (Redrawn after Kuntz, R. E. and
Chandler, A. C., *J. Parasitol.,* 42, 625, 1956. With permission.)

Philophthalmus gralli, and thus there is the possibility of tegumental secretion of hy-
drolytic enzymes which function in the lysis of surrounding snail tissues. This has been
confirmed in a recent histochemical study[121] of the rediae of *Fasciola hepatica* in which
evidence was provided that both the redial tegument and the cecum are enzymatically
active with probable absorptive and digestive functions. More information on this subject was elaborated upon by Erasmus.[43]

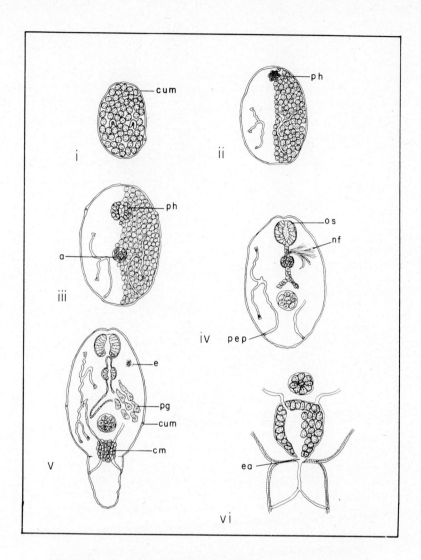

FIGURE 1-25.   Stages in the development of the excretory system of the cercaria of *Heterophyes aequalis* in Egypt. (i) Cercaria after appearance of flame cells, (ii) Young cercaria with two pairs of flame cells, early differentiation of pharynx, (iii) Older cercaria, early differentiation of acetabulum, (iv) Later stage with two pairs of flame cells but moderately well developed intestine, (v) Excretory tubes in region of mesodermal mass prior to fusion, (vi) Posterior end of cercaria, excretory tubes fused and surrounded by loose mass of cells, excretory atrium well developed. a, acetabulum; cm, cells of mesodermal mass; cum, cercarial vitelline membrane; e, pigmented "eye" spots; ea, excretory atrium; nf, nerve fibers; os, oral sucker; pep, primary excretory pore; ph, pharynx; and pg, penetration glands. (Adapted after Kuntz, R. E. and Chandler, A. C., *J. Parasitol.*, 42, 626, 1956. With permission.)

specific layer of the future cyst wall of the metacercaria. Cercariae which encyst in the open have batonnet cells in their cystogenous glands and these unroll to form ply after ply of a laminated innermost cyst layer in several species such as *Fasciola hepatica*, *Notocotylus attenuatus*, and *Parorchis acanthus*.

Mucoid gland cells are present under the tegument and have been reported in several gymnocephalous cercariae, in particular, fasciolids, notocotylids, paramphistomatids,

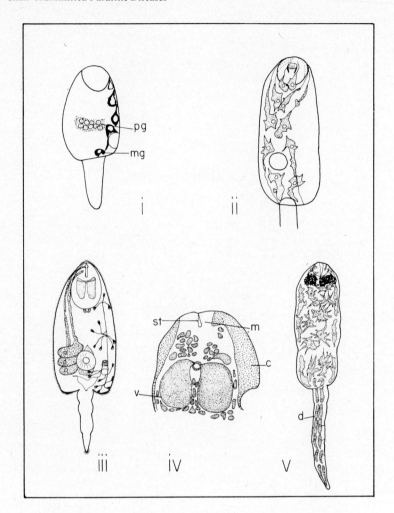

FIGURE 1-26.   Mucoid glands in cercariae. (i) and (ii) Stages in
the development of the cercariae of the heterophyid *Euryhelmis
monorchis* showing mucoid glands, (iii) Emerged virgulate cercaria
showing internal features of this cercarial type, (iv) Frontal section
through the anterior end of a nonemerged virgulate cercaria showing the relation of the virgula to other structures of the anterior
end, (v) Notocotylid cercaria of *Notocotylus urbanensis*, showing
mucoid glands. c, cuticula (tegument); d, duct in tail; m, mucus;
mg, mucoid gland; pg, penetration gland; st, stylet; and v, virgula.
(After Kruidenier, F. J., *Am. Midl. Natl.,* 1951, 1953.)

and echinostomes, and in the xiphidiocercaria group as in the paragonimids. Their
secretion is mainly acid polysaccharide in nature, and in the virgulate xiphidiocercaria
the secretion accumulates in the virgula (Figure 1-26).

Penetration gland cells are a common feature of several cercarial types. The cells
are situated near the acetabulum, lateral, posterior, and anterior to it. Their ducts are
long and open at the oral sucker, or the penetration organ, in the strigeids and schistosomes, two groups of cercariae in which these gland cells have been studied in some
detail, histochemically and by means of the light and electron microscopes. In histochemically and by means of the light and electron microscopes. In strigeid cercariae,
the penetration gland cells are in two groups, one anterior to the acetabulum and one

55

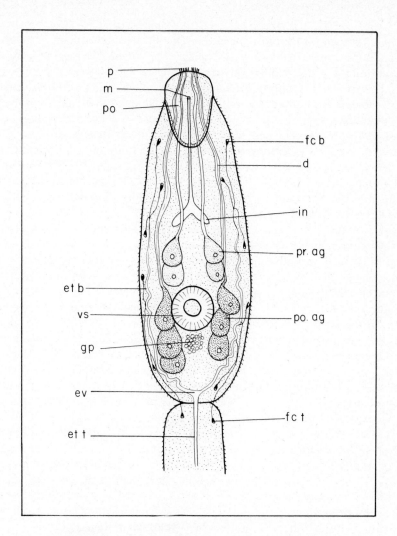

FIGURE 1-27. Cercarial body of *Schistosoma mansoni*.
d, ducts of penetration glands; et b, excretory tubule in
body; et t, excretory tubule in tail; ev, excretory vesicle;
fcb, flame cell in body; fct, flame cell in tail; gp, genital
primordium; in, intestine; m, mouth; p, papillae at anterior
tip of penetration organ; po, penetration organ; po. ag,
postacetabular gland; pr. ag, preacetabular gland; and vs,
ventral sucker.

posterior to it, but the cells appear similar. In *Posthodiplostomum minimum* there is
of one pair anterior to the acetabulum and two pairs posterior to it. The secretions of
these glands are an acid mucopolysaccharide and a carbohydrate-protein complex. The
penetration glands of the schistosomes have been thoroughly studied by Stirewalt and
associates, and by Kruidenier.[156] There are two pairs of preacetabular, and three pairs
of postacetabular glands (Figure 1-27). The preacetabular cells are largest and acido-
philic or eosinophilic with large granules in the cytoplasm, while the postacetabular
are smaller basophilic and their cytoplasm contains small granules. The preacetabular
and postacetabular glands of the schistosomes can be differentiated histologically and
histochemically. The preacetabular glands are PAS-negative but can be stained with

the alizarin red present in Purpurin, suggesting the presence of calcium. The postacetabular glands, however, are PAS-positive and are associated with mucoid secretion.[159] The secretion of the postacetabular glands is believed to be used by the cercaria to adhere to the skin surface during the location and initial penetration into the skin, whereas the alkaline secretion of the preacetabular glands probably helps in the penetration of the keratagogenous layer. The ultrastructure of these acetabular glands in the case of *Schistosoma mansoni*[39] shows them to consist of enlarged aboral areas (funduses) and their oral extensions as ducts. The glands are morphologically similar except for their shape and secretory globules. In the funduses of the postacetabular glands the globules were of a single type, spherical to irregular in shape, with numerous electron-dense areas. Some of the preacetabular secretory globules were of uniform density, while others showed electron-lucid areas.

Enzyme-containing preacetabular gland secretion of *S. mansoni* cercariae, which is emitted exhaustively under the horny layer during skin penetration, is thought to modify skin so as to facilitate movement through it of the infective larvae during their invasion of the mammalian hosts. It may also have other functions. Campbell et al.,[20a] studying further the properties of the enzyme(s) secreted from the preacetabular glands, found that the enzyme is relatively storage stable in water or glycine-NaOH buffer. It is highly sensitive to the buffer system, glycine-NaOH buffer providing the highest proteolytic activity against Azocoll and gelatin substrates. In the glycine buffer the pH optimum for the enzyme is 8.5 to 8.8; the temperature optimum is 51°C. Enzymatically active secretion can be collected over a nonpenetrable lipidized surface. The stimulus in use is skin surface lipid, which does not lend itself to quantitative application. Thus, in an attempt to quantitate the protease-containing secretion, an improved method of secretion was used, in which two commercially available free fatty acid fractions of skin lipid active in stimulating cercariae to penetrate skin, linolenic, and linoleic acids were substituted for skin surface lipid.[157]

Several types of sense organs have been reported in cercarial bodies. In many cercariae a pair of eye spots are light receptors and in some, such as the notocotylids, there might be three. Photoreceptors which are closely associated with the brain and consist of separate sensory and pigment-containing cells are among the features common to both free-living and many parasitic flatworms. Although certain adults in the Monogenea and Aspidogastrea have pigmented photoreceptors, they are limited in the Digenea to immature stages: the miracidium, cercaria, and, rarely, the germinal sacs in the molluscan host. The photoreceptors of the cercaria disintegrate as it develops to the adult. Probably the majority of the Digenea lack pigmented photoreceptors at any stage, and when present, they may occur in either the miracidium or cercaria of a sepcies, or both. In some families their presence or absence seems to be correlated with the extent to which those larvae are free living, but in others no such correlation exists.

A possible photoreceptor in cercariae of *Schistosoma mansoni* has been reported for the first time by Short and Gagné.[150] At least two structures near the anterior end of the cercaria appear by electron microscopy to be ovoid and composed of a thin cytoplasmic wall surrounding a cavity which contains lamellae; the latter are extensions of modified cilia which arise from the inner wall of the cavity. These structures with a membrane-lined cavity and lamellae resemble structures interpreted as possible gravity and photoreceptors in trematode miracidia, and as photoreceptors in a monogenean larva, a ctenophore, and certain molluscs.

Several types of so-called papillae (Figure 1-28), stainable with silver nitrate, which project from the surface of the tegument or capsules embedded within the tegument, have been reported in light microscope studies from cercariae and other stages of a wide variety of trematodes.[89,138,149,151] These papillae have been generally interpreted as being sensory in nature because of their distribution, morphology, and innervation,

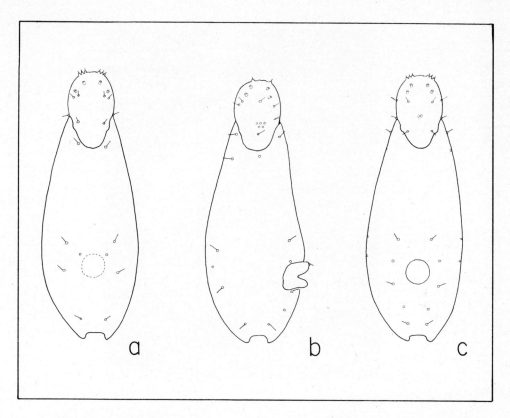

FIGURE 1-28. Argentophilic "papillae" on the body of *Schistosoma mansoni* cercariae. (a) Dorsal view, (b) Right lateral view, (c) Ventral view. (Redrawn after Short, R. B. and Cartlett, M. L., *J. Parasitol.*, 59, 1041, 1973.)

and because they resemble similar structures shown to be sensory in other invertebrates. Some investigators have indicated that patterns of papillae may be of systematic value.[89,138,149,53,151] Not all the papillae appear to be projections or elevations, as the word papilla implies, and it has been suggested to use the term "argentophilic cuticular structures" for these sensory structures.[62] The number, distribution, and structure of these papillae on the cercaria of *Schistosoma mansoni* were studied by both light and scanning electron microscopy.[149] It was found that the basic number of body papillae (excluding those at the anterior tip) is 62: 18 ventral, 4 acetabular, 20 dorsal and 20 lateral (10 on each side). At the anterior tip there are 14 papillae in two groups of 7 (Figure 1-29). Each group of papillae is associated with a group of, apparently, 7 gland duct openings. Some of the papillae are raised, uniciliated (the most numerous), some are in the form of pits, while others are in the form of a disklike elevation with a small knob in the center. The schistosome (*S. mansoni*) cercariae do not show sexual differences in their papillar pattern. However, as to species of schistosome cercariae, Richard[137,138] was able to differentiate cercariae of *S. mansoni*, *S. rodhaini*, *S. haematobium*, and *S. bovis*, and Short and Kuntz[151] were also able to separate cercariae of *S. mansoni* and *S. rodhaini* on the basis of their papillar patterns. That cercariae of *S. mansoni* and *S. rodhaini* can be differentiated by their papillar patterns could be useful in the field because these two species are closely related, comprising the *S. mansoni* group or complex in Africa, and in certain localities they occur together in snails of the same species, *Biomphalaria pfeifferi* or *B. sudanica*.

Many cercariae possess a well-developed tail which varies in length, may possess a fin fold, and may be forked. The length of the furcae also varies among the forked-

A

FIGURE 1-29.   Scanning electron micrographs of *Schistosoma mansoni* cercariae. (A) Ventral view showing acetabulum with papillae. (Magnification × 5500.) (B) Anterior end, right side. Note anterior tip papillae (3) opening I (arrow), pit (D_1), and ciliated papillae (D_3, 4, and L_1). (Magnification × 4200.) (C) Anterior tip, ventral view. Note pits (V_1, V_2) and bulbous anterior tip papillae with collars. (Magnification × 4200.) (D) Papillae and gland duct openings (G) at anterior tip. (Magnification × 7000.) (Courtesy of Drs. Short and Cartlett.)

tail cercarial types, and they may bear fin folds. Tactile hair-like processes extend from the tail stem of many strigeids, and in some of these the tail stem may contain caudal bodies in the form of vesicles. Strong muscular components, both circular and longitudinal, are common in the cercarial tail.

2. Cercarial Types

(Figure 1-30) Cercariae of Digenea are of various types and their classification is usually based on their external morphology. It should be noted that the cercarial types discussed below are in general descriptive and do not indicate phylogenetic relationships among the forms producing them. Exceptions to this are: the Gymnocephalous-Echinostome type, the Xiphidiocercariae type, the Furcocercous type, the Pleurolophocercous type, and the Amphistome type.

Early morphological designations were based on the number and position of the suckers present in the cercarial body; the following types were, and some of them still are in use: monostome, distome, amphistome, and gasterostome. The designations are similar to those used with the adult flukes.

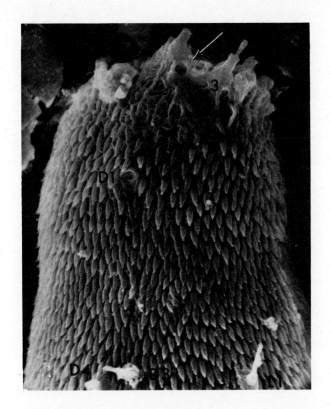

FIGURE 1-29B

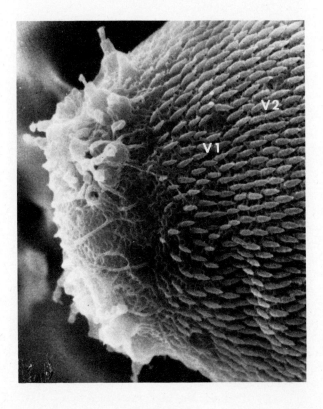

FIGURE 1-29C

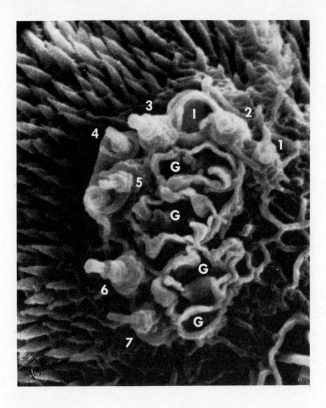

FIGURE 1-29D

Another classification of cercarial types takes into consideration the shape and relative size of the tail, and the presence of accessory structures on the tail, such as folds. The following types are recognized.

Cercariaeum cercaria — This type of cercaria lacks a tail, it does not usually leave the molluscan host, and the adult stage is a member of one of the families Brachylaemidae (Figure 1-31), Monorchiidae, Zoogonidae, and Cyclocoelidae.

Leucochloridium cercaria — This cercaria develops inside branched sporocysts in terrestrial or amphibious snails. The sporocysts develop pigmented or banded brood sacs into which the cercariae migrate for encystment. The adults are flukes of the genus *Leucochloridium* of the family Brachylaemidae.

Microcercous cercaria — This type has a short, knoblike, triangular or cup-shaped tail. This short tail may be covered with spines and may or may not be distinct from the body of the cercaria. This cercaria is represented in several unrelated families, among which are: Nanophyetidae, Paragonimidae, some members of the Dicrocoeliidae, and Brachylaemidae. The microcercous cercariae of the families Nanophyetidae and Paragonimidae will also be considered later under another important type, the xiphidiocercaria.

Pleurolophcercous cercaria — The tail is provided with a dorso-ventral fin fold; the body has pigmented eyespots; and the cercaria develops in a redia in prosobranch (operculate) snails and encysts in fishes. This type is represented in the families Heterophyidae and Opisthorchiidae.

Parapleurolophocercous cercaria — This cercaria is similar to the pleurolophocercous except that the tail has a lateral fin fold in addition to the dorso-ventral fin fold. The cercaria is represented in the family Heterophyidae.

Furcocercous cercaria — This cercaria is characterized by a forked tail, some species

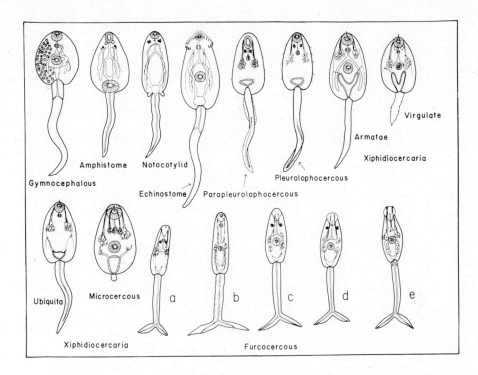

FIGURE 1-30.   Various cercarial types of digenetic trematodes included in this book.

may possess eyespots; and a pharynx may be present or absent. Several subtypes are recognized and the division is mainly based on the relative length of the furcae to the tail stem and the presence or absence of a pharynx and eyespots:

1.   *Brevifurcate-apharyngeate* cercaria of the families Schistosomatidae and Spirorchiidae — the body does not possess a dorso-median fin fold; and cercariae develop in simple sporocysts and penetrate the final host directly, hence two-host life cycle, they may or may not have eyespots.
2.   *Brevifurcate-pharyngeate clinostomatid* cercaria of the family Clinostomatidae — there is a dorso-median fin fold on the body; ventral sucker vestigial; and develops in rediae in aquatic snails and encysts in second intermediate host.
3.   *Longifurcate-pharyngeate* cercaria of the superfamily Strigeoidea — dorso-ventral fin folds on furcae may or may not be present; and develops in filiform sporocysts in aquatic snails and encysts in a second intermediate host.

Another classification of cercariae takes into consideration the presence of specialized body structures. Some of these types are the following:

Echinostome cercaria — This cercaria possesses a collar of large spines around the oral sucker. The cercaria develops in collared rediae, with stumpy appendages in aquatic snails and encysts in the open or requires a second intermediate host: a snail, an amphibian, or a fish. Adult flukes are members of the family Echinostomatidae. The cercaria in related families may show modifications in time of appearance of collar and collar spines and in degree of development of these structures (families Psilostomatidae, Philophthalmidae, Cathaemasiidae, Rhopaliasidae, and Fasciolidae).

Gymnocephalus cercaria of the family Fasciolidae — It is a basic distome cercaria, similar to the echinostome cercaria, but lacks a collar and collar spines.

Xiphidiocercaria— This cercaria possesses a stylet in the oral sucker and there are well-developed penetration glands and mucoid cells.

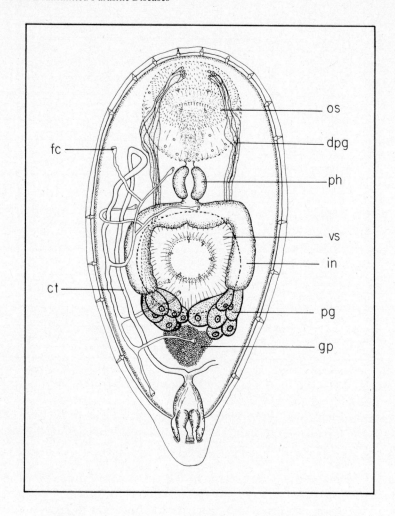

FIGURE 1-31.   Mature emerged cercaria of the brachylaemid *Pos-tharmostomum helicis*, dorsal view. Excretory system shown only on one side. Note presence of vitelline membrane, although nuclei of membrane are not visible. ct, collecting tubule; dpg, ducts of penetration glands; fc, flame cell; gp, genital primordium; in, intestine; os, oral sucker; pg, penetration gland; ph, pharynx; and vs, ventral sucker. (From Ulmer, M. J., *Trans. Am. Microsc. Soc.,* 70, 319, 1951).

The troglotrematids have this type of cercaria, but the tail is stumpy, knob-like, and microcercous in type. The latter develops in rediae. The xiphidiocercariae of the microphallids, gorgorerids, plagiorchiids, lecithodendriids, and dicrocoelids develop in sporocysts. There are several subtypes of the Xiphidiocercaria:

1.   *Virgulate* cercaria is that of the majority of trematodes of the family Lecithodendriidae. In this cercaria there is a bilobed or pyriform virgula organ located in the region of the oral sucker. The material secreted by the virgula organ aids the cercaria in attachment to the host and may also provide some protection for the cercaria. The tail of this cercaria is usually shorter than the body and the ventral sucker is smaller than the oral sucker.

2.   *Ubiquita* cercaria is without a virgula organ and with or without a vestigial ventral sucker. This cercaria is represented in the family Microphallidae.

3. *Armatae* cercaria is the subtype in which the virgula organ is absent and the suckers are of equal size, or the ventral sucker may be larger than the oral one. Trematodes of the families Plagiorchiidae and Telorchiidae produce this type of cercaria.

3. Physiology and Histochemistry

Investigations have been conducted on the carbohydrate, lipid, and protein metabolism of the cercaria. Glycogen, lipids, and amino acids have been demonstrated in many cercarial species and are no doubt obtained from the molluscan host via the redia or the sporocyst. Infected snails placed in water with[14] C-labeled glucose produce labeled cercariae in a few days, and the label continues to be detected in the cercariae for at least 2 months after the beginning of the experiment.[87] The amino acid content of certain cercarial species studied resembles to a great extent those of the molluscan host. Lipids, consisting of neutral fats and fatty acids, have been demonstrated histochemically in the excretory system and parenchymal cells, and on the surface of developing cercariae.

Cercariae of various trematodes have been studied histochemically and a number of enzyme systems have been detected. Acid phosphatase activity was demonstrated in the cercaria and other larval stages of *Fasciola hepatica*. A recent histochemical study of the cercariae of *F. hepatica*[121] showed that eserine-resistant esterase activity was present in the primordial intestinal cells of free-swimming cercariae; however, similar reactivity was noticeably absent from cercariae which were still enclosed within the rediae. Germinal clusters inside rediae, and the intestinal cells of developing and mature cercariae, stained for acid phosphatase and hexosaminidase; the reaction was frequently associated with discrete cytoplasmic granules.

Among schistosome cercariae cholinesterase, esterase, and aminopeptidase activity has been detected. The aminopeptidase activity is usually demonstrated in the intestinal ceca, cholinesterase, and esterase activity in the nervous elements.

4. Biology

The cercariae, after being produced as germinal embryos in rediae or sporocysts, may encyst inside these larval forms, may leave the germinal sacs, in certain species, and mature in the molluscan host's tissues before they exit from the snail, or, in other species they may leave the snail shortly after they depart from the rediae or sporocysts. Cercarial production varies from one group of Digenea to the other, and usually depends also on the molluscan host's size and its nutritional status. Small numbers were reported for the paramphistomes as compared to the echinostomes and clinostomes; in the latter case a figure of 25,000 cercariae was given as the probable mean total cercarial output resulting from single miracidia of certain echinostome species, and 30,000 to 40,000 for some clinostome species. The strigeids are known to produce a much larger number of cercariae, and a figure of 60,000 as the daily cercarial output is not uncommon. The schistosome group is also known to produce large numbers of cercariae. However, varying figures are found in the literature relative to the mean daily cercarial output, as seen in Table 1-3. The highest daily output recorded from an individual *Biomphalaria glabrata* infected with *Schistosoma mansoni* in Saint Lucia was 16,000, and from one *Bulinus (Physopsis) globosus* infected with a Nigerian strain of *S. haematobium* was 10,525.

Certain environmental factors have been found to affect cercarial emergence as they also affect the rate of development of the larval stage inside the snail. It is quite possible that external environmental factors may also influence the rate of development of the parasite and the emergence of the cercariae indirectly, through their effect on the physiological processes of the snail. Among the factors influencing cercarial emerg-

Table 1-3
OUTPUT OF CERCARIAE OF *SCHISTOSOMA MANSONI* AND *S. HAEMATOBIUM* IN VARIOUS ENDEMIC AREAS, AND BY EXPERIMENTALLY INFECTED SNAILS

Schistosoma species	Snail species	Cercarial output	Remarks	Ref.
S. mansoni (Pernambuco, Brazil)	*Biomphalaria glabrata*	Mean number of cercariae produced per day per snail: (7 to 10 mm diameter) = 154.5 cercariae (13 to 16 mm diameter) = 381.9 cercariae	Snails exposed to 10 miracidia; temperature of exposure and maintenance, 26°C. Weekly counts done: 10 for 1st group of snails (7 to 10 mm diameter, and 13 for 2nd group of snails (13 to 16 mm).	8
S. mansoni (Rhodesia)	*Biomphalaria pfeifferi*	Average number of cercariae per infected snail per 2 hr varied from 0 to 1415 (mean 277.4)	100 snails collected from a habitat east of Salisbury, othe same day of every alternate week, at approximately the same time of day for an entire year (April 1963 to March 1964)	148
S. haematobium (Rhodesia)	*Bulinus (Physopsis) globosus*	0 to 502 (mean 129.8)		
S. mansoni	*Biomphalaria straminea* (several Brazilian strains)	For 2 experimental groups a) median 144.0 maximal 1930 b) median 10.5 maximal 1480	The higher the susceptibility of the snail strain is, the longer the snails will live and more cercariae they will shed.	7
S. haematobium (Tanzania)	*Bulinus (Physopsis) nasutus*	Mean daily output of individual *naturally* infected snail varied between 14 and 4119	While a rise in temperature from 23 to 30°C stimulated emergence of cercariae, the effect of heat was subsidiary to that of illumination. The pattern of output being modified in response to alterations in the cycle of illumination.	

Table 1-3 (continued)
OUTPUT OF CERCARIAE OF *SCHISTOSOMA MANSONI* AND *S. HAEMATOBIUM* IN VARIOUS ENDEMIC AREAS, AND BY EXPERIMENTALLY INFECTED SNAILS

Schistosoma species	Snail species	Cercarial output	Remarks	Ref.
S. mansoni	*Biomphalaria glabrata* (from Jabuticatubus, Minas Gerais, Brazil).	It was estimated that naturally infected snails shed cercariae for an average of 7 days; the total number of cercariae for each snail being about 16,000		74
S. mansoni	*Biomphalaria salinarum* (South-West Africa).	Range from less than 200 to about 1600 cercariae	Snails kept in outdoor aquaria from the time they were exposed to miracidia. Counts of cercariae done from first shedding until death of snail. There was day-to-day irregularity of numbers of cercariae shed throughout all seasons.	132
S. haematobium	*Bulinus (Physopsis)* sp. (Transvaal)	Range from less than 200 to about 1700 cercariae	Counts of cercariae not in text but in graphs. Shedding of cercariae done without disturbing snails.	
S. mansoni	*Biomphalaria alexandrina* (Near Alexandria, Egypt).	Daily output from infected snails in the field estimated at 957.7 cercariae.	Output of cercariae is snail-size specific. When the size-specific output is adjusted to the size-composition of infected snails taken from the field the estimate of 957.7 in the previous column is arrived at. However, the number may vary with the season.	27

Table 1-3 (continued)
OUTPUT OF CERCARIAE OF *SCHISTOSOMA MANSONI* AND *S. HAEMATOBIUM* IN VARIOUS ENDEMIC AREAS, AND BY EXPERIMENTALLY INFECTED SNAILS

Schistosoma species	Snail species	Cercarial output	Remarks	Ref.
S. haematobium (from Nigeria)	Bulinus (Physopsis) globosus	Range of mean number cercariae production of single snails examined biweekly during entire life span (snail exposed to 7 miracidia) = 47—320 (mean value 147)	57% of 1031 daily counts being less than 201 cercariae; 78.4 less than 401; 86.9% less than 601; 93% less than 1001, and 97% less than 2001. One specimen, however, shed 10,525 cercariae	177
S. haematobium (from Dezful, Iran)	Bulinus (Bulinus) truncatus (from Dezful, Iran)	Range of mean number cercariae, etc. as above = 56—417 (mean value 198)	Of 323 daily counts 75.2% less than 201, 92.2% less than 401, 98.4% less than 601, 99.3% less than 800 Only two daily counts recorded at levels of between 1000 and 2000 cercariae.	
S. mansoni	Biomphalaria straminea (several Brazilian strains)	For 2 experimental groups a) median 144.0 maximal 1930 b) median 10.5 maximal 1480	The higher the susceptibility of the snail strain is, the longer the snails will live and more cercariae they will shed.	7
S. mansoni	Biomphalaria glabrata (several Brazilian strains).	For 3 experimental groups Median Maximal 2015 9580 1818 9150 462 1900		
S. mansoni	Biomphalaria glabrata (St. Lucia strain)	Mean/Snail/day = 1780	Highest daily output recorded from an individual snail was over 16,000.	161

ence are temperature, light, pH, salinity, and CO_2 tension. Temperature seems to be a significant factor. In several species cercarial production is arrested below 10°C, and in others below 15°C. The cercariae of the schistosomes *Heterobilharzia americana* and *Schistosoma mansoni* do not emerge below 16°C; optimum temperatures are usually between 24 and 28. The upper limit of the temperature range is more variable and depends mainly on the thermal death point of the snail. There is a correlation between light intensity and the emergence of ocellate cercariae, but exceptions exist. It is the experience in many laboratories that isolating the snails singly and changing the water in which the snails are maintained will stimulate cercarial emergence of several trematode species. This does not seem to be correlated with other environmental factors.

It has often been observed that excessive emergence of cercariae from the snails might affect the latter's life span. Examples are the snails *Bulinus (Physopsis) globosus* and *Bulinus (Bulinus) truncatus* infected with *S. haematobium*, a Nigerian strain and an Iranian strain, respectively. It was shown recently by Webbe and James[177] that snails with a high cercarial shedding capacity are not as long-lived as snails producing fewer cercariae. The results are shown in Figure 1-32.

There is a marked diurnal periodicity in the emergence of cercariae of several species of digenetic trematodes. This has been reported not only on the basis of experimental infections of laboratory-reared snails but also on the basis of field observations in the case of the human species of schistosomes. Methods used in the field were filtration of water from natural snail habitats and detection of the cercariae quantitatively during different periods of the day and night. The phenomenon of diurnal periodicity of the emergence of cercariae of certain trematodes and the exact mode of action of some stimuli which trigger the emergence are still not clear. It has been repeatedly reported that such a periodicity corresponds to active periods of the definitive hosts, mammals and birds, or in some cases to the activity of the second intermediate hosts. It has also been suggested that the physiology of the snail, or certain snail hormones, might have some effect on the periodic emergence of the cercariae. The cercariae of the lung flukes *Paragonimus kellicotti*, and *P. mexicanus* start emergence in the late afternoon and early evening,[112] and this might correspond to the activity of the crayfish hosts (*P. kellicotti*) or the crab hosts (*P. mexicanus*). Among the schistosomes, cercariae of certain species of bird schistosomes show peak emergence in the early hours of the morning, corresponding to the activity of their bird definitive hosts. It has been shown in the case of the bovine species, *Schistosoma bovis* and *S. mattheei* in Africa and the Middle East, *S. haematobium* infecting humans in Africa and the Middle East and *S. mansoni* infecting humans in Africa and some parts of the Americas that, although the cercariae commence their emergence in the morning (about 8 or 9 a.m.), they have a peak emergence about the middle of the day. The cercariae of the Oriental *S. japonicum* on the other hand have a peak emergence in late afternoon and early evening, and although they infect humans, they also infect a long list of wild and domestic animals. Most of the cercariae of the African rodent schistosome *S. rodhaini* emerge in the early hours of the morning. Moreover, cercariae of the rodent schistosome in the U.S. and Canada, *Schistosomatium douthitti*, also exhibit diurnal periodicity. Under laboratory conditions, and presumably in nature, cercariae of *S. douthitti* leave the snail during the first few hours after sunset. They move freely for a few minutes and then adhere to the surface film of the water. They remain as such until they are apparently disturbed by the activity of the mammalian hosts. Under natural conditions there must also be some cercariae which emerge during the early hours of the morning, judging from their ability to cause dermatitis during the daytime. The emergence after sunset, until the early morning, corresponds to the period when the rodent hosts are active. The writer[109] has shown, however, that the cercariae of *S. douthitti* can be induced to emerge twice instead of once daily by subjecting them to darkness during

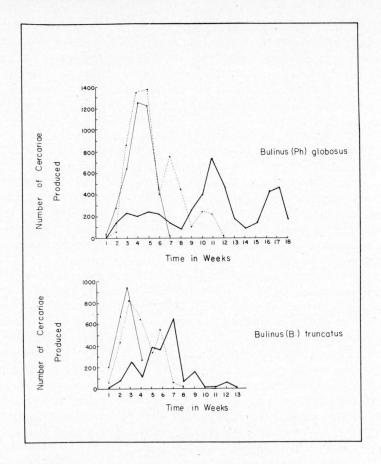

FIGURE 1-32. Showing the individual cercarial outputs of three specimens of each of *Bulinus (Bulinus) truncatus* and *Bulinus (Physopsis) globosus* infected with *Schistosoma haematobium* during their entire lifespan. (Redrawn from Webbe, G. and James, C., *J. Helminthol.*, 46, 185, 1972.)

the morning and evening. Six laboratory-infected *Lymnaea (Pseudosuccinea) columella* shed a total of 472 cercariae between 8:30 a.m. and 12 noon, at 25°C in a dark cabinet, and a total of 320 cercariae after they were isolated again in the dark after sunset.

With some schistosome cercariae which show peak emergence during the day, reversal of their emergence was accomplished. Those of *S. mansoni* emerging from the snail *Biomphalaria glabrata* were induced to show a peak emergence in the evening by reversing the light cycle. The diurnal periodicity of some schistosome cercariae is illustrated by the daytime hourly production of cercariae of *S. haematobium* from the snails *Bulinus (Bulinus) truncatus* and *Bulinus (Physopsis) globosus* under laboratory conditions. The daytime hourly production of cercariae was recorded over a period of 8 hr following the application of light and heat stimuli, after a period in complete darkness, compared with similar cercarial production from snails kept under normal aquarium conditions with the onset of some natural light at or about 06.00 hr.[177] The peak cercarial production of *B. (Ph.) globosus* which were covered occurred at 4 and 5 hr after application of the light and heat stimuli at 09.00 hr when 33% and 29%, respectively, of the total cercarial production of the ten snails took place. A prominent peak in the number of cercariae was produced by *B. (B.) truncatus* 5 hr after application of the light and heat stimuli, when 50% of the total day's production from the ten snails occurred, followed by 20% and 16% of the total production 6 and 7 hr, respectively, afterwards. After complete darkness a more concentrated output of cercariae takes place during the 4th, 5th, and 6th hr in the case of both species studied. With an initial period of natural light the cercarial production is spread over a longer period, commencing at the 2nd hr.

The active swimming period of the cercariae in the water after it leaves the snail varies from one species to the other. In some cases encystment is very rapid, while in others the cercariae have a prolonged swimming period before they encyst in a second intermediate host or penetrate the final host directly. Relative to age of the cercariae, it has been demonstrated that the percentage of cercariae of the schistosomes *(S. mansoni)* which are able to penetrate and infect the mammalian host decreases with increasing age. This was confirmed by the writer,[107] and it was found that a small percentage of the cercariae were still able to infect mice 24 hr after their emergence from the snail; some were alive after 48 hr. With *Schistosomatium douthitti* there was no significant difference in the percent of worm return resulting from cercariae 1 to 12 hr of age, but the return decreased considerably during the next 12 hrs, when it was 10.8%. Some cercariae were still infective during the second day, because 6.9% worm return was obtained using 36-hr old cercariae.[109]

The swimming behavior of some cercarial species is very characteristic and is helpful for species identification. The cercariae of *Heterobilharzia americana* swim toward light, thus aggregate in the water toward the light source, usually close to but not at the surface of the water. The cercariae of *Schistosomatium douthitti*, another American mammalian schistosome, are apparently not affected by light and become attached to the surface film soon after emergence. The cercariae of *Schistosoma mattheei*, after emergence from the snail, become more or less evenly distributed throughout the water, while those of *S. intercalatum* become concentrated near the surface film.[189] The majority of the cercariae of *S. mansoni* are located near the surface of the water, but some are halfway to the bottom. They swim actively with intermittent periods of rest. Among the bird schistosomes the cercariae of *Trichobilharzia ocellata* swim actively towards the light but tend to attach to wall of the container. Those of *T. physellae* swim less actively and not as directly toward light and attach to wall of container on the light side or bottom, while those of *T. stagnicolae* swim toward light, and do not attach but are suspended motionless in water close to wall of container.

During their free-living existence in the water several environmental factors influence the behavior and the survival of the cercaria. Some of these physical and chemical factors are summarized in Table 1-2. To those referred to in the table is added the information that the swimming activity of the cercariae of *S. mansoni* is influenced by salt solutions of differing ionic composition.[12]

Not all cercariae emerge from their molluscan hosts. The majority of cercariae of certain Digenea which utilize land snails and slugs as intermediate hosts do not leave the snail, and these are in general members of the families Brachylaemidae, Leucochloridiidae, and Brachycoeliidae. Examples of these are the flukes *Postharmostomum helicis, P. gallinum, Brachylaema virginiana, Ectosiphonus rhomboideus, Entosiphonus thompsoni, Hasstilesia tricolor*, and *Leucochloridium* spp. Cercariae of some of these trematodes remain inside the snail until it is eaten by the definitive host; others leave the sporocysts in which they developed, and become encysted in various organs of the gastropod. There are some species, however, which leave the snail and become encysted in another snail, of the same or different species, or in a slug. For example in the case of *Brachylaema virginiana* and *Panopistus pricei* the cercariae, which are produced in branched sporocysts, leave the sporocyst, creep over the surface of their land snails or slugs, and are transferred to another snail or slug upon contact. In several members of the family Dicrocoeliidae, the sporocyst with its content of cercariae leaves the terrestrial snail and becomes covered with mucus from this host, which keeps it moist until it is ingested by the second intermediate host, usually an arthropod but in some cases a lizard (in the case of *Platynosomum fastosum*). In the cyclocoelid group the cercariae also do not emerge out of the snail, which may be terrestrial or may be aquatic. Eggs of *Cyclocoelum elongatum* (*C. dollfusi*), are passively ingested by land snails, and the metacercariae are found encysted within the rediae producing them. However, eggs of *Cyclocoelum obscurum* are released into the water, and emerged miracidia (each containing a redia) attach to the freshwater snail, *Gyraulus hirsutus*. The rediae bore into the molluscan host, leaving the empty miracidia behind. The developing cercariae encyst within the rediae as early as the 14th day postexposure.

g. The Metacercaria

The cercarial stage of many trematodes is followed by a metacercarial stage, usually as a result of encystment of the cercaria, but in some cases this stage remains unencysted (Figure 1-33). The body of this stage usually resembles that of the cercaria. However, some of the cercarial adaptive structures (such as stylet, spines, and mucoid and penetration and cystogenous glands) are lost. The metacercaria in some cases does not increase in size, while others expand and assume a characteristic shape. Within the body, however, are found some of the organ systems of the cercaria, such as the digestive, excretory, and nervous systems. The excretory system may remain the same, as in the fasciolids, echinostomes, and microphalids, or there may be a slight increase in the number of flame cells, as in the opisthorchids. The extent of development of the reproductive elements varies among the Digenea. In some species they might remain in the form of the genital anlagen of the cercaria until their full development in the definitive host when the metacercaria is ingested, or they may show slight or pronounced development, in some species to the extent of formation of viable eggs. Metacercariae of the strigeids have an entirely different body from that of the cercaria, and sometimes they are not encysted.

Encystment in the open, on vegetation, or on external surfaces takes place among the paramphistomatids, fasciolids, notocotylids, and philophthalmids. Cercariae of philophthalmids readily encyst in a pipette when this is used to draw them up. In nature some notocotylids encyst on snail shells. Among the echinostomes, encystment of some cercarial species occurs in the open, others encyst in snails (same or different individual) or in bivalves or tadpoles, while in others it takes place in fish, thus resembling another group, namely the heterophyids and opisthorchids. In several members of the xiphidiocercaria and the furcocercous cercaria groups, encystment takes place in various invertebrate and vertebrate hosts. In the brachylaemids encystment may take place in the same sporocyst, in the same snail or in a different individual. The life cycle of the spirorchids (blood flukes of turtles) and the schistosomes (blood flukes of birds and mammals) does not have a metacercarial stage. Precocious development of the metacercaria in the second intermediate host is well known among certain groups such as the plagiorchids, lecithodendriids, allocreadiids, and clinostomatids.

In the majority of cysts, the wall is formed of more than one layer. The nature and arrangement of the layers varies among different groups of Digenea, and mainly depends on whether encystment is in the open, in the tissues of invertebrates, or in those of vertebrates. However, differences also do occur among representatives of each of these types of encystment. It has been demonstrated by several investigators that the cyst of those forms encysting in the open has more layers than those which penetrate a second intermediate host before encysting. The cyst wall of *Fasciola hepatica* consists of four layers: the outer consists of tanned protein, followed by a thin fibrous layer of mucoprotein and acid mucopolysaccharide, then a third layer divided into three

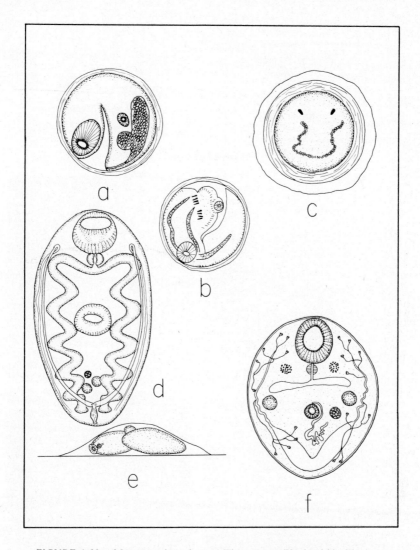

FIGURE 1-33. Metacercariae of some Digenea. (a) Plagiorchid (*Plagiorchis* sp.), from tissues of aquatic pulmonate snails, (b) Echinostome (*Echinoparyphirum flexum*), from tissues of aquatic pulmonate snails, (c) Amphistome (*Paramphistomum microbothrium*), on vegetation, (d) Brachylaemid (*Postharmostomum helicis*), from tissues of terrestrial pulmonate snails, (e) Philophthalmid (*Parorchis avitus*), having a hemispherical shape, on vegetation, (f) Lecithodendrid, from aquatic insects. Note in (d) and (f) the advanced development of the metacercaria, especially the reproductive organs. Some original, others adapted after various authors.

sublayers of muco-protein, acid mucopolysaccharide and neutral mucopolysaccharide. The fourth (innermost layer) is made of lamellae of protein in a protein-lipid matrix. The metacercarial cyst wall of *Notocotylus attenuatus* consists of three layers but the chemical composition of these layers is basically similar to those of the cyst wall of *F. hepatica*.

The structure and composition of the cyst wall of metacercaria of the philophthalmid *Cloacitrema narrabeenensis* has recently been investigated by Dixon.[38] This is another fluke whose cercaria encysts in the open, and it was found that its cyst, therefore, is similar to others encysting in the open, especially that of *Fasciola hepatica*. The cyst of *C. narrabeenensis* (Figure 1-34) is composed of four layers: an outermost layer of

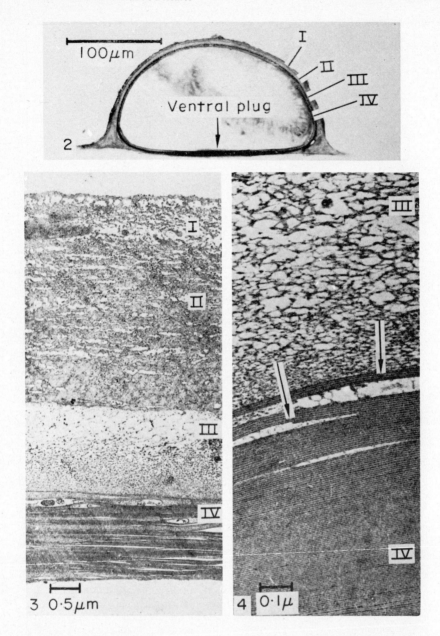

FIGURE 1-34.   Metacercarial cyst of the phylophthalmid *Cloacitrema narrabeenensis.* (2) Light micrograph of section through the metacercarial cyst showing the morphology of the cyst and the four layers making up the cyst wall. The metacercaria has been displaced within the cyst during sectioning. (3) Electron micrograph of section through a portion of the cyst wall surrounding the metacercaria showing the appearance of the four layers of the cyst wall. (4) Electron micrograph of section through part of the inner cyst wall. The arrows indicate favorable sections through the lamellae of layer IV where the triple nature of the lamellae can be resolved. (Photograph after Dixon, K. E., *Int. J. Parasitol.,* 5, 113, 1975. With permission of Pergamon Press and Dr. Dixon.)

acid mucopolysaccharide, a layer of protein which is presumed to be tanned, a layer of neutral mucopolysaccharide, and an innermost layer of keratinized protein. In the center of the ventral side of the inner cyst wall, the keratinized layer is incomplete and this ventral plug region is composed of neutral mucopolysaccharide. The only difference between this cyst wall and that of *F. hepatica* metacercariae is that the order of the two layers of the outer cyst is reversed.

The metacercarial cyst wall of certain members of the xiphidiocercaria group consists of two layers, an outer of carbohydrate-protein complex and an inner of protein. The cyst wall of *Nanophyetus salmincola* (Family Nanophyetidae) consists of two layers. A thick, hyaline, noncellular inner layer surrounds the metacercaria; histochemical tests revealed that it is composed of a carbohydrate-protein complex. The outer layer is cellular in appearance, and these cells look like modified kidney cells of the fish host (where the cyst is located) that have become oriented around the metacercaria, constituting a well-defined layer. The capsulated appearance created in the formation of the outer layer suggests a tissue reaction of the fish host to the presence of the parasite, but it appears that it also forms a part of the cyst wall.

As has been mentioned above, the strigeid metacercaria has a different and more complex body than the cercaria, and it may be encysted or remain free. The shape of these metacercariae is of various types, the excretory system is more developed, and becomes associated with a newly formed lymph system, or "reserve bladder system" of some authors. In addition a tribocytic organ, or adhesive organ, develops and is usually prominent; the digestive system is well developed, and the strigeid metacercariae which do not encyst continue feeding until their death, or until they are ingested by the definitive host. Strigeid metacercariae are found in fish, frogs, snails, or leeches. They are usually well developed and so large in size that they had been given generic names, especially before some life cycles were elucidated. Not only that they are host specific but also organ specific, for example, brain cavity, eye, muscles, etc. The body of the strigeid metacercaria is generally divided into a forebody and a hindbody, but the hindbody may be small in some, while in others the divisions of the body are inconspicuous and unclear (Figure 1-35). The strigeid metacercarial types are the Diplostomulum, the Neascus, the Tetracotyle, and the Prohemistomulum. In the Diplostomulum there is a clear division of the body; this larva is free, unencysted. The forebody is still the prominent portion, the hind part being conical; in the adult stage it will accomodate the reproductive organs. There is usually a pair of lappets at the sides of the oral sucker. The diplostomulum is found in fish, amphibians, and snails, and is the metacercaria of the genera *Diplostomum, Fibricola, Alaria, Pharyngostomoides*, and others. The forebody and hindbody are both well developed in Neascus. Lappets are absent; there is a cyst wall of parasite origin, and this metacercaria parasitizes fish eaten by the definitive hosts where it develops into the genera *Crassiphiala, Posthodiplostomum*, and others. In the Tetracotyle the hindbody can hardly be seen; it is a large and thick metacercaria, usually with additional suckers, and is contained in a distinct cyst wall of parasite origin. The hosts for this larva are fish, amphibians, leeches, and snails, and it is the larval stage of such genera as *Cotylurus, Apharyngostrigea*, and *Apatemon*. The Prohemistomulum is found in fish, snails and leeches, and it is the metacercaria of species of the genera *Mesostephanus* and *Prohemistomum*. There are no lappets, but the larva is usually contained in a cyst wall formed by the parasite.

Some information is available on the histochemistry of strigeid metacercariae. Recently histochemical and thin layer chromatographic studies were conducted on adults and Tetracotyles of *Cotylurus* sp., from the snail *Physa heteropha* in New Jersey.[49] Oil Red O staining demonstrated neutral lipids in the intestinal lumina, eggs and vitellaria of adults, and the excretory system of metacercariae. As determined by TLC, encysted metacercariae incubated in a nonnutrient salt solution excreted free fatty acids, sterols, and sterol esters into the medium. Moreover, TLC analysis detected free sterols, free fatty acids, and sterol esters in metacercariae and adults. Metacercariae

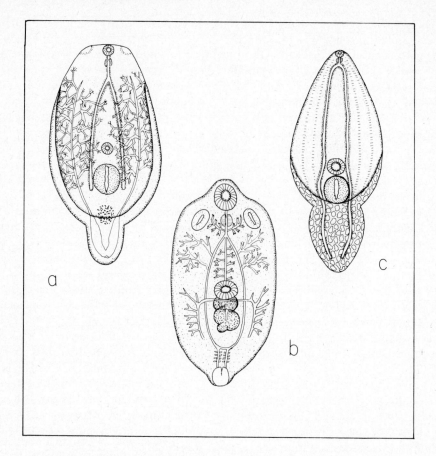

FIGURE 1-35.   Metacercarial stages of the strigeoid flukes. (a) *Diplosto-mulum*, (b) *Tetracotyle*, (c) *Neascus.*

of other groups have also been studied histochemically. In the metacercaria of *Leu-cochloridiomorpha constantiae* the excretory system is lipid-negative histochemically but the intestine is positive. Cholesterol detected in the incubate medium is presumably excreted from the mouth.[50]

Metacercariae of Digenea vary as to the minimum time required for their develop-ment and infectivity. These variations depend on whether the metacercariae encyst in the open, on herbage and surfaces, and whether or not they require a period of growth and metamorphosis inside the second intermediate hosts. Those which encyst in the open are infective to the final host in a very short period: after a few hours (notocotyl-ids), on the first day (paramphistomatids), and up to 5 days at a temperature of about 22°C (*Fasciola hepatica* and other fasciolids). Those metacercariae which encyst in a second intermediate host, but do not grow within this host, need at least several days (5 to 15) until they become infective. On the other hand, the metacercariae which grow inside their second intermediate hosts, whether they undergo encystment or stay free, require a longer period than those mentioned above to become infective to the defini-tive hosts.

h. The Mesocercaria

The mesocercaria is a larval stage in the life cycle of certain strigeid flukes (e.g., *Alaria* spp.), placed between the cercaria and the metacercaria. It resembles the body of the cercaria, utilizes frogs as second intermediate host, but is found in a wide range

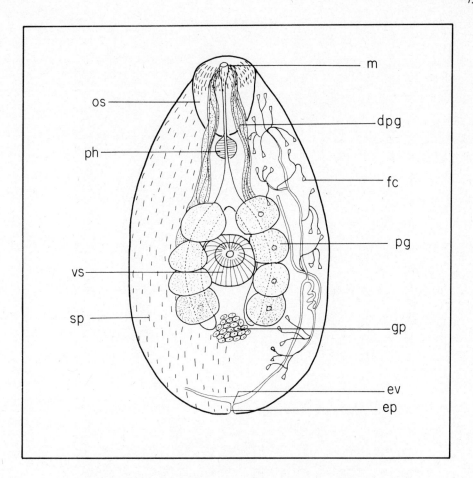

FIGURE 1-36.   Mesocercaria of *Alaria* sp. from connective tissue of cotton-
mouth snake in Louisiana. dpg, ducts of penetration glands; ep, excretory
pore; ev, excretory vesicle; fc, flame cell; gp, genital primordium; m, mouth;
os, oral sucker; pg, penetration gland; ph, pharynx; sp, spine; and vs, ventral
sucker.

of paratenic hosts, including frogs, snakes, birds, and mammals. These animals become infected by eating parasitized frogs. The mesocercariae have also been reported
in humans. The mesocercaria develops into a metacercaria in an additional intermediate host before it is ingested by the definitive host.

   The body of the mesocercaria is larger than that of the cercaria, and the excretory
system is more developed than that of the cercaria (Figure 1-36). In species of *Alaria*
there are usually four penetration gland cells in the region of the ventral sucker, and
their ducts, as in the cercaria, open at the oral sucker. The number of these cells appears to vary from species to species, but their cytoplasm always stains with Neutral
Red and Nile Blue sulfate. In the frog host the mesocercariae are usually free shortly
after the penetration of the cercariae, but later they become enclosed in a thin layer
of fibrous tissue. In paratenic hosts a varying degree of host response occurs, in some
cases leading to the formation of a capsule whose cavity is filled with fluid and the
mesocarcariae.

   Humans are among the paratenic hosts (see Volume II, Chapter 14 on Strigeids).
Only a few cases were reported but one has been fatal.[46] In this case several thousand
mesocercariae were estimated to be present in several locations: in the peritoneal cav-

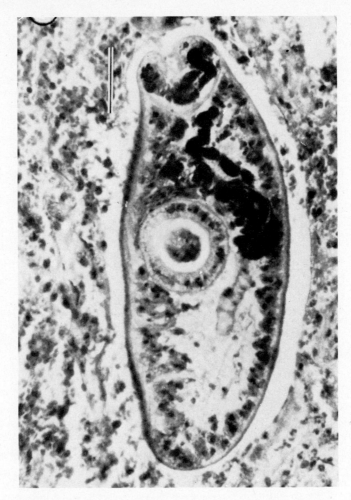

FIGURE 1-37.   Mesocercaria of *Alaria americana.* Longitudinal section
of mesocercaria in lung of a patient who died as a result of the infection.
Note lack of leucocytic response. Scale = 50 μm. (After Freeman R. S.,
et al., *Am. J. Trop. Med. Hyg.,* 25, 805, 1976. By courtesy of the *Am.
J. Trop. Med. Hyg.* and Dr. Freeman.)

ity, bronchial aspirate, brain, heart, kidneys, liver, lungs (Figure 1-37), lymph nodes,
pancreas, spleen, and stomach. The victim, from Canada, probably ate inadequately
cooked frog legs while hiking. The mesocercariae, about 300μ long, penetrated through
the stomach wall and spread to the various organs both directly and via the circulatory
system. Death occurred 9 days from the onset of symptoms, resulting from asphyxia-
tion due to extensive pulmonary hemorrhage, probably caused by immune-mediated
mechanism.

Other human cases are on record, where a mesocercaria was found in the retina of
the eye of a girl in Ontario, Canada, and mesocarcariae 500μ in length were found in
two areas of intradermal swelling removed from the upper thigh and iliac crest of a
man in Louisiana.[11] In the latter case there was much pathology due to the presence
of the mesocercariae.

i. *The Schistosomulum*

The schistosomes do not have a metacercarial stage in their life cycle, on the other
hand; the cercaria penetrates directly into the skin of its bird or mammalian host.

Shortly after that it is transformed into a "schistosomulum", which migrates in the host's body, eventually reaches the liver in which it grows to maturity, then moves on to its habitat location, the mesenteric venules or the vesico-prostatic and uterine plexuses of veins. The changes in environment which the cercaria passes through are quite remarkable. After exiting from the sporocyst where it develops, it settles in the snail host's tissues and is bathed in its hemolymph, then emerges from the snail to the fresh water or sea water environment, depending on the schistosome species. It swims freely in the water for a short period, up to 2 or 3 days (under laboratory conditions this period may be a matter of minutes), after which it penetrates the skin tissue. A remarkable feature of the cercaria is the swiftness with which it adapts to its different environments. It has been demonstrated that the free-living cercaria in the water and the newly arrived larva in the skin are within a short time different organisms, serologically, metabolically, and structurally.

The elimination of the acetabular secretory apparatus of the cercariae of schistosomes is one of the earliest events occurring in the transformation to schistosomulum. The space left within the cercaria is eventually filled with small undifferentiated cells resulting from an ensuing rapid cellular multiplication. The trigger for in vivo glandular secretion is the direct contact of the cercaria with the host skin, whereas the in vitro glandular secretion has been reported to be stimulated by many substances.

Morphological changes take place in the tegument of the schistosomulum. These involve the loss, interruption, or modification of the glycocalyx and replacement of the cercarial trilaminate membrane by the schistosomular heptalaminate one, a condition which is noted in 7-day old schistosomula recovered from the lung. Both changes occur quickly, that of the glycocalyx more so than that of the surface membrane. The changes in the surface membrane occur by the loss of the trilaminate membrane through casting off of microvilli and by binding to the inner aspect of the tegumentary surface membrane of coiled membranous inclusions. These are produced in Golgi regions of the perikaryons and moved through the cytoplasmic processes into the surface syncytium. This appears to be a continuous means of maintenance of the surface membrane of the vertebrate stages of the schistosomes.

In a review of the subject of changes from cercaria to schistosomulum Stirewalt[156] discussed the differential characteristics of cercariae (c) and schistosomula (s) of *Schistosoma mansoni*; these are summarized as follows: Tail present (c), absent (s); water-adapted (c), water-intolerant (s); serum complement sensitive (c), insensitive (s); oral sucker alternately protruded and retracted (c), permanently protruded (s); pre- and postacetabular glands full (c), evacuated or nearly so (s); locomotion by alternate attachment of suckers (c), restricted, worm-like (s); no ingestion (c), ingestion (s); glycocalyx intact (c), lost, interrupted or modified (s); surface membrane trilaminate, not pitted, no surface microvilli (c), heptalaminate pitted transient surface microvilli (s); Cercarienhüllen Reaktion positive in antiserum (c), negative (s); surface PAS and Alcian blue-positive (c), negative (s); membranous inclusions in tegument few (c), numerous (s); surface unstable in selected chemicals (c), stable (s); amino acid uptake low (c), high (s); enzyme activity in extract high (c), absent or low (s); and metabolism for energy (c), for synthesis (s).

B. Taxonomy of Digenea and Digenean Families

The classification of the Digenea has always been a difficult matter. Diverse opinions have been expressed over the last century, and various systems were recommended. The diversity in the taxonomic treatment of the Digenea resulted from differences of opinion concerning the significance and systematic importance of certain morphological and developmental features. Early systematics and grouping of the Digenea were based on merely morphological features of the adult flukes, especially the number and the position of the suckers, the shape of the body, whether it is divided into fore- and hindbody or not, and whether a head collar provided with spines is present or absent, etc. Thus there was the grouping into distomes, monostomes, amphistomes, holostomes, and echinostomes. Such features were realized later to be only of an adaptive nature and not of phylogenetic significance. However, the terms listed above are still being used, but not for taxonomical purposes: rather, in a descriptive way.

Life history studies later on revealed discrepancies and inconsistencies in the old grouping. Investigations on the morphology of larval stages helped in revealing some internal relationships among the various flukes. Of the features used in systematics were the morphological type of the cercaria, the miracidia, and the excretory system of these larval stages, especially in the cercaria. Thus whether there are one or two pairs of flame cells in the miracidium, the type and shape of bladder, the extent of the main collecting ducts, and the distribution and grouping of the flame cells, and whether the cercariae develop in rediae or sporocysts were taken into consideration. Making use of the homologies of the excretory systems in cercariae, and also the arrangement of the flame cells in the miracidia, La Rue first showed the probable relationship between the Strigeidae and the Schistosomatidae, families which could scarcely be linked on the basis of adult characters only. On the same bases are added to these two families the Clinostomatidae and the Cyathocotylidae, as closely related to the Strigeidae, and even the Bucephalidae. However, it has been emphasized later that the excretory system may also be misleading. Further life history studies indicated that the characteristics of the excretory system are not sufficiently consistent to indicate close relationships in some groups. Stunkard showed that three genera in the Heterophyidae show differences in their excretory system. McMullen demonstrated differences in the excretory system of members of the families Plagiorchiidae, Lecithodendriidae, Dicrocoeliidae, and Microphallidae; all had been grouped in the Plagiorchioidea. In these latter families the excretory bladder varies from sac-like to an I-, Y-, or V-shape.

One of the systems which is most favored by many parasitologists at present is that of La Rue.[82] The system takes as its primary considerations the formation of the cercarial excretory bladder, i.e., whether it has an epithelial lining or not. Thus in the superorder Anepitheliocystidia (a new name) comprising the orders Strigeatoidea, Echinostomida (a new name) and Renicolida (a new name), the thin-walled bladder of the early stages persists as a nonepithelial structure, although in some families, for example Fasciolidae, the bladder may become covered by muscle fibers and other tissues from the body wall. However, in the superorder Epitheliocystidia (a new name), comprising the orders Plagiorchiida (a new name) and Opisthorchiida (a new name), there is an epithelial bladder formed as a result of the migration of cells to the early embryonic bladder, the latter disintegrating and being replaced by a cellular wall.

La Rue, in the above system, also took into consideration certain morphological features of the miracidia (number of flame cells) and the cercaria, such as whether the latter develops in a sporocyst or in a redia and whether it encysts in the open or in a second intermediate host.

Trematode Families

In this book, the main basis for the order by which the digenean families are arranged is the significance and extent of geographical distribution of the infections or diseases which their member species produce in their hosts. Thus the schistosomes are considered first, followed by the paragonimids. However, it was found more practical in some cases to adhere to phylogenetic relationships, and thus the less significant forms are considered following, or in association with, one of the major disease producing agents. An example is the placement of the Nanophyetidae and other troglotrematids, following the Paragonimidae. Moreover, toward the end of the section on the

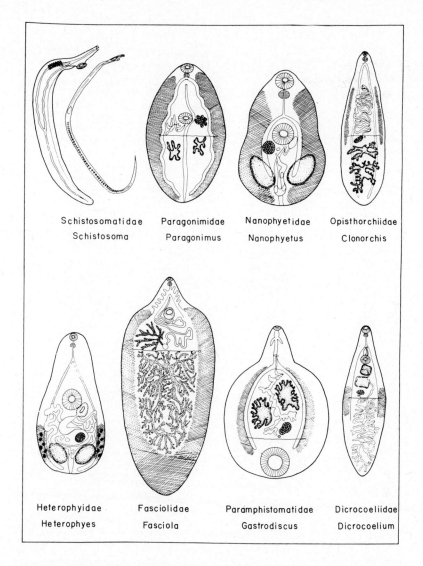

FIGURE 1-38.   Representatives of some digenean families considered in this book, for comparison of adult fluke morphology.

trematodes, the less pathogenic species and their families are treated in groups, in some cases based on their phylogenetic relationship. The diagnostic features of the families to follow, are based on the information in Yamaguti[193] Dawes[32] and the writer's trematode collection (Figures 1-38 and 1-39).

1. Family Schistosomatidae Poche, 1907

Digenea in which the sexes are separate. Pharynx absent, esophagus short. Ceca joining together to form a single slender limb terminating near posterior extremity. Suckers present or absent. Acetabulum, when present, anterior to genital pore. Body of male may be widened to form gynaechophoric canal. Testes four or more, anterior or posterior to cecal union. Cirrus pouch present or absent. Female more slender than male. Ovary elongate, sometimes spirally curved, anterior to cecal union. Uterus more or less winding, in intercecal field; eggs not operculated, embryonated when leaving the host, with terminal or lateral spine, or spineless. Vitellaria extending from ovarian, or pre- or postovarian, zone to posterior extremity. Parasitic in blood vessels of birds and mammals.

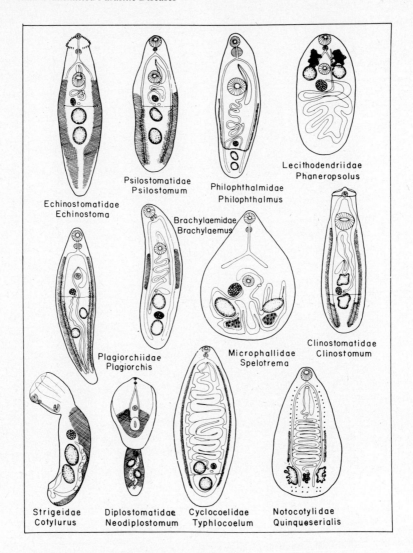

FIGURE 1-39.   Continuation of Figure 1-38 to compare morphological features of digeneans.

a. Subfamily Schistosomatinae Stiles and Hassall, 1898

Suckers present. Gynaecophoric canal well developed. Common cecum without lateral dendritic branches. Testes situated anterior to cecal union. Genital pore postacetabular. Female body cylindrical or nearly so, though attenuated anteriorly.

Belonging to this subfamily are all genera from mammals: *Schistosoma* Weinland, 1858; *Orientobilharzia* Dutt and Srivastava, 1955; *Bivitellobilharzia* Vogel and Minning, 1940; *Schistosomatium* Tanabe, 1923; *Heterobilharzia* Price, 1929.

Some genera occurring in birds also belong to this subfamily, and these genera are: *Ornithobilharzia* Odhner, 1912; *Austrobilharzia* Johnston, 1917 (syn. *Microbilharzia* Price, 1929).

The above genera comprise the following species: *Schistosoma*: *S. haematobium* (Bilharz, 1852) Weinland, 1858; (Syn. *Bilharzia magna* Cobbold, 1859; *B. capensis* Harley, 1864); *S. intercalatum* Fisher, 1934; *S. mattheei* Veglia and LeRoux, 1929; *S. bovis* (Sonsino, 1876) Blanchard, 1895; *S. leiperi* LeRoux, 1955; *S. spindale* Montgomery, 1906; *S. indicum* Montgomery 1906; *S. incognitum* Chandler, 1926 (Syn. *S. suis* Rao and Ayyer, 1933); *S. nasale* Rao, 1933; *S. mansoni* Sambon, 1907; *S. rodhaini* Brumpt, 1931; *S. japonicum* Katsurada, 1904; *S. margrebowiei* Le Roux, 1933; *S. hippopotami* Thurston, 1963.

Bivitellobilharzia: B. loxodontae Vogel and Minning, 1940.

Schistosomatium: S. douthitti (Cort, 1915) Price, 1929 (Syn. *S. pathlocopticum* Tanabe, 1923).

Heterobilharzia: H. americana Price, 1929.

Orientobilharzia: O. turkestanicum (Skrjabin, 1913); *O. datti* (Dutt and Srivastava, 1952); *O. nairi* (Mudalair and Ramanujachari, 1945); *O. bomfordi* (Montgomery, 1906); *O. harinasuti* (Kruatrachue et al., 1965). *Ornithobilharzia* (in birds): *O. odhneri* Faust, 1924; *O. canaliculata* (Rudolphi, 1819) Odhner, 1912; *O. pricei* (Wetzel, 1930).

Austrobilharzia (in birds): *A. variglandis* (Miller and Northrup, 1926) Penner, 1953.

There are three other subfamilies of the family Schistosomatidae and all members of these subfamilies parasitize birds.

b. Subfamily Bilharziellinae Price, 1929

Suckers present. Gynaecophoric canal well or poorly developed, or absent. Common cecum without lateral dendritic branches. Testes situated posterior to cecal union. Cirrus pouch present. Male genital pore usually away from acetabulum. Body of male cylindrical or flattened. Body of female distinctly flattened in region of common cecum; posterior extremity may be enlarged. Ovary equatorial. Female genital pore immediately postacetabular. Genera include *Bilharziella* and *Trichobilharzia*.

Bilharziella Looss, 1899: *B. polonica* (Kowalewski, 1895) Looss, 1899.

Trichobilharzia Skrjabin and Zakharow, 1920: *T. physellae* (Talbot, 1936) McMullen and Beaver, 1945 (Syn. *Pseudobilharzia quarquedulae* McLeod, 1937); *T. ocellata* (LaValette, 1854) McMullen and Beaver, 1945; *T. stagnicolae* (Talbot, 1936) McMullen and Beaver, 1945; *T. yokogawai* (Oiso, 1927) McMullen and Beaver, 1945; *T. anatina* Fain, 1955; *T. berghei* Fain, 1955; *T. nasicola* Fain, 1955; *T. spinulata* Fain, 1955; *T. schoutedeni* Fain, 1955; *T. maegraithi* Kruatrachue et al., 1968.

c. Subfamily Dendrobilharziinae Mehra, 1940

Body of male flattened. Gynaecophoric canal poorly developed. Common cecum with short lateral dendritic branches. Suckers absent. Testes situated posterior to cecal union. Cirrus pouch present. Male genital pore in anterior part of body. Body of female flattened nearly throughout. *Dendrobilharzia* Skrjabin and Zakharow, 1920: *D. pulverulenta* (Braun, 1901) Skrjabin, 1924; *D. loossi* Skrjabin, 1924; *D. anatinarum* Cheatum, 1941.

d. Subfamily Gigantobilharziinae Mehra, 1940

Body cylindrical or nearly so in both sexes. Suckers absent. Gynaecophoric canal poorly developed or absent. Testes situated posterior to cecal union. Cirrus pouch absent. Male genital pore at anterior end of gynaecophoric canal. Female with lateral lobe-like projections at posterior extremity. *Gigantobilharzia* Odhner, 1910: *G. acotylea* Odhner, 1910; *G. huttoni* Leigh, 1953; *G. sturniae* Tanabe, 1948; *G. gyrauli* (Brackett, 1940) Brackett, 1942; *G. huronensis* Najim, 1950; *G. egreta* Lal, 1937; *G. tantali* Fain, 1955; *G. monocotylea* Szidat, 1930.

2. Family Paragonimidae Dollfus, 1939

Lung flukes of mammals. Large, thick, and ovoid distomes, with spinose tegument. Acetabulum comparatively small, near midbody. Pharynx present; ceca long, undulating. Testes usually symmetrical, lobed; cirrus pouch absent. Ovary lobed, submedian, pretesticular. Vitellaria dendritic, lateral and dorsal, extending whole length of body. Uterus short, but strongly convoluted. Genital pore immediately postacetabular.

Miracidium with one pair of flame cells; cercariae microcercous xiphidiocercaria, develop in rediae, without appendages, in prosobranch (operculate) snails, and encyst in crustaceans. Genus *Paragonimus* Braun, 1899: *P. westermani* (Kerbert, 1878) Braun, 1899. Several other species will be discussed in Volume II, Chapter 1.

3. Family Nanophyetidae Dollfus, 1939

Small intestinal flukes, body pyriform or elongate; oral sucker subterminal, well developed; acetabulum, large in middle third of body. Testes symmetrical, in hindbody, sometimes occupying nearly whole lateral fields of hindbody (*Sellacotyle*). Cirrus pouch present (*Nanophyetus*) or absent (*Sellacotyle*); posterodorsal to acetabulum when present, enclosing bipartite seminal vesicle. Genital pore midventral, postacetabular. Ovary median, pretesticular. Uterus winding in postacetabular median area between testes. Excretory vesicle elongate saccular. Vitellaria extending dorsally from level of pharynx or esophagus to posterior extremity. Cercariae encyst in fish.

Genus *Nanophyetus* Chapin, 1927: *N. salmincola salmincola* (Chapin, 1927) Filimonova, 1968; *N. salmincola schikhobalowi* (Skrjabin and Podjapolskaja, 1931) Filimonova, 1968.

Genus *Sellacotyle* Wallace, 1935: *S. mustelae* Wallace, 1935.

4. Family Achillurbainiidae Dollfus, 1939

Large flat flukes inhabiting mammalian tissues with oral sucker and pharynx; acetabulum in anterior half of body. Intestinal ceca undulating. Testes in the form of numerous follicles, scattered mostly in hindbody; seminal vesicle tubular, winding; cirrus and cirrus pouch absent. Ovary slightly submedian, behind acetabulum; genital pore median at level of bifurcation of ceca. Uterus winding in front of ovary and between ceca. Vitellaria dense, follicular in fore- and hindbody.

Genus *Achillurbainia* Dollfus, 1939: *A. nouvelli* Dollfus, 1939; *A. recondita* Travassos, 1942.

5. Family Opisthorchiidae Braun, 1901

Members are parasites of several classes of vertebrates, where they mainly inhabit the bile ducts or gall bladder. Body small, to medium-sized, flat, translucent, usually elongate. Suckers usually rather weakly developed. Ceca simple, may or may not reach to posterior extremity. Testes diagonal or tandem, rarely opposite, usually in posterior part of body; they may be branched or ovoid; cirrus pouch usually absent. Ovary submedian, usually pretesticular. Vitelline follicles extracecal, in the form of narrow bands, varying in extent. Uterus with an ascending limb, with numerous coils, winding in the area between ovary and genital pore, and containing many small embryonated eggs.

Miracidia with one pair of flame cells; cercariae pleuro- or parapleurolophocercous, develop in rediae in prosobranch aquatic snails, and usually encyst in fish.

a. Subfamily Opisthorchiinae Looss, 1899

Genus *Clonorchis* Looss, 1907: *C. sinensis* (Cobbold, 1875) Looss, 1907.

Genus *Opisthorchis* Blanchard, 1895: *O. felineus* (Rivolta, 1884) Blanchard, 1895; *O. viverrini* (Poirier, 1886), Stiles and Hassall, 1896.

Genus *Amphimerus* Barker, 1911: *A. pseudofelineus* (Ward, 1901).

Genus *Cyclorchis* Luhe, 1908: *C. campula* (Cobbold, 1876) Luhe, 1909.

Genus *Pachytrema* Looss, 1907: *P. calculus* Looss, 1907.

Genus *Diasiella* Travassos, 1949: *D. diasi* (Travassos, 1928) Travassos, 1949.

b. Subfamily Metorchiinae Lühe, 1909

Genus *Metorchis* Looss, 1899: *M. albidus* (Braun, 1893) Looss, 1899; *M. conjunctus* (Cobbold, 1860) Looss, 1899.

Genus *Parametorchis* Skrjabin, 1913: *P. complexus* (Stiles and Hassall, 1894) Skrjabin, 1913.

6. Family Heterophyidae Odhner, 1914

Members are intestinal parasites of birds and mammals. Small or very small ovoid or pyriform spined flukes with weak suckers, the ventral usually enclosed in a ventro-genital sac. The latter variously modified containing one or more gonotyles and genital pore. Cirrus pouch absent, but male and female ducts fused to form a hermaphroditic duct with a common genital ejector opening into the genital sac. Seminal vesicle usually large, distinct and free in parenchyma. One or two testes. Uterus having transverse folds extending posterior to ovary, and even posterior to testes in some genera. Eggs not numerous. Vitellaria, follicular arranged on the lateral or the median sides of the ceca; generally in the posterior region, but variable in extent. Excretory vesicle Y-, V- or even T-shaped.

In general larval characteristics similar to those of Opisthorchiide. Several subfamilies among which are

a. Subfamily Heterophyinae Ciurea, 1924

Ventral sucker not enclosed in genital sinus; gonotyl postero-lateral to the ventral sucker, bearing a row of chitinous rodlets. Genus *Heterophyes* Cobbold, 1866, example *H. heterophyes* (Siebold, 1852) Stiles and Hassall, 1900.

b. Subfamily Metagoniminae Ciurea, 1924

Ventral sucker lateral, enclosed in genital sinus; gonotyl inconspicuous, in the form of one or two papilla-like bodies. Genus *Metagonimus* Katsurada, 1912, example *M. yokogawai* (Katsurada, 1912).

7. Family Fasciolidae Railliet, 1895

Members are parasites of herbivorous mammals and humans, where they inhabit the bile ducts, liver or intestine. Large flat, leaf-like distomes; anterior end of body cone-shaped in some genera. Suckers usually close to each other, may or may not be equal in size. Cuticle spined or not. Intestinal ceca simple or with numerous diverticula. Testes tandem, usually branched, but may be without branches or lobes. Cirrus pouch present and well developed. Ovary pretesticular, branched or entire; seminal receptacle reduced or absent. Vitellaria of small numerous follicles in a major part of the body, usually confluent posteriorly; follicles may be situated dorsal and ventral to intestinal ceca or only ventral to them. Uterus with relatively few coils, anterior to ovary, and with large operculate eggs. Genital pore preacetabular.

Miracidia with one pair of flame cells; cercariae gymnocephalous, develop in rediae with collar and appendages (procrusculi), in aquatic or amphibious pulmonate snails; cercariae encyst in the open on vegetation. Three subfamilies are

a. Subfamily Fasciolinae Stiles and Hassall, 1898

Body with (*Fasciola*) or without (*Fascioloides*) cephalic cone. Ceca dendritic, long; testes profusely branched; ovary branched; vitellaria dorsal and ventral to ceca (*Fasciola*), or only ventral (*Fascioloides*).

Genus *Fasciola* Linnaeus, 1758: *F. hepatica* Linnaeus, 1758; *F. gigantica* Cobbold, 1856.

Genus *Fascioloides* Ward, 1917: *F. magna* (Bassi, 1875) Ward, 1917.

b. Subfamily Fasciolopsinae Odhner, 1910

With or without cephalic cone; ceca simple, long; testes branched; vitellaria extending whole length of ceca, or confined to lateral fields of hindbody; uterus long or short. In *Fasciolopsis* cephalic cone absent; cirrus pouch extending far back of acetabulum; vitellaria confined to hindbody; uterus long. In *Parafasciolopsis* cephalic cone present; cirrus pouch preacetabular; vitellaria extending whole length of ceca; uterus short.

Genus *Fasciolopsis* Looss, 1899: *F. buski* (Lankester, 1857) Stiles, 1901.
Genus *Parafasciolopsis* Ejsmont, 1932: *P. fasciolaemorpha* Ejsmont, 1932.

c. Subfamily Protofasciolinae Skrjabin, 1948.

Intestinal ceca simple; testes round, median, tandem, at about middle of body; ovary round.

Genus *Protofasciola* Odhner, 1926: *P. robusta* (Lorenz, 1881) Odhner, 1926.

8. Family Paramphistomatidae Fischoeder, 1901

Members are intestinal parasites of several classes of vertebrates. Thick muscular body; with or without ventral pouch. External surface may carry papillae, but not spines. Acetabulum large, usually terminal, may be subterminal. Oral sucker terminal, with or without posterior pouch-like muscular diverticula. Pharynx absent; esophagus with a posterior muscular bulb in some genera. Ceca simple, usually sinuous, extending to posterior end of body near acetabulum. Testes tandem or oblique, may be lobed; usually in middle third of body. Ovary usually post-testicular. Vitellaria acinous or follicular, lateral. Uterine coils mostly intercecal, in the form of one ascending limb. Genital pore anterior with or without muscular genital sucker. A well-developed "lymph" system present.

Miracidia with one pair of flame cells; amphistome cercariae develop in rediae without collar or appendages, in aquatic pulmonate snails, and encyst in the open, usually on vegetation, a few species on skin of amphibians. Several subfamilies include

a. Subfamily Paramphistomatinae Fischoeder, 1901

Genus *Cotylophoron* Stiles and Goldberger, 1910: *C. cotylophorum* (Fischoeder, 1901) Stiles and Goldberger, 1910. Genus *Paramphistomum* Fischoeder, 1901: *P. microbothrium* Fischoeder, 1901.

b. Subfamily Brumptiinae Stunkard, 1925

Genus *Brumptia* Travassos, 1921: *B. bicaudata* (Poirier, 1908).

c. Subfamily Cladorchiinae Lühe, 1909

Genus *Stichorchis* (Fischoeder, 1901) Looss, 1902: *S. subtriquetrus* (Rudolphi, 1814).
Genus *Wardius* Barker and East, 1915: *W. zibethicus* Barker and East, 1915.

d. Subfamily Gastrodiscinae Monticelli, 1892

Genus *Gastrodiscus* Leuckart, 1877: *G. aegyptiacus* (Cobbold, 1876) Railliet, 1893; *G. secundus* Looss, 1907.
Genus *Gastrodiscoides* Leiper, 1913: *G. hominis* (Lewis and McConnel, 1876) Leiper, 1913.

e. Subfamily Pseudodiscinae Näsmark, 1937

Genus *Pseudodiscus* Sonsino, 1895: *P. collinsi* (Cobbold, 1875) Stiles and Goldberger, 1910. Genus *Watsonius* Stiles and Goldberger, 1910: *W. watsoni* (Conyngham, 1904) Stiles and Goldberger, 1910.

9. Family Dicrocoeliidae Odhner, 1910

Members are parasites of birds and mammals; some in reptiles, where they inhabit the liver, bile ducts, gall gladder, or pancreatic duct. Delicate, elongate, leaf-like, translucent distomes; small or medium size, with well-developed suckers which vary in relative size. Ceca long. Testes opposite, tandem or oblique, close behind ventral sucker; cirrus and cirrus pouch present, small. Ovary behind testes. Vitellaria well developed, lateral to ceca in the middle region of body. Uterus having descending and ascending limbs; uterus coils filling most of body posterior to gonads. Genital pore anterior to ventral sucker.

Eggs small, thick shelled, embryonated; ingested by land snails; xiphidiocercariae develop in sporocysts, and encyst in arthropods.

Two subfamilies; those parasitic in birds and mammals are in the subfamily Dicrocoeliinae Looss, 1899.

Genus *Dicrocoelium* Dujardin, 1845: *D. dendriticum* (Rudolphi, 1819) Looss, 1899; *D. hospes* Looss, 1907.

Genus *Athesmia* Looss, 1899: *A. foxi* Goldberger and Crane, 1911.

Genus *Eurytrema* Looss, 1907: *E. pancreaticum* (Janson, 1889) Looss, 1907.

Genus *Platynosomum* Looss, 1907: *P. fastosum* Kossack, 1910.

Genus *Zonorchis* Travassos, 1944: *Z. allentoshi* (Foster, 1930) Travassos, 1944.

10. Family Echinostomatidae Looss, 1902

Members are intestinal parasites of birds and mammals; a few in reptiles. Spinous elongate distomes with head collar surrounding oral sucker, and the collar usually provided with a single or double crown of large spines; larger than body spines. Oral sucker and pharynx present; acetabulum in anterior third of body, usually much larger than oral sucker. Ceca long, reaching to hindbody. Testes tandem, sometimes diagonal, variable in shape, in hindbody. Cirrus pouch present, well developed. Ovary median or submedian, pretesticular; seminal receptacle absent. Vitellaria follicular, lateral, posterior to level of ventral sucker. Uterus in intercecal field anterior to ovary, containing large eggs. Genital pore median, preacetabular.

Miracidia with one pair of flame cells; echinostome cercariae developing in aquatic snails, in rediae having a collar and appendages (procrusculi); cercariae may encyst in the open, or in snails, in amphibians or in fish.

Several subfamilies include:

a. Subfamily Echinostomatinae Looss, 1899

Genus *Echinostoma* Rudolphi, 1809:
E. revolutum (Froelich, 1802) Looss, 1899; *E. lindoense* Sandground and Bonne, 1940; *E. malayanum* Leiper, 1911.

Genus *Echinoparyphium* Dietz, 1909: *E. flexum* (Linton, 1892) Dietz, 1910.

Genus *Euparyphium* Dietz, 1909: *E. melis* (Shrank, 1788). Genus *Hypoderaeum* Dietz, 1909: *H. conoideum* (Bloch, 1872) Dietz, 1909.

Genus *Paryphostomum* Dietz, 1909: *P. sufrartyfex* (Lane, 1915) Bhalerao, 1931.

Genus *Petasiger* Dietz, 1909: *P. nitidus* Linton, 1928.

b. Subfamily Echinochasminae Odhner, 1910

Genus *Echinochasmus* Dietz, 1909: *E. donaldsoni* Beaver, 1941.

Genus *Patagifer* Dietz, 1909: *P. bilobus* (Rudolphi, 1819) Dietz, 1909.

c. Subfamily Himasthlinae Odhner, 1910

Genus *Himasthla* Dietz, 1909: *H. muehlensi* Vogel, 1933.

11. Family Psilostomatidae Odhner, 1911

Parasites of birds, rarely of reptiles and mammals. Anatomy similar to Echinostomatidae but lacking head collar and collar spines. Several subfamilies and genera include:

Genus *Psilostomum* Looss, 1899: *P. marilae* Price, 1942.
Genus *Sphaeridiotrema* Odhner, 1913: *S. globulus* (Rudolphi, 1819) Odhner, 1913.
Genus *Psilorchis* Tharpar and Lal, 1925: *P. hominis* Kifune and Takao, 1973.

12. Family Cathaemasiidae Fuhrmann, 1928

Intestinal parasites of birds. Body elongate, usually spinous. Suckers well developed, the ventral larger than oral. Oral sucker may form a pentagonal hood-like expansion. Internal anatomy silimar to Echinostomatidae but head collar and circumoral spines absent. Uterus entirely anterior to ovary. Genital pore preacetabular. Esophagus may have a pair of lateral diverticula. Several subfamilies and genera include:

Genus *Ribeiroia* Travassos, 1939: *R. ondatrae* (Price, 1931) Price, 1942.

13. Family Philophthalmidae Looss, 1899

Parasitic distomes commonly known as eye flukes since they occur in conjunctival sac of birds, occasionally in humans. Also parasites of cloaca and intestine of birds. Body elongate or oval, usually spinous. Acetabulum in middle third of body, larger than oral sucker. Pharynx well developed; intestinal ceca reaching to posterior border. Testes in posterior half, tandem or opposite, lobed or oval; cirrus pouch well developed. Ovary pretesticular; vitellaria U- or V-shaped in pretesticular lateral fields; uterine coils between testes and acetabulum, but may extend beyond these structures, uterine eggs containing oculate miracidia with redia. Genital pore at or near intestinal bifurcation, that is, anterior to ventral sucker. Excretory bladder of various shapes but has long, wide lateral arms. Eggs usually embryonated when leaving the host; miracidium with a fully developed redia; cercariae encyst in the open, almost immediately.

Genus *Philophthalmus* Looss, 1899: *P. gralli* Mathis and Leger, 1910.
Genus *Parorchis* Nicoll, 1907: *P. acanthus* (Nicoll, 1906) Nicoll, 1907.
Genus *Cloacitrema* Yamaguti, 1935: *C. michiganense* McIntosh, 1938.
Genus *Ophthalmotrema* Sobolev, 1943: *O. numenii* Sobolev, 1943.

14. Family Notocotylidae Lühe, 1909

Members are parasites of ceca and gut of birds and intestine of mammals. Small, elongate or oval flukes where ventral sucker absent (monostomes); oral sucker small; pharynx absent. Characteristic longitudinal rows of glands or ridges usually found on ventral surface of body. Two symmetrical testes close to posterior border; cirrus pouch well developed and long. Ovary median, usually intertesticular, may be posttesticular. The follicular or tubular vitellaria lateral to intestinal ceca, mostly anterior to testes. Uterus extensive and exhibits transverse coils in intercecal field between ovary and genital pore; numerous eggs with long bipolar filaments. Genital pore median, usually close to bifurcation of intestinal ceca, but may be far posterior to this level. Excretory bladder small but possesses two long arms which unite anteriorly. Cercariae monostomate; pharynx lacking, with a pair of adhesive organs present at posterior end of body, cercariae encyst in the open, on surfaces.

Genus *Notocotylus* Diesing, 1839: *N. attenuatus* (Rudolphi 1809); *N. urbanensis* (Cort, 1914) Hanah, 1922; *N. stagnicolae* Herber, 1942; *N. skrjabini* Ablassov, 1953.
Genus *Catatropis* Odhner, 1905: *C. verrucosa* (Froelich, 1789) Odhner, 1905.
Genus *Quinqueserialis* Skwortzow, 1935: *Q. quinqueserialis* Barker and Laughlin, 1911.

Genus *Ogmocotyle* Skrjabin and Schulz, 1933: *O. pygargi* Skrjabin and Schulz, 1933; *O. ailuri* Price, 1954; *O. indica* (Bhalerao, 1942) Ruiz, 1946.

15. Family Plagiorchiidae Lühe, 1901

Parasites of several classes of vertebrates, where they occur in intestine, gall bladder, ureters, or cloaca. Shape of body variable, elongate, pyriform, or oval. Intestinal ceca simple, unbranched, variable in length, reaching about the level of the midbody, or extending to posterior extremity. Testes in posterior third of body, oblique or opposite, rarely tandem; cirrus pouch well developed, usually dorsal to ventral sucker. Ovary is median, or submedian and pretesticular; vitellaria mostly lateral in posterior part of body; uterus with a descending branch between the two testes, reaching posterior part of body, and an ascending branch, also between the testes; genital pore median or submedian between oral and ventral suckers. Excretory bladder usually Y-shaped. Eggs embryonated when laid. Xiphidiocercariae develop in sporocysts in aquatic pulmonate snails, encystment usually in insects.

Genus *Plagiorchis* Lühe, 1899: *P. javensis* Sandground, 1940; *P. micracanthus* Macy, 1931; *P. muris* Tanabe, 1922.

16. Family Prosthogonimidae Nicoll, 1924

Distome flukes of birds, rarely occurring in mammals. In birds inhabit cloaca, bursa Fabricii, and oviduct; in mammals they occur in liver. Body flat, translucent, rounded posteriorly but tapering anteriorly; with acetabulum in anterior half of body. Intestinal ceca reaching close to posterior border. Two symmetrical testes, postacetabular; cirrus pouch well developed, elongated, located between the suckers. Ovary pretesticular; vitellaria in lateral fields, usually in groups in middle third of body; most of posterior half of body filled with uterine coils, containing small numerous eggs. Genital pore lateral to oral sucker or ventral to pharynx. Excretory bladder Y-shaped.

Genus *Prosthogonimus* Lühe, 1899: *P. ovatus* (Rudolphi, 1803) Lühe, 1899; *P. macrorchis* Macy, 1934.

Genus *Mediogonimus* Woodhead and Malewitz, 1936: *M. ovilacus* Woodhead and Malewitz, 1936.

17. Family Microphallidae Travassos, 1920

Adults resemble closely members of the family Heterophyidae. Parasites of intestine of several vertebrate classes, mainly birds; a few species in fishes and mammals, including humans and subhuman primates. Small to very small flukes, usually pyriform, spinose. Oral and ventral suckers small; suckers equal or ventral small; ventral sucker in posterior half of body. Ceca short, widely divergent. Testes opposite or slightly diagonal, cirrus sac transverse, anterior to ventral sucker. Ovary submedian, usually pretesticular. Coils of uterus filling space between posterior extremity and testes. Eggs small, numerous. Genital atrium near acetabulum, containing male papilla or accessory suckers in some genera. Vitellaria in small lateral clusters, variable in position. Excretory bladder V-shaped. Life cycle more or less similar to family Plagiorchiidae; xiphidiocercariae of ubiquita or armatae type, usually develop in sporocysts and encyst in crustaceans. Four subfamilies.

Genus *Spelotrema* Jägersköld, 1901: *S. brevicaeca* Africa and Garcia, 1935) Tubangui and Africa, 1939 (Syn. *Heterophyes brevicaeca* Africa and Garcia, 1935).

Genus *Microphallus* Ward, 1901: *M. nicolli* (= *Spelotrema nicolli*) Cable and Hunninen, 1938; *M. minor* Ochi, 1928.

18. Family Lecithodendriidae Odhner, 1910

Members parasitic in several vertebrate classes; but primarily parasites of bats. Some species in humans. Delicate flukes, spherical or elongate; spinous or nonspinous. Digestive system with pharynx; esophagus and intestinal ceca vary considerably in length, but usually short. Acetabulum small at or near middle of body. Testes usually symmetrical at varying levels in body; usually lateral to ventral sucker. Ovary submedian in anterior or posterior part of body. Cirrus pouch usually present, but may be absent; genital pore usually in forebody but variable in position, median, submedian, or lateral. Vitellaria follicular and in groups on each side, usually in anterior part of body, may be in posterior part. Several lobes of uterus in posterior part of body, containing numerous small eggs. Excretory bladder usually V-shaped. Xiphidiocercariae develop in sporocysts in aquatic pulmonate snails; cercariae usually encyst in insects; metacercariae advanced in development. Several subfamilies.

Genus *Lecithodendrium* Looss, 1896: *L. linstowi* Dollfus, 1931.

Genus *Allassogonoporus* Olivier, 1938: *A. marginalis* Olivier, 1938.

Genus *Cephalophallus* Macy and Moore, 1954: *C. obscurus* Macy and Moore, 1954.

Genus *Phaneropsolus* Looss, 1899: *P. bonnei* Lie-Kian Joe, 1951; *P. simiae* Yamaguti, 1954.

Genus *Prosthodendrium* Dollfus, 1931: *P. glandulosum* (Looss, 1896); *P. lucifugi* Macy, 1937.

19. Family Strigeidae Railliet, 1919

Intestinal parasites of mammals and birds. Body of many of the fluke members of this family divided into two regions, a forebody and a hindbody, by means of a transverse constriction. Forebody usually cup shaped or bulb shaped and contains a small oral sucker with or without pseudosuckers on each side; a small pharynx usually present, a small acetabulum, and a characteristic two-lobed foliaceous tribocytic organ occupying most of the forebody. Hindbody usually cylindrical containing reproductive organs. Two testes, tandem or diagonal; cirrus pouch absent; copulatory bursa well developed; ovary pretesticular; vitellaria follicular in fore- and hindbody or confined to either fore- or hindbody; genital pore terminal or near posterior extremity. Well-developed lymph or reserve bladder system present in the form of a prominent network and reservoirs throughout body. Eggs large, few. Miracidium with two pairs of flame cells; cercariae furcocercous, developing in sporocysts in aquatic pulmonate snails; metacercariae (Tetracotyle) in fish, snails, leeches. Subfamilies Strigeinae and Bolbocephalodinae in birds and Duboisiellinae in mammals.

Genus *Strigea* Abildgaard, 1793: *S. elegans* Chandler, 1947; *S. falconis* Szidat, 1928.

Genus *Apharyngostrigea* Ciurea, 1927: *A. bilobata* Olsen, 1940; *A. ibis* Azim, 1935; *A. pipientis* (Faust, 1918).

Genus *Parastrigea* Szidat, 1928: *P. cincta* (Brandes, 1888) Szidat, 1928.

Genus *Cotylurus* Szidat, 1928: *C. cornutus* (Rudolphi, 1809); *C. flabelliformis* (Faust, 1917) Van Haitsma, 1931.

Genus *Apatemon* Szidat, 1928: *A. gracilis* (Rudolphi, 1819) Szidat, 1928.

20. Family Diplostomatidae Poirier, 1886

Members are intestinal parasites of mammals and bids. Body divided into distinct forebody and hindbody. Forebody foliate or spatulate, with or without pseudosuckers, with an acetabulum and pharynx and with a well-developed round or elliptical tribocytic organ and compact gland below. Hindbody cylindrical, containing two tandem testes, and a pretesticular ovary. Vitellaria follicular in forebody or in hindbody or in both. Uterus short; eggs few, large. Genital pore usually dorsoterminally. Reserve bladder system well developed as in Strigeidae. Cercaria furcocercous, metacercaria

(Diplostomulum, or Neascus) in fish, amphibians, and snails. Two subfamilies, Diplostominae and Alariinae.

Genus *Diplostomum* v. Nordmann, 1832: *D. spathaceum* (Rudolphi, 1819) Braun, 1893; *D. flexicaudum* (Cort and Brooks, 1928) Van Haitsma, 1931.

Genus *Mesophorodiplostomum* Dubois, 1936: *M. pricei* (Krull, 1934) Dubois, 1936.

Genus *Neodiplostomum* Railliet, 1919: *N. americanum* Chandler and Rausch, 1947.

Genus *Fibricola* Dubois, 1932: *F. texensis* Chandler, 1942.

Genus *Uvulifer* Yamaguti, 1934: *U. ambloplitis* (Hughes, 1927) Dubois, 1938.

Genus *Alaria* Schrank, 1788: *A. alata* (Goeze, 1782); *A. canis* La Rue and Fallis, 1934.

21. Family Cyathocotylidae Poche, 1926

Parasites of birds and mammals (rarely in fishes). Body rounded, oval, or pyriform, division into two regions usually not distinct. Tribocytic organ circular, elliptical or saucer shaped. Oral sucker and pharynx present; acetabulum present or absent. Testes variable in position, usually tandem. Cirrus sac present near posterior end of body. Ovary intertesticular, or opposite anterior testis. Genital pore terminal. The follicular vitellaria variable in extent and position. Uterus short; eggs large, few.

Genus *Cyathocotyle* Mühling, 1896: *C. prussica* Muhling, 1896; *C. orientalis* Faust, 1922.

Genus *Prohemistomum* Odhner, 1913: *P. vivax* (Sonsino, 1892).

Genus *Mesostephanus* Lutz, 1935: *M. fajardesis* (Price, 1934) Lutz, 1935; *M. appendiculatoides* (Price, 1934) Lutz, 1935; *M. longisaccus* Chandler, 1950.

22. Family Clinostomatidae Lühe, 1901

Members include parasites of birds and mammals, in addition to reptiles. They inhabit the mouth and esophagus. Usually large or medium-size flukes. Oral sucker small or vestigial; anterior end of body may be retractile and oral sucker becomes surrounded by collar-like fold of body wall. Acetabulum very large in anterior third of body. Esophagus short; ceca simple or sinuous, or with conspicuous lateral branches; ceca long reaching to posterior extremity. Testes tandem in hindbody; cirrus pouch present. Ovary intertesticular, submedian. Uterus with ascending and descending lobes in intercecal fields, posterior to acetabulum. Vittelaria in lateral fields of hindbody, extending into intercecal area. Genital pore in posterior third of body. Cercariae brevifurcate; pharyngeate; oral sucker replaced by extensible penetration organ as in Schistosomatidae; acetabulum rudimentary; penetration glands as in Strigeidae and Diplostomatidae; eyespots pigmented; cercariae develop in rediae; three host life cycle. Three subfamilies.

Genus *Clinostomum* Leidy, 1856: *C. attenuatum* Cort, 1913; *C. marginatum* Rudolphi, 1809.

23. Family Brachylaemidae Joyeaux and Foley, 1930

Parasites of mammals and birds, rarely of amphibians; with well-developed oral sucker and pharynx; with well-developed or small acetabulum; very short esophagus and intestinal ceca reaching posterior border. Body usually elongate, rarely oval, with or without spines. Testes tandem or oblique in posterior part of body; cirrus pouch present, variable in position. Ovary between or anterior to testes or opposite anterior testis; vitellaria lateral, variable in extent; uterine coils (ascending and descending) between intestinal ceca, mainly anterior to testes and ovary, may or may not reach intestinal bifurcation; genital pore median or submedian usually near posterior border of body but may be a little further anteriorly. Excretory vesicle small, but has two long lateral arms. Embryonated eggs ingested by land snails and slugs; cercariae tailless, microcercous or obscuromicrocercous, develop in branched sporocysts. Cercariae may develop to metacercariae without encystment, in same snail or another snail of same or different species. However, *Leucochloridiomorpha constantiae* utilizes aquatic snails (*Campeloma decisum*); furcocercous cercariae develop in branched sporocysts; cercariae encyst in another snail of same species.

Genus *Brachylaema* Dujardin, 1843: *B. migrans* Dujardin, 1845; *B. helicis pomatiae* Diesing, 1850; *B. virginiana* Dickerson, 1930.

Genus *Postharmostomum* Witenberg, 1923: *P. gallinum* (Witenberg, 1923); *P. helicis* (Leidy, 1874) Robinson, 1949.

Genus *Entosiphonus* Sinitzin, 1931: *E. thompsoni* Sinitzin, 1931.

Genus *Ectosiphonus* Sinitzin, 1931: *E. rhomboideus* Sinitzin, 1931.

Genus *Leucochloridium* Carus, 1835: *L. problematicum* Magath, 1920.

Genus *Leucochloridiomorpha* Gower, 1938: *L. constantiae* (Müller, 1935).

24. Family Cyclocoeliidae Kossack, 1911

Members parasitic in body cavity, air sacs, trachea, or nasal cavity of birds. Medium size to large. Oral and ventral suckers vestigial or absent, but mouth terminal or subterminal. Alimentary tract with well-developed pharynx, short esophagus; ceca simple or with lateral diverticula, and ceca unite near posterior extremity to form a characteristic ring (cyclocoel). Testes intercecal, posterior, oblique, or opposite; cirrus pouch small. Ovary intertesticular or opposite anterior testis. Uterus with close transverse coils, occupying almost all of intercecal area; may extend laterally beyond ceca, and contains embryonated eggs without filaments. Vitellaria extracecal, along entire length of ceca; long, may unite posteriorly. Genital pore median, close to anterior end of body. Miracidium with redia; may hatch and penetrate aquatic snails in some species, or is ingested by land snails in others. Cercariae encyst inside the redia. Two subfamilies.

a. Subfamily Cyclocoelinae Stossich, 1902

Members have simple ceca, without lateral diverticula.
Genus *Cyclocoelum* Brandes, 1892: *C. macrorchis* Harrah, 1922.

b. Subfamily Typhlocoelinae Harrah, 1922

Members have ceca with lateral diverticula.
Genus *Typhlocoelum* Stossich, 1902. Genus *Tracheophilus* Skrjabin, 1913.

C. Infections with Digenea as Zoonoses

Digenetic trematodes are of world-wide distribution, where they parasitize a wide variety of vertebrate hosts. In several areas they are parasites of both animals and humans, and thus the infections represent typical zoonoses. Zoonoses have been defined by WHO and by FAO as "those diseases and infections which are naturally transmitted between vertebrate animals and man." Some felt that this definition is too wide, in that it includes infections that animals acquire from man that are merely incidental and of no public health importance. However, in spite of this and other drawbacks, the definition has been widely accepted, on condition that there should be either proof or strong circumstantial evidence that there is transmission between animals and man. Some classifications of the zoonoses have been proposed; they may indicate the direction and degree of infection. It has been argued that terms used in the classification of the zoonoses are difficult to pronounce, not to mention comprehend. In this brief account of trematode infections as zoonoses such terms are avoided.

There are several examples of trematode infections which fall under the category of zoonoses. Details of the zoonotic significance of each infection will be included, when applicable, in subsequent chapters of this book. It is the purpose of this present account to treat the trematodes together as a group of parasites producing infections that constitute zoonoses in some areas.

Some of the infections with Digenea occur primarily in animals but can be transmitted to man, whereas others occur primarily in humans but can be transmitted to animals. The extent of human infections, however, varies from one trematode to the other, especially the first category. Although it is hard to draw sharp lines in grouping the infections, the following is an attempt to do so.

1. Trematodes which are primarily parasites of animals, but on occasion are transmitted to man and thus do not constitute important zoonoses — (a) Incidental trematodes of man, which are very rare and occur mainly by chance or in unusual circumstances: examples are the opisthorchids *Metorchis conjunctus* and *Pseudamphistomum truncatum*; the echinostomatids *Echinochasmus perfoliatus*, *Himasthla Muehlensi*, *Paryphostomum sufrartyfex*, and *Echinoparyphium paraulum*; the clinostomid *Clinostomum complanatum*; the philophthalmid *Philophthalmus* sp.; the plagiorchid *Plagiorchis javensis*; the cyathocotylid *Prohemistomum vivax*; the mesocercariae of the diplostomatid *Alaria* spp.; the amphistome *Watsonius watsoni*; and the dicrocoelids *Dicrocoelium dendriticum* and *Dicrocoelium hospes*, and (b) Same as (a) but of comparatively higher prevalences only in certain limited areas: examples are the echinostomes *Echinostoma ilocanum*, *Echinostoma malayanum*, *Echinostoma hortensis*, and *Hypoderaeum conoideum*; the microphalid *Spelotrema brevicaeca*; the lecithodendriids *Phaneropsolus bonnei*, and *Prosthodendrium molenkampi*; and the schistosomes *Schistosoma bovis*, *S. rodhaini*, *S. indicum*, and *S. margrebowiei*.

2. Trematodes which are primarily parasites of animals, but high prevalences occur in man in certain areas, and constitute important zoonoses: examples are the fasciolids *Fasciola hepatica*, *F. gigantica*, and *Fasciolopsis buski*; the opisthorchids *Opisthorchis felineus* and *O. viverrini*; the schistosome *Schistosoma mattheei*; the heterophyids *Heterophyes heterophyes* and *Metagonimus yokogawai*; the amphistome *Gastrodiscoides hominis*; and the nanophyetid *Nanophyetus salmincola schikhobalowi*.

3. Trematodes which are of high prevalence in both animals and man, and both of these hosts are probably maintenance hosts of the flukes: examples are the opisthorchid *Clonorchis sinensis*; the paragonimids *Paragonimus* spp.; and the schistosome *Schistosoma japonicum*.

4. Trematodes which are primarily parasites of humans but can be transmitted to animals: examples are *Schistosoma mansoni*, *S. haematobium*, and *S. intercalatum*.

Some of the above trematodes and their zoonotic significance are discussed below.

Fasciola hepatica and *F. gigantica* are parasites of herbivorous animals, mainly sheep and cattle, in many parts of the world. In ordinary circumstances, human infection is rare, but outbreaks have often been reported, and have mainly been caused by the custom in some countries of eating watercress. Human infections can also be acquired by drinking water containing freely floating metacercariae. Sizeable outbreaks have occurred in some European countries, for example, England, France, Holland, and Italy; on Hawaii; and in several countries of Central and South America. Infections in France involved about 500 persons in 1956 to 1957. In England there were outbreaks in 1960, 1970, and 1972. In Peru 9% (range, 4.5 to 34.2%) of 1,011 children from six villages were infected in 1975 and had eggs in their stools. Although human infections are usually contracted by eating fresh watercress, it is believed that in some places such as Peru lettuce and alfalfa cause the infection, and not watercress. Only a few cases have been reported in Africa and Asia due to *Fasciola gigantica,* and it is believed that the infections are more common, but not adequately reported. In Ruwanda and Burundi ingestion of watercress caused some human infections, but chewing infected grass or eating green rice have also been responsible.

Certain animal schistosome species such as *Schistosoma bovis, S. indicum, S. rodhaini,* and *S. margrebowiei* produce a patent infection in man, but their occurrence has been relatively rare. Their zoonotic significance, however, stems from the fact that their cercariae in the water infect humans, and confer a certain degree of immunity, called heterologous immunity, against the more pathogenic and common schistosomes of man. This phenomenon, for which the term zooprophylaxis has been coined, was demonstrated in laboratory experiments using combinations of several species and strains of animal and human schistosomes and is also based on field observations.

Fasciolopsis buski is an intestinal fluke and is highly endemic in China (mainly in Chekiang Province), India, Southeast Asia, the Philippines, Taiwan, Hong Kong, and Indonesia. Metacercariae of this fluke are found encysted on several aquatic plants and their fruits. When the nuts are peeled off with the teeth before they are eaten, the cysts are released and are ingested. In addition to man other animals, especially the pig, are naturally infected. In considering the zoonotic nature of infections with *F. buski,* one of the problems is that the prevalence of infection in humans does not usually coincide with the prevalence in animals. In some areas in China there is a very high prevalence of infection among humans, whereas the parasite does not exist in pigs. In other areas infection rates are almost equal, and in still others infection rates may be moderate among pigs and very low among humans. The latter case can be explained by the fact that humans in these areas cook the plants before they eat them, rather than explaining the difference on the basis of the occurrence of different strains of the parasite, viz., a pig strain and a human strain.

Schistosoma mattheei is a common parasite among cattle in South, Central, and East Africa, but in this geographic area it also infects sheep, goat, zebra, horse, wild antelopes, baboons, and humans. In some parts of South Africa, especially in the Transvaal, infection rates up to 59% have been found among inhabitants of some villages. The eggs of *S. mattheei* are found in the feces or urine of the patients, who are usually also found infected with either *S. mansoni* or *S. haematobium.* It has been suggested that the importance of this zoonotic species will increase with the increase in livestock development in Africa.

Several members of the family Heterophyidae are intestinal parasites of carnivores and piscivorous mammals and birds. These hosts become infected by eating raw fish containing the metacercariae of the parasites. Normally when heterophyidiasis is well established in wild and domestic fish-eating mammals the population is at great risk of infection with the flukes. Evidence that human heterophyidiasis is of extensive world distribution is the frequent finding, at autopsy, of small spined flukes deep in the intestinal mucosa, in virtually all areas of the world where fish are eaten raw or inadequately cooked. At least nine species, belonging to six genera, have been reported as human parasites, mainly in the Orient, Southeast Asia, some Pacific islands, Australia, the Mediterranean area, southern and eastern Europe, and in West Africa. These species include: *Heterophyes heterophyes, H. heterophyes nocens, Metagonimus yokogawai, Haplorchis taichui, Haplorchis yokogawai, Haplorchis calderoni, Haplorchis vanissima, Stellantchasmus falcatus, Stamnosoma armatum,* and *Cryptocotyle lingua.* Among the commonest of the above species as parasites of man are *Heterophyes heterophyes* in areas bordering the Mediterranean, and *H. heterophyes nocens* and *Metagonimus yokogawai* in the Orient, where they also parasitize dogs, cats, and foxes. Next to these in the rate of occurrence in man are *Haplorchis taichui, H. yoko-gawai,* and *Stellantchasmus falcatus.*

Gastrodiscoides hominis is an amphistome fluke which infects humans mainly in India, Vietnam, Kazakhstan in Russia, and the Philippines. Animals are found infected with *G. hominis* in these areas in addition to Malaysia, Thailand, Java, and Burma, and among the important animal hosts are pigs, monkeys, Napu mouse deer, and rats. However, the role which these animals play in the transmission of the infection to man is not always known with certainty. Also not known is the true relationship between the human and animal flukes. It has been observed that in some areas where the fluke is endemic and of public health importance, its occurrence in pigs had not been noticed, and vice versa. It has been suggested that the flukes from these two hosts are different strains, and some have even named the pig fluke *Gastrodiscoides hominis* var. *suis,* but this has not been generally accepted.

Members of the genus *Nanophyetus* are restricted to two enzootic areas, namely, the northwestern region of the U.S. where the etiologic agent is *Nanophyetus salmincola,* and areas in Siberia, U.S.S.R., where the etiologic agent is *Nanophyetus salmincola schikhobalowi.* Infection is mainly prevalent among dogs, foxes, coyotes, and bears. However, in the enzootic area of Siberia the infection is also very common among humans. Although no human infections were ever reported in the enzootic area of the U.S. experimental infection in one person became patent.

Among the important zoonotic liver flukes are *Clonorchis sinensis, Opisthorchis felineus,* and *O. viverrini,* which are of wide occurrence in some parts of the world. *Clonorchis sinensis* is the most important and it was estimated that about 19 million people are infected in the Orient and Southeast Asia, with the highest prevalences being in mainland China, Japan, Korea, and Taiwan. In China the main endemic area of clonorchiasis is Kwantung province. *Opisthorchis felineus* is a liver fluke which infects humans in central Siberia, certain other parts of Asia, and Eastern Europe. *Opisthorchis viverrini* is frequently encountered in humans in Southeast Asia, especially in northeast Thailand and Laos. The infection is contracted through eating raw fish, a common food in all the areas where the parasites are encountered. The infection is thus shared with a number of animal reservoir hosts, principally dogs and cats, and a number of wild animals such as badgers, martins, and the Korean and Japanese otters and others have been found infected and play a role in the maintenance of the infections. Fish culture in the endemic areas favor the transmission of infection, because fresh grass, straw, pig excreta, and human nightsoil are used as fish food.

Paragonimiasis is a lung fluke infection in many parts of the Orient, Southeast Asia, West Africa, and South America. *Paragonimus westermani* is the principal human lung fluke in Japan, Korea, China, Taiwan, The Philippines, Asiatic U.S.S.R., and other parts of the Far East. In these endemic areas several mammalian reservoir hosts are found, among which are dogs, cats (both domestic and wild), panther, fox, wolf, leopard, wolves, tigers, pigs, and mongoose. In some West African countries, namely Cameroon, Liberia, Nigeria, and Zaire, the etiologic agents of the infections are *Paragonimus africanus* and *P. uterobilateralis.* Animals which were found naturally infected in West Africa include the dog, the mongoose, and certain subhuman primates. In South America the infections are endemic in Peru and Ecuador, and isolated human cases have been found in Central America. The taxonomic status of the etiologic agent(s) of the human infections in the Americas is not known with certainty but could be one or more of several species parasitizing animals such as cats, opossums, and raccoons. Among these lung fluke species are *Paragonimus peruvianus, P. amazonicus, P. mexicanus,* and *P. caliensis.* Infections with *Paragonimus* spp. are contracted through eating raw or improperly cooked freshwater crabs and crayfish in the Orient, and crabs in West Africa and South and Central America. In Japan, it was recently found that wild boars are naturally infected with the immature flukes and are usually eaten raw by hunters and other local inhabitants, and so are believed to be a source of human infections. Moreover, it seems that in nature several kinds of animals could serve as paratenic hosts of *Paragonimus*, and when eaten by wild carnivores, such as leopards and tigers, cause the natural infection in these hosts. Some of the factors which help in the human infections are the beliefs of certain communities in the Orient in the medicinal value of juices prepared from raw crustaceans and the consumption by women of certain tribes in the Cameroon of raw crabs in order to increase their fertility.

One of the important zoonotic aspects of paragonimiasis is that immature lung flukes are often encountered in ectopic sites in the human body. These are regarded by many parasitologists to be due to either infection by the lung fluke species which normally matures in the human lungs, or to represent a nonpatent infection by one of the several animal species occurring in these areas.

Schistosoma japonicum is an Oriental schistosome which occurs in Japan, China, Taiwan, the Philippines, the island of Sulawesi in Indonesia, Thailand, Laos, Cambodia, and Malaysia. It has been suggested by some parasitologists that *S. japonicum* evolved as a parasite of small mammals, possibly rodents, and subsequently became adapted to other mammals including man. A wide range of domestic and wild animals have been found infected, among which are dogs, cats, cattle, water buffalo, rodents, pigs, horses, donkeys, sheep, goat, and monkeys. Although some of these animals probably do not play an important role in transmission, others have been demonstrated to be of significance as maintenance hosts. It was shown by Pesigan et al.,[207] who determined the relative transmission index for the different hosts in Palo, Leyte, the Philippines, that humans contribute almost 75% of the infection and the rest is contributed by the reservoir hosts. This index, expressed as a percentage of the total role played by all the hosts involved is: man, 75.7; dog, 14.4; cow, 5.7; rat, 1.5; pig, 1.5; water buffalo, 1.2; and goat, 0.04. If the hosts are arranged according to their miracidia-producing capacity, not taking into consideration their relative population strengths, they are as follows: cow, 82.2; dog, 10.4; man, 5.6; water buffalo, 1.1; pig, 0.5; goat, 0.2; and rat, 0.01. Recently, Cabrera[196] showed that rats are important in transmission on Leyte, the Philippines. Eggs of *S. japonicum* were detected in livers of 405/557 (72.7%); in intestines of 270/550 (49.1%), and in the feces of 48/385 (12.5%). Eggs in the feces of the infected rats were determined by hatching techniques to be viable in 58% of 19 field rats examined. In Japan, where the prevalence of the disease in humans has declined in the last decade, it was also found that rats are probably important in transmission. Infection rates in the Tone River Basin, were 16.8% in rats, 0.4% in humans, and 7.6% in cows. One of the features of schistosomiasis japonica is the existence of several strains of the parasite and their infectivity and pathogenicity to various animals, factors which complicate the problem of determination of the relative role played by various animals in transmission. There is a zoophilic strain on Taiwan which infects a wide variety of domestic and wild animals but will not develop to maturity in humans. Moreover, although dogs, in addition to humans, are infected on Khong Island in Laos, it seems that the water buffalo (*Bubalus bubalis*) is not susceptible, as no natural infection was found in this animal, although it frequents the same waters where humans are believed to contract the infection. The schistosome species which infects humans on Khong Island was named *Schistosoma japonicum*-like schistosome by some investigators, while others regard it only as a strain of the *Schistosoma japonicum* occurring in the main foci, China, Japan, the Philippines, and Sulawesi. However, because its eggs are smaller than the parasite in other countries, its snail host is different (the aquatic *Lithoglyphopsis aperta*, and not the amphibious *Oncomelania* spp.), and the clinical picture manifested by it is different, it has recently been considered as a different species, *S. mekongi*.

Man is probably the most important host of *Schistosoma mansoni*. However, several animals have been found naturally infected, especially in South America and East Africa, and the question has always existed whether such animals are of significance in transmission of the infection among themselves and to man. There were some reviews[201,205,206] of this subject some 20 years ago. Among the hosts known at that time for *S. mansoni* were: monkeys, *(Cercopithecus sabeus)* on St. Kitts Island (imported to the island), and *Cercopithecus aethiops* in South and East Africa; baboons (*Papio doguera*) in East Africa; dogs in East Africa; opossums in Brazil; shrews in Egypt and Zaire (formerly the Belgian Congo); a wide variety of rodents in various parts of Brazil; and one rodent, the gerbil, in Egypt. More information about additional hosts has been added since then and experiments and field observations were made in an attempt to determine the role, if any, played by these animals in transmission. Cattle were found naturally infected in Brazil, but eggs were seen in the intestinal mucosa of only one.[195] It was also demonstrated that calves can be experimentally infected, and the worms develop to maturity with the production of a large number of eggs in the feces. Additional species of subhuman primates were found naturally infected; these included the squirrel monkey (*Saimiri sciurea*) in Surinam, *Cercopithecus* sp. in Tanzania and Ethiopia, and Patas monkey (*Erythrocebus patas*) in the region of Kano, Nigeria.[200] Baboons were found to be commonly infected in the Lake Manyara National Park, Tanzania. Snails were found infected in the nearby river, and 15 of 17 visitors to the area contracted *S. mansoni* and became sick after bathing in the river; no local or other human contacts were evident at the time, or 18 months later at the termination of the study.[199] Thus it was demonstrated that, at least in the study area, baboons can act as reservoir hosts of *S. mansoni* and transmit the infection to man, but it is still not known if they are true maintenance hosts of the parasite. It is possible that the baboons initially became infected as a result of human contamination, of one part of the river which is frequented by the baboons, by one of the several laborers who, shortly before the study and the outbreak of the disease among the visitors, had helped to construct a camp for a game research officer. The camp is situated by the bank of the river.

Natural infection with *S. mansoni* was also encountered in 5/5 wild-caught raccoons (*Procyon cancrivorus nigripes*), in the state of Minas Gerais in Brazil.[197]

Relative to rodents and their natural infection with *S. mansoni*, the rat *Nectomys squamipes squamipes* is often found infected in Brazil. It was also demonstrated experimentally that infection can be transmitted among infected and uninfected animals if they are placed together in an enclosure containing a ditch with snails. Two field rats were also found in 1976 in Venezuela passing viable eggs in their feces. The Nile rat, *Arvicanthis niloticus*, was found naturally infected with both *S. mansoni* and *S. haematobium* in Egypt.[204] It should be pointed out that the Nile rat is common in cultivated areas of Egypt, and it is possible that it becomes infected through contacts with cercariae-infected waters, or by ingesting infected snails when they become stranded on the mud as the water recedes in irrigation canals and drains. Recent investigations on Guadeloupe showed high infection rates among rats, *Rattus rattus* and *Rattus norvegicus*. Studies have been carried out since 1975 to determine the significance of such infections on transmission of human schistosomiasis and the results were summarized in 1978.[208] The conclusions were that in foci where there are human habitations, the role of rats is extremely slight, compared to that of man, and rats are incapable of maintaining the cycle there. In other foci where no humans live, where the water bodies containing the snails are in the form of large ponds, and where nearly all the rats are infected, they can maintain the cycle, but with very low infection rates among the snail *Biomphalaria glabrata*, and with very low cercarial densities (from 0 to 2.6 cercariae per liter of water as compared to 48 to 230 cercariae per liter in the other habitats inhabited by humans).

Relative to the natural infection rates among wild animals, the author[202] examined a small number of animals in the endemic foci on the island of St. Lucia, and found them negative for *S. mansoni*; these included eight rats, *Rattus norvegicus*, from Soufriere, Cul de Sac, and Castries; six mice, *Mus musculus*, from Castries; five mongoose, *Herpestes mungo*, from Castries and Dennery; and four opossums, *Didelphis marsupialis insularis*, from Dennery and Cul de Sac. However, experimental infections with *S. mansoni* cercariae of some wild rodents from Louisiana (a nonendemic area), some of which (or related species) are common to other parts of the Americas, indicated high susceptibility rates. Of those which passed numerous viable eggs in the feces were: 4 out of 4 rice rats (*Oryzomys palustris*); 2 out of 2 white-footed mice (*Peromyscus leucopus*), and 11 out of 14 nutrias (*Myocastor coypus*). Moreover, five out of eight experimentally exposed opossums (*Didelphis virgineanus*) passed viable eggs in the feces, and in 6 raccoons (*Procyon lotor*), also from Louisiana, although the worms developed to maturity and some eggs were found in the tissues, none were found in the feces.[203]

It is generally agreed, based on our present knowledge, that man is the only significant maintenance host of *Schistosoma haematobium*. It is realized that some animals have been found naturally infected with this schistosome, but the number of animals found infected of each species was very small. Among the animals found naturally infected are baboons and monkeys in East Africa, rodents in South Africa and Kenya, and pigs in Nigeria. Moreover, baboons (*Papio rhodesiae*) were found infected in Rhodesia, near Salisbury, and a chimpanzee (*Pan satyrus*) which had originated in Sierra Leone, West Africa and was imported into the U.S.[198] In Egypt the Nile rat (*Arvicanthis niloticus*) was found infected with *S. haematobium* in association with *S. mansoni*.

Schistosoma intercalatum is a member of the *S. haematobium* complex with terminally-spined eggs. It causes a human intestinal form of schistosomiasis in Gabon, Zaire, Cameroon, and the Central African Republic. Reports of natural infections in animals have so far been rare, and it is believed that in nature, like *S. haematobium*, humans are the only significant maintenance hosts. However, a wide range of laboratory hosts has been determined.[209] It is of interest to note that the first report of a schistosome in a primate was probably due to *S. intercalatum*. "Bilharzia magna" was described by Cobbold in 1859, from a sooty mangabey (*Cercocebus atys*), and although many authors have speculated that Cobbold's worm was *S. haematobium*, Wright et al.,[209] based on their studies on *S. intercalatum*, inferred that it could have been this species rather than *S. haematobium*.

D. The Molluscan Intermediate Hosts
1. Introduction and Classification of the Mollusca

The Phylum Mollusca comprises the snails, slugs, clams, oysters, squids, octopuses, and nautili. The word Mollusca is derived from the Latin *mollis*, meaning soft, and the term originated in reference to squids and cuttlefish, a group of the Mollusca of which the shell is always (except in the pearly nautilus) either reduced and covered by the soft flesh of the animal, or is completely absent. Later, the name extended to the more familiar snails, slugs, and bivalves as well. It is an appropriate term anyway on account of the very soft nature of the bodies of these animals. Molluscs have certain characteristics in common: a foot used in snails for creeping over surfaces, in clams for plowing, burrowing, or boring, and modified in squids for seizing the prey. All molluscs possess a structure called the "mantle" which envelops internal organs, namely those located in the pallial and visceral mass. The mantle is by far the most unique structure in members of this phylum. Besides enveloping internal organs, the principal function of the mantle is secretion of the protective shell, in those forms which possess one. A mantle collar is the free edge of the mantle which hangs down like a shirt's collar from the pallial region, covering, to a varying extent, the cephalic region. Actually it is a specialized groove in the mantle collar which performs the function of secreting the shell.

There are other characteristics of the Mollusca. The type of cleavage in the egg is similar in many forms. The cleavage and resulting veliger larva in many marine snails, bivalves, and chitons resemble the cleavage and trochophore larva of marine annelids. In general there is bilateral symmetry in adult molluscs; however, the viscera and shell are coiled in the gastropods and cephalopods. In addition there are three germ layers and no segmentation; the epithelial covering consists of one layer, mostly ciliated and with mucous glands lying underneath and opening at the surface. The digestive gland is complete, often U-shaped or coiled. The buccal cavity is provided with a buccal mass containing a food rasping organ, the radula, which carries transverse rows of minute chitinous teeth but the radula is absent in bivalves. There is usually an extensive digestive gland and often salivary glands. The circulatory system has a heart, with one or two auricles and a ventricle (usually in a pericardial cavity), an aorta, and a few other vessels, and numerous blood connective tissue sinuses (open system). Respiration is by one to many gills (ctenidia) or a "lung" which is a modified part of the mantle cavity and veins. The coelom is reduced to cavities of nephridia, gonads, and pericardium. The sexes are usually separate, but there are some forms which are hermaphroditic, and a few are protandric. Fertilization is external or internal. They are mostly oviparous; the egg cleavage is determinate and unequal. There is a veliger (trochophore) larva, or parasitic stage (Unionidae), or development is direct (Pulmonata, Cephalopoda).

Molluscs are of diverse form and size. Chitons vary in length from 1 to 18 cm. Snails vary from less than 1 mm to 24 cm or longer as in the conch, *Strombus*, but most snails are less than 5 cm in diameter or height. The shells of bivalves vary from 1 cm long up to 90 cm in the giant clam, *Tridacna*. Molluscs are of wide distribution in both time and space, having a continuous fossil record since Cambrian time. It is one of the largest phyla, next only to the Arthropoda in the number of known species. Estimates of about 80,000 are very modest, with the snail group making about three quarters of this number. Members of the Mollusca live in a variety of habitats: the littoral zone, deep sea, coral reefs, mudflats, rivers, lakes, man-made fresh-water bodies, forests, and in the deserts.

The Mollusca is evidently a very old group of animals, and the ancestral mollusc must have lived on earth many millions of years ago. This naturally is of significance from the point of view of the evolution of the Digenea and the involvement of molluscs in their life cycle. The nearest approach to the ancestral mollusc is the monoplacophoran genus *Neopilina*, discovered not long ago (in 1952) by the Danish Galathea expedition in the Pacific Ocean off the coast of Mexico at a depth of 3570 m. The group Monoplacophora was previously established for some fossils of Palaeozoic age which were thought to be primitive limpet-like gastropods. However, with the discovery of *Neopilina galathcae* and *N. ewingi* it is clear that the Monoplacophora are more primitive than the gastropods. *Neopilina* has a saucer-shaped shell, more or less circular in outline, the largest specimen measuring 37 mm long, 33 mm in width, and 14 mm high. *Neopilina* exhibits clear metamerism, more particularly in the gills, the excretory organs, the gonads, and the musculature. Thus it seems that it has closer relationships with arthropods, in view of the presence of a digestive gland, an open circulation, a reduced coelomic cavity, and the type of excretory organ known as mixonephridium in both molluscs and arthropods. It is possible that the two groups, molluscs and arthropods, together with annelids, might have evolved from animals possessing short segmented bodies. This view is now favored over an older one of an early turbellarian (flat-worm) ancestor.

The gastropods, comprising most of the intermediate hosts of the Digenea, have become modified structurally from the ancestral type by an alteration in the arrangement of the mantle cavity, an enlargement of the head and foot, an enlargement of the visceral mass in order to accomodate not only the viscera, but also a mantle cavity big enough to shelter the gills at all times and to allow the head to be withdrawn into it when the animal is disturbed or exposed to unfavorable circumstances.

Further development has resulted in a structure coiled in a spiral, the familiar helicoid spiral of the snail, but this is different from the process called "torsion". Torsion is independent of spiral coiling, and it occurs during the embryonic development of all gastropods. Some gastropods are not coiled, but all gastropods, at one time in their phylogenetic history, have undergone torsion. Torsion results in bringing the mantle cavity to the front of the body, while the visceral and pallial organs are altered in position by twisting through 180° in relation to the head and foot. This is the present situation with the soft parts of the gastropods. The majority of the intermediate hosts of the Digenea are gastropods; but some parasitize bivalves, and there are records of a few scaphopods being involved.

a. Classification of the Mollusca

The living molluscs are divided into six classes: Monoplacophora, Amphineura, Scaphopoda, Bivalvia (Pelecypoda), Cephalopoda, and Gastropoda.

1. Monoplacophora

Many old fossil monoplacophorans are on record. The only living representative is the genus *Neopilina*, with a few species, and it has been stated above that *Neopilina* was discovered only recently, off the Pacific coast of Mexico. Its body is unique among the Mollusca, being segmented, and the gills and many internal organs are paired and metamerically repeated. Such features have been taken into consideration in determining the affinities of the Phylum Mollusca.

2. Amphineura

Members of this class have an elongate body in which the head is reduced and the tentacles are absent. The shell, when present, has eight plates. There is a nerve ring around the mouth and two pairs of nerve cords. Amphineurans are divided into the Aplacophora (Solenogasters) and the Polyplacophora (Loricata). The Aplacophora are wormlike, without a shell, but with a thick integument bearing minute calcareous spicules, and the foot is rudimentary. They are bottom dwellers, occurring at 60 to 6000 ft, in ooze or on corals and hydroids, on which they feed. Examples are *Neomenia* and *Chaetoderma*. The Polyplacophora are the chitons, in which the body is elliptical and the foot is large and flat; gills, in pairs varying in number but usually numerous, are situated in a groove around the foot. The sexes are separate with one gonad, and there is a trochophore larva in the life hisory. The shell is in the form of a mid-dorsal row of eight broad plates. Examples are *Chiton, Cryptochiton*, and *Tonicella*.

3. Scaphopoda (Greek *skaphe*, boat + *podos*, foot)

In the Scaphopoda the shell and mantle are slenderly tubular, slightly curved and open at both ends. The body has a conical foot, delicate tentacles around the mouth, but gills are absent. The sexes are separate and there is only one gonad; there is a veliger larva in the life history. Scaphopods are the "Tooth Shells", or "Tusk Shells"; they are marine, living in sand or mud in shallow waters or down to 15,000 feet. Examples are *Dentallium* and *Siphonodentalium*.

4. Bivalvia or Pelecypoda (Greek *pelekys,* hatchet + *podos,* foot)

The shell is made up of two lateral valves, hence the common name bivalves. The valves are usually symmetrical, with dorsal hinge and ligament, and are closed by one or two adductor muscles. The mantle consists of right and left lobes, its margin commonly forming posterior siphons to control flow of water in and out of the mantle cavity. Bivalves lack a head, jaws and radula, but the mouth is provided with labial palps. The foot, often hatchet shaped, extends between the mantle lobes when in motion. The gills are usually platelike. Bivalves are usually dioecious, some are hermaphroditic. The gonad opens into the mantle cavity. In their life history, some have a veliger larva, while others have a glochidial stage. In the latter case, the eggs deposited in brood chambers (marsupia) in the gills of the female, develop, by total but unequal cleavage, into the glochidium larva. This larva has two valves which are closed by an adductor muscle and a long larval thread, and in some (*Anodonta*, and others) the valves may bear hooks ventrally, but others (*Unio, Quadrula,* and *Lampsilis*) lack hooks. When shed into the water through the female's exhalent siphon, the parasitic glochidia attach to fish, are surrounded by a host capsule, feed and grow, and later escape from the capsule, become detached and fall on the bottom to become free living. Bivalves are mostly marine, but some are in fresh water.

5. Cephalopoda (Greek *kephale,* head + *podos,* foot)

These are the most highly developed molluscs, and include the squids, octopuses, and nautili. The shell may be external or internal, or may be absent. The body of the cephalopod has a large head, with conspicuous and complex eyes. The mouth is provided with horny jaws and is surrounded by eight or ten arms or many tentacles. A radula is present in the buccal cavity. The nerve ganglia are grouped in the head, and are enclosed in a cartilage-like covering. Cephalopods are all marine, are dioecious, and their development is direct. There are two orders: the Tetrabranchia has shell external, coiled in one plane, divided by internal septa; two pairs of gills; and several tentacles. Examples are the ammonites, *Ammonites,* and the nautili, of which there is only one living genus of about four species, *Nautilus pompilius,* the pearly nautilus, found in the eastern Pacific and Indian Oceans at depths to 1800 ft. The order Dibranchia's shell is internal and reduced, or absent and the body cylindrical or globose, often with fins. There are eight or ten arms, with suckers, and one pair of gills and nephridia. There are eyes with lens, and there is an ink sac. Among the Dibranchia, the Decapoda have ten arms; examples are *Spirula,* in deep tropical seas, occasionally in the North Atlantic, *Sepia,* the cuttle-fish in Atlantic waters, *Loligo,* the squid, and *Architeuthis,* the giant squid. Another group of the Dibranchia, the *Octopoda,* have eight arms; examples are *Argonauta,* the paper nautilus and *Octopus,* the common octopus or devilfish.

6. Gastropoda (Greek *gaster,* belly + *podos,* foot)

These are the snails, slugs, limpets, whelks, conchs, etc. Gastropods are very abundant in salt and fresh water and on land. They constitute the largest molluscan class, and they occur from the tropics to subpolar regions, to depths of 17,000 ft in the sea and at heights to 18,000 ft in the Himalayas. Most marine gastropods produce large numbers of eggs that develop into a veliger larval stage before metamorphosing to the adult form. Land snails occur from moist tropical regions to temperate climates and also in arid deserts. Fresh-water snails have occupied the majority of waterbodies on the earth.

In the gastropods, the visceral mass is usually asymmetrical in a spirally coiled shell, the latter being either right- or left-handed (shell in some is conical, reduced, or absent). In the body of a gastropod, the head is united with the foot to form the head-foot, bearing one or two pairs of tentacles and one pair of eyes. They are monoecious or dioecious, with 1 gonad, usually with special ducts. They are mostly oviparous, some are ovoviviparous. The class Gastropoda is usually divided into three subclasses (considered by some as orders).

a. Opisthobranchiata

The shell is small or absent. There are two pairs of tentacles, with the eyes at the base of the posterior pair. They possess gills (except the nudibranchs), and the gills are posterior to the heart. Opisthobranchs are monoecious, and the genital pores are separate. They are all marine. Examples are *Haminea*, with shell; *Tethys*, sea hare, reaching up to 30 cm long, shell vestigial and internal; and *Doris, Aeolis* and *Polycera*, the nudibranchs or sea slugs, in which the shell is absent, and respiration is by skin, or by papilla-like gills.

b. Prosobranchiata

The gills are anterior to the heart, and both gills and mantle cavity are facing anteriorly. The shell is often large, usually with a calcareous or horny operculum, hence the name "operculates". The head part of the head-foot is often snout-like, with two nonretractile tentacles and two eyes. They are dioecious, oviparous, and the eggs are usually in capsules; a few are ovoviviparous. Prosobranchs are mostly marine, but some representatives are in fresh water, and a very few are terrestrial (land operculates). The prosobranchs are subdivided into three orders.

Order Archaeogastropoda (the Diotocardia) — This order comprises the oldest prosobranchs. They are all marine, herbivorous or deposit scrapers, and are characterized by having two auricles, and usually two gills, each with two rows of lamellae. They include the true limpets, the Patellacea, *Patella, Patina, Haliotis, Fissurella, Acmae*, etc. and also comprises the Neritacea.

Order Mesogastropoda — This order contains the majority of the prosobranchs. Members have a single auricle and a single gill with a single row of lamellae. The gills are comb-like and are attached to the mantle along their entire length. The radular ribbon is long, with seven teeth in each row. They live over sandy or muddy bottoms, and some forms burrow into the substratum. Representative marine families are Littorinidae, Potamididae (= Cerithiidae), Turritellidae, and Scalidae. Their shells are top-shaped spire, or fusiform, or spindle-shaped. Some limpets are also represented in this order, e.g., Capulidae. Representative families from fresh water are: Hydrobiidae, (*Oncomelania, Pomatiopsis, Amnicola, Aroapyrgus, Bithynia, Parafossarulus*), Pleuroceridae (*Pleurocera, Goniobasis, Semisulcospira*), Thiaridae (*Thiara, Brotia, Cleopatra*), Viviparidae (*Campeloma, Viviparus*), and Pilidae (= Ampullaridae) (*Pila, Pomacea, Marisa, Lanistes*).

Order Neogastropoda — This order comprises marine snails which are surface dwellers on rocky shores, as well as burrowing forms. The general morphology of the body is similar to the Mesogastropoda, but members are characterized by one to three transverse rows of strong radular teeth. Examples are the families Muricidae, Buccinidae (the true whelks), Nassariidae, Fasciolariidae, Galeodidae, Volutidae, Olividae, and Conidae. Members of the Conidae have two lanceolate teeth in each row of the radula and are associated with a poison gland.

c. Pulmonata

These are the fresh water and land snails and slugs and are mostly small, with a simple spiral shell, or it may be reduced or absent. There are one or two pairs of tentacles. Gills are lacking; the mantle cavity is anterior and its vascular lining forms an air-breathing lung, which opens on the right side by a contractile pore, the pneu-mostome. They are monoecious (hermaphroditic), with a single gonad, are mostly oviparous, and their development is direct. The Pulmonata is divided into three orders.

Order Basommatophora — The members of this order are mostly fresh-water snails, but there are some representatives on land and in salt water. They have one pair of contractile tentacles with eyes at the bases, hence the designation Basommatophora. The male and female genital tracts open separately at the exterior, but the male and female genital pores are close, with the male pore near the tentacle. Representatives in fresh water include Families Planorbidae (*Biomphalaria, Bulinus, Segmentina, Anisus, Planorbis*, etc.); Physidae (*Physa, Aplexa*); Ancylidae (*Ferrissia*); and Lymnaeidae (*Lymnaea, Fossaria, Stagnicola*, etc.). Representatives in salt or brackish water include Families Siphonaridae (*Siphonaria*) and Gadiniidae (*Gadinia*). Representatives on land include family Carychiidae (*Carychium*). Members of the Ellobiidae (*Melampus, Phytia*) are terrestrial or semiamphibious.

Order Stylommatophora — These are the terrestrial snails and slugs. The head-foot has two pairs of tentacles, with the eyes at the tips of the posterior (dorsal) tentacles. The tentacles can be everted and inverted. Slugs may possess a rudimentary and concealed shell or none at all. Representative families of land snails include Helicellidae (*Helicella, Hygromia*); Polygyridae (*Polygyra*); Endodontidae (*Anguispira, Helicodiscus*); Cioneilidae (*Cionella*); Succineidae (*Succinea*); and Helicidae (*Helix*). Representative families of slugs include Philomycidae (*Philomycus, Pallifera*); Limacidae (*Deroceras, Limax*); and Arionidae (*Arion*).

Order Systellommatophora — The members of this order are tropical slugs, which possess two pairs of contractile tentacles with eyes at the tip of the posterior (dorsal) pair. The shell is usually absent. A representative family is Veronicellidae (*Veronicella*) (Figure 1-40).

b. Gross Morphology of the Gastropods

Because the majority of the molluscan intermediate hosts of the Digenea fall under the class Gastropoda, a brief consideration of the members of this class is included in this chapter, rather than a treatise on the morphology of all other molluscan classes. The soft parts (animal) of the gastropod are usually enclosed in a shell. Slugs of the subclass Pulmonata may or may not have a shell, and when it is present it is rudimentary, flat, and concealed beneath the mantle.

1. The Shell

The shell consists of three layers: the periostracum, the prismatic layer, and the nacreous layer. The periostracum, or epidermis, is a thin outer protective layer comprised of conchiolin, which is a protein-polysaccharide complex secreted by a groove in the mantle edge, or collar. The periostracum protects the underlying layers from the effect of acidic elements in the water. The prismatic layer, constituting the main part of the shell, lies underneath the periostracum and consists of crystals (prisms) of calcium carbonate in the form of either argonite or calcite or both. These two types, when present, occur in separate layers, and the crystals are embedded in a matrix or lattice of conchiolin. The mantle edge also secretes the crystalline layer, but from a region which is situated more to the interior than that producing the periostracum, in order to protect the crystals from the effect of the environment. It has been demonstrated that the calcium utilized in shell formation may enter the mantle directly, as well as through other parts of the mollusc with transfer to the mantle by the hemo-lymph. The innermost layer of the shell, that is, the layer which borders on the mantle surface, is the nacreous layer which is smooth, composed of calcium carbonate, and

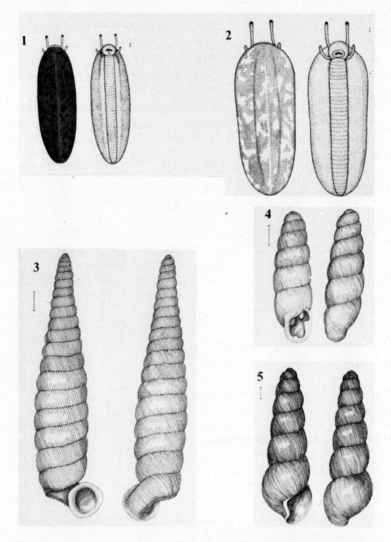

FIGURE 1-40.   Some slugs and terrestrial pulmonate snails, illustrated. (1) and
(2) are tropical slugs of the family Veronicellidae, order Systellomatophora, of
the subclass Pulmonata. (1) *Veronicella floridana* (Leidy), dorsal (left) and ventral (right) views, (2) *Vaginulus occidentalis* (Guilding), dorsal (left), and ventral
(right) views. Note in (1) and (2) the two pairs of contractile tentacles with the
eyes at the tip of the dorsal pair. (3), (4), and (5) are terrestrial pulmonate snails.
(3) *Brachypodella portoricensis* (Pfeiffer), (4) *Gulella bicolor* (Hutton); an
Asiatic snail introduced into various countries. Note apertural teeth. (5) *Subulina octona* (Bruguiere); has a circumtropical distribution (found in Malaysia,
Hawaii, South America, Puerto Rico, St. Lucia, etc.). (Illustrated with the help
of Mrs. Linda Stoner.)

is secreted by epithelial cells from almost the entire mantle surface. In the bivalve
molluscs these are the cells which are stimulated by foreign objects, for example inert
material or dead digenean metacercariae, form invaginations in the mantle, and secrete
concentric layers of nacre to surround the object and in this way form a pearl.

   Some gastropod shells which are common in certain parts of the world are shown
in Figure 1-41. The shell of the gastropod mollusc is usually in the form of a simple,
regular spire of various shapes, and the shell aperture in the prosobranchs only is
provided with a calcareous or horny operculum, which is attached to the foot of the

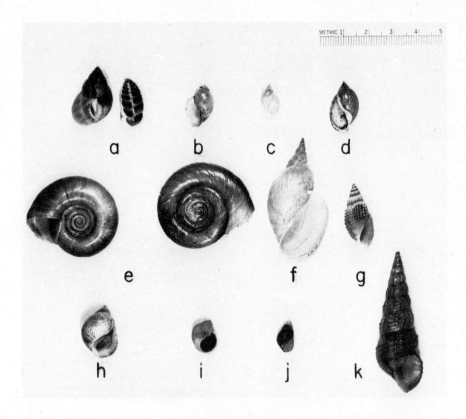

FIGURE 1-41.   Various shells of fresh-water and marine snails, to show shell features. (a) two shells cut longitudinally to show columella and whorls, (b) *Bulinus (Physopsis)* sp., (c) *Lymnaea gedrosiana*, Iran, (d) *Physa anatina*, Louisiana, (e) *Biomphalaria glabrata*, Brazil, (f) *Lymnaea stagnalis*, Minnesota, (g) *Thiara granifera*, Philippines, (h) *Littorina irrorata*, Gulf of Mexico, (i) *Littorina littorea*, Scotland, (j) *Bithynia goniomphala*, Thailand, (k) *Brotia* sp., Thailand.

animal. The growth lines of the operculum are of taxonomic importance. These may be circular lines revolving around a central point (concentric operculum), or they may be spirally arranged; when they are present in large numbers, it is a multispiral operculum, and when only few are present it is a paucispiral operculum. The shape of the operculum is also diagnostic; it may be oval, round, or spindle shaped. The spiral coiling is characteristic of shells, and the segments forming the body of the shell, known as whorls, revolve around a centrally situated pillar, the columella, oriented along the longitudinal axis inside the shell, and cannot be observed from the exterior. Its basal part, however, can be seen, and it may be straight, twisted, or abruptly terminated (cut off); in the latter case the columella is designated as being truncate, as in species of the subgenus *Physopsis* of the planorbid genus *Bulinus*. The spiral coiling of the shell may be right-handed (dextral shell: lymnacids, most of freshwater and marine prosobranchs), or left-handed (sinistral shell: physids, bulinids). The general shape of the shell varies, and it may be flat discoidal (planorbids, such as *Biomphalaria, Planorbis, Helisoma, Indoplanorbis, Segmentina, Hippeutis*, and *Anisus*); elevated (the majority of shells, and they may be conical, globose, or turrete); or patelliform (fresh water and marine limpets) Figure 1-42. Shell markings on the surface are in the form of both sculpture and color. Figure 1-43 and 1-44. A shell surface may be smooth or it may possess obvious or fine striations, both axial (growth lines) and/or

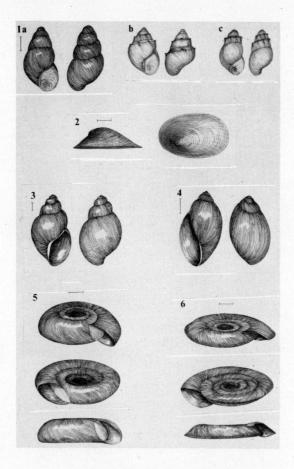

FIGURE 1-42.   Some fresh-water snails of the Caribbean island of St. Lucia, representing one prosobranch family, the Hydrobiidae, and the four families of the freshwater Basommatophora, showing some shell features. (1) *Pyrogophorus parvulus* (Guilding), Hydrobiidae; a, smooth form; b and c, coronate forms. Note paucispiral operculum. (2) *Gundlachia radiata* (Guilding); (Ancylidae); lateral and dorsal views showing characteristic radiating lines of the shell, of this freshwater limpet. (3) *Fossaria cubensis* (Pfeiffer) (Lymnaeidae); Dextral shell. (4) *Physa cubensis* Pfeiffer (Physidae); Sinistral shell. 5. *Drepanotrema lucidus* (Pfeiffer) (Planorbidae); upper (right, or umbilical) side, lower (left, or spire) side, and lateral view. 6. *Drepanotrema depressissimum* (Moricand) (Planorbidae); same views as in (5). (Illustrated with the help of Mrs. Linda Stoner.)

spiral. Certain structures may be present on the surface of some shells, and these include ribs or costae (*Bulinus (Bulinus) forskalii, Oncomelania hupensis*); tubercles or beads (*Thiara granifera, Thiara tuberculata, Brotia* spp, *Nassarius obsoletus, Cerithidea cingulata*); spines (many members of the marine family Muricidae); papillae; and hairs (especially in terrestrial gastropods). In some shells, the axial and spiral sculpture of the surface forms a reticulate pattern (*Viviparus* spp). Raised spiral ridges or keels characterize some shells, such as those of *Parafossarulus*. The aperture of the shell may be drawn to form a canal at the base, such as in the fresh-water family Pleuroceridae (*Goniobasis, Pleurocera*), and the marine family Cerithiidae (*Cerithidea*).

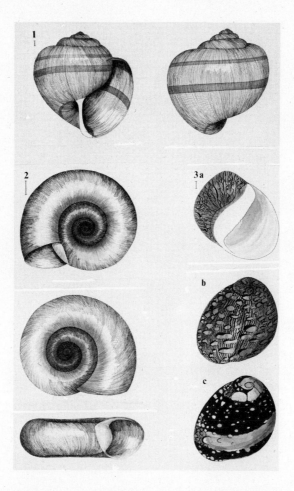

FIGURE 1-43.   Some fresh-water snails of St. Lucia (illustrated). (1) *Pomacea glauca* (Linnaeus), note spiral bands
on body whorl, (2) *Biomphalaria glabrata* (Say); three
views; lower (left or spire) side; upper (right or umbilical)
side; lateral view showing shell height, (3) *Neritina punctulata* Lamarck, showing different color patterns. (Illustrated
with the help of Mrs. Linda Stoner.)

2. The Soft Parts (Animal)

The body of a gastropod mollusc is divided into a head, foot, mantle region (pallial
region), and a visceral mass (see Figures 1-45 and 1-46.) The head and foot are united
to form a head-foot region which bears the tentacles and eyes, and on which are situated the mouth and the genital apertures; the anal opening is situated at its demarcation
from the pallial region. The mantle embraces the "neck" of the snail, and covers the
pallial region and visceral mass. Although its basic structure is the same, the mantle
varies at its edge, the mantle collar, in the area covering the pallial region and that
covering the visceral mass. In some prosobranchs an elongation from the mantle
known as the siphon fits into the siphonal, or anterior, canal of fusiform or spindle-
shaped shells. The osphradium is a chemoreceptor present in both prosobranchs and
pulmonates, and situated at the mantle edge or at the base of the siphon, when the
latter is present. The osphradium varies in size in different forms, and is believed to
function in detecting food substance at a distance, and in testing the quality of the
water entering the mantle cavity.

FIGURE 1-44.   Some terrestrial (1 and 2) and fresh-water snails of the Caribbean island of St. Lucia. (1) *Pleurodonte orbiculata* (Férussac), (Camaenidae), adult specimens. Note closed umbilicus, and thick and reflected peristome, (2) *Pleurodonte orbiculata*, medium size, and immature specimens. The immature specimen in the center shows an open umbilicus, (3) *Biomphalaria glabrata* (Say), (Planorbidae); different views, (4) *Pomacea glauca* (Linnaeus), (Ampullaridae). Note operculum mounted separately, (5) *Neritina punctulata* Lamarck, (Neritidae). Note callus, and operculum.

3. The Digestive System

The digestive system of gastropods starts at the mouth opening, which is situated anteriorly or antero-ventrally on the head-foot. The mouth leads into a buccal cavity containing the buccal mass which encloses the radular ribbon. Leading from the buccal cavity is an esophagus, which is followed by the stomach, then an intestine which makes a distinct loop and turns anteriorly to open at the anus, which is situated in the region of the mantle collar. There are two jaw-like plates at the mouth opening, opposed to which is a chitinous rasping ribbon, the radula, resting on a tongue, or odontophore. The radula shows a great diversity among gastropods, and the number, shape, size, and position of the cusps on the central, lateral, and marginal teeth are important taxonomic features. In pulmonates the rows of teeth on the lingual ribbon may be V-shaped, as in members of the Physidae, or in straight horizontal lines as in the other families. In members of the Planorbidae the central tooth is bicuspid, the lateral teeth are large and either bi- or tricuspid, and the marginals are long, narrow and multicuspid or serrated. In the Lymnaeidae the central tooth is unicuspid. The number of teeth in each row in the radulae of prosobranch gastropods is much less than that in radulae

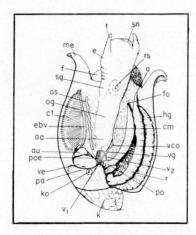

FIGURE 1-45(A).   *Littorina littorea*: animal removed from shell and mantle cavity opened middorsally to display its contents. Some other structures are seen by transparency. a, anus; aa, anterior aorta; au, auricle; cm, columellar muscle; ct, ctenidium; e, eye on eye stalk; ebv, efferent branchial vessel; f, foot; fo, female opening; hg, hypobranchial gland; k, kidney; ko, kidney opening; me, mantle edge; og, esophageal gland; os, osphradium; pa, posterior aorta; po, pallial oviduct; poe, posterior esophagus; r, rectum; rs, radular sac; sg, salivary gland; sn, snout; t, tentacle; v_1, nerve to heart and kidney; v_2, genital nerve; vco, visceral connective; ve, ventricle; and vg, visceral ganglion. (Redrawn after Fretter, V. and Graham, A., *British Prosobranch Molluscs*, Royal Society of London, 1962.)

of pulmonate gastropods. In the prosobranchs basal denticles may or may not be present on the central tooth. In the Hydrobiidae, for example, the central tooth is multicuspid, with basal denticles; the laterals are hatchet shaped and multicuspid, and the marginals are slender and multicuspid. In the Thiaridae, the central tooth is multicuspid and without basal denticles, and there are only one lateral and two marginal, each with few or many cusps. In all gastropods a "radular formula" depicts the number of lateral and marginal teeth in each row, on either side of the central (1).

The esophagus, especially in terrestrial snails, is widened posteriorly to form the crop, in which food is stored and digested. In fresh-water pulmonates (Figure 1-46) a gizzard, strongly muscular, forms the main portion of the stomach; it is absent in the terrestrial forms. The gizzard leads posteriorly to the pylorus of the stomach, into which empties the main collecting duct from the voluminous digestive gland. In the prosobranchs there is a ciliated region of grooves and ridges in the anterior globular portion of the stomach. In addition, in most prosobranchs there is an evaginated style sac containing an acellular, semitransparent rod known as the crystalline style. The sac lies alongside the anterior portion of the intestine, with which it may communicate in a few prosobranchs. The style sac and style are confined to herbivorous prosobranchs; the style releases amylolytic enzymes. In carnivorous prosobranchs, however, the style sac and style are absent and the stomach is reduced in size.

Associated with the digestive system of gastropods are a number of secretory glands. In addition to the salivary glands, there are the esophageal glands and the midgut or

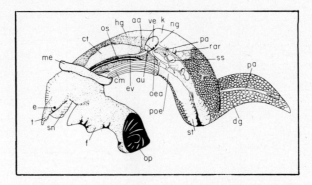

FIGURE 1-45(B).   *Littorina littorea*: animal removed from shell and seen from left side. aa, anterior aorta; au, auricle; cm, columellar muscle; ct, ctenidium; dg, digestive gland; e, eye on eye stalk; ev, efferent branchial vessel; f, foot; hg, hypobranchial gland; k, kidney; me, mantle edge; ng, nephridial gland; oea, esophageal artery; op, operculum; os, osphradium; pa, posterior aorta; poe, posterior esophagus; rar, renal artery; sn, snout; ss, style sac region of stomach leading forward to intestine; st, stomach; t, tentacle; and ve, ventricle. (Redrawn after Fretter, V. and Graham, A., *British Prosobranch Molluscs,* Royal Society of London, 1962.)

digestive gland which opens into the stomach. The digestive gland is a multipurpose gland and in addition to secretion of enzymes it performs other functions such as absorption, phagocytosis, nutrient storage, and excretion.

4. The Circulatory System

The circulatory system of gastropods consists of a heart, a few arteries, and a few veins, but mainly an open system of hemolymph (blood) sinuses, or lacunae, with connective tissue walls. In the primitive prosobranchs, the Archaeogastropoda, the heart consists of one ventricle and two auricles, hence the designation Diotocardia. In the Mesogastropoda and Neogastropoda (the Monotocardia) and the pulmonates, there is only one auricle and one ventricle (Figure 1-46). The hemolymph is forced into the sinuses via arteries, and after bathing the organs and tissues is returned to the auricle via the branchial vein in prosobranchs, or via the pulmonary and renal veins in the pulmonates. Both of the latter veins in the pulmonates pass along the lateral sides of the kidney and unite near the blind end of the saccular portion of the kidney before entering the auricle.

5. The Respiratory System

Although provided with gills (prosobranchs) or lungs (pulmonates), exchange of gases also takes place throughout the surface of the mantle. The lung, in pulmonates, is a modified part of the mantle cavity, and opens to the exterior through a respiratory pore, the pneumostome. The gills are made up of filaments which vary in extent and number in different forms, and are situated in the mantle cavity, with full access to the environmental water. They rarely project freely on the back or at the sides. Some groups, like the Ampullaridae and Siphonaridae, possess both lungs and gills. Among the pulmonate fresh water basommatophorans, the Planorbidae and Ancylidae possess, in addition to the lung, a structure known as the pseudobranch, with a strongly folded surface, which is also respiratory in function.

109

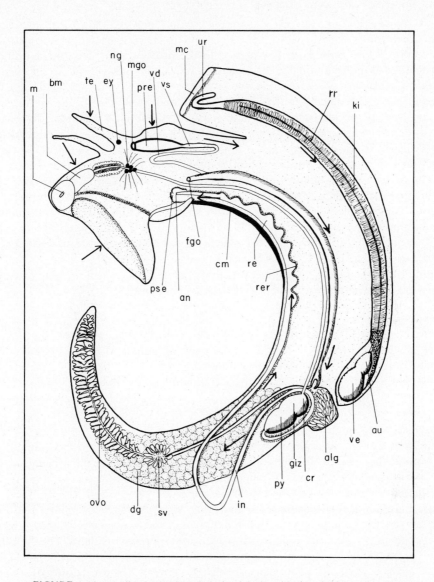

FIGURE 1-46. A dissected *Biomphalaria glabrata*, showing the internal organization of planorbid, basommatophoran snails. alg, albumen gland; an, anal opening; au, auricle; bm, buccal mass; cm, columellar muscle; cr, crop of stomach; dg,
digestive gland; ey, eye; fgo, female genital opening; giz, gizzard of stomach; in,
intestine; ki, kidney; m, mouth; mc, mantle collar; mgo, male genital opening; ng,
nerve ganglia; ovo, ovotestis; pre, preputium; pse, pseudobranch; py, pylorus; re,
rectum; rer, rectal ridge; rr, renal ridge; sv, seminal vesicles; te, tentacle; ur, ureter;
vd, vas deferens; ve, ventricle; and vs. vergic sac. Arrows indicate sites of penetration of the miracidia of *Schistosoma mansoni* (and other trematode species); the
route of migration of the daughter sporocysts to the digestive gland and ovotestis
region; the migration of the cercaria in the opposite direction (anteriad) and their
emergence to the external environment through a thin epithelial layer near the mantle collar.

6. The Excretory System

The kidney is the main excretory organ. In the prosobranchs it is found near the pericardial sac at the beginning of the visceral mass. The kidney empties to the exterior anteriorly, at a point on the mantle roof close to the anal opening. There are two kidneys in the Archaeogastropoda (diotocardians), and one kidney in the Mesogastropoda and Neogastropoda (monotocardians). In the pulmonates the kidney is located on the roof of the lung, on almost the entire length of that sac. Its posterior (distal) part is adjacent to the pericardium. In both prosobranchs and pulmonates the cavity of the kidney is connected posteriorly to the pericardial cavity via renopericardial canals. The cavities of the kidney and the pericardium, in addition to the canals, especially in terrestrial snails, are favorite sites for the cercariae and metacercariae of some digeneans, for example the brachylaemids. The wall of the kidney, as well as the mantle in the basommatophorans, are favorite sites for the early sporocyst stage of some groups, such as the fasciolids, and for metacercariae of others, such as certain echinostomes whose cercariae encyst in snails.

In addition to the kidney, excretory functions are also performed to some extent by the digestive gland and by the amoebocytes of the hemolymph.

7. The Nervous System

The nervous system of gastropods consists of two cerebral ganglia, paired pedal and visceral ganglia, and two or three additional pairs of ganglia. The ganglia are united by commissures to form one structure, situated anteriorly in the head-foot, referred to sometimes as the "brain"; nerve fibers emanating from the ganglia and commissures innervate the various organs.

8. The Reproductive System

Gastropods exhibit variation in their reproductive system. The sexes are separate in the prosobranchs, whereas the pulmonates are hermaphroditic. In the prosobranchs a gonad (testis or ovary) is situated in the posterior (distal) part of the snail in the region of the digestive gland, close to the top of the spire in elevated shells (Figure 1-47). In the male snail, out of the testis there is a male duct which has at its beginning a number of seminal vesicles, then usually receives a duct from a distinct sausage-shaped prostate gland. A part of the duct, in some forms, functions as a prostate gland. The duct then proceeds along the floor of the mantle cavity until it is connected with the penis (Figure 1-47), or opens at the genital pore when the penis is absent. The genital pore is situated dorsally, on the neck of the snail. In the Viviparidae the right tentacle is short and includes the penis. The penis is of taxonomic importance since it exhibits variations in structure and shape among various prosobranchs. In the female prosobranch snail the ovary is also embedded in the digestive gland, and the oviduct which arises from it shows two portions: one, the visceral oviduct, is in the visceral mass, and the other, the pallial oviduct, lies along the mantle cavity until it opens at the genital pore, which is usually near the anus. One to several small sacs, the seminal receptacle or spermathecae, open into the oviduct at the demarcation between the visceral and pallial oviducts. In viviparous prosobranchs the pallial oviduct is enlarged and is modified to form a brood pouch, which usually contains a large number of eggs and developing embryos.

The genitalia of some pulmonate planorbids are shown in Figure 1-48. In the pulmonates, an ovotestis, in which both eggs and sperm are formed, is located at the top of the visceral mass, its anterior end being overlapped by the digestive gland. A hermaphoditic duct arises from the ovotestis, has a number of outgrowths, the seminal vesicles, and then at the level of the stomach divides into a male and a female tract which proceed parallel to each other. The female tract receives secretions from a num-

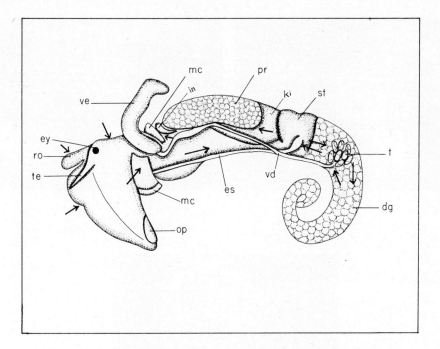

FIGURE 1-47.   A dissected specimen of a male *Oncomelania* sp., showing internal organs. dg, digestive gland; es, esophagus; ey, eye; mc, mantle collar; op, operculum; pr, prostate gland; ro, rostrum; st, stomach; t, testis; te, tentacle; vd, vas deferens; and ve, verge (penis). Arrows indicate the sites of penetration of the miracidium of *Schistosoma japonicum* (and certain other digenetic trematodes); the route of migration of the daughter sporocysts to the digestive gland and testis, and the migration of the cercariae in the opposite direction and their emergence to the outside environment through a thin epithelial layer near the mantle collar.

ber of associated glands (muciparous and albumen glands), and is differentiated into an oviduct, a uterus, and a vagina which opens at the female genital opening in the basommatophorans, and at the common genital opening in the stylommatophorans. A duct from a spermathecal sac opens into the vagina shortly before the genital pore. The male tract is differentiated into a sperm duct, which receives the secretion of a prostate gland, and then tapers to form a vas deferens, which joins the penial complex. The latter is differentiated into a vergic sac that contains the verge or penis, and a preputium that opens to the exterior at the male genital pore in fresh-water pulmonates, or at the common genital opening in terrestrial snails and slugs. Among the genitalia of pulmonates considerable variations exist, especially relative to the ovotestis, the spermathecal duct and sac, the prostate gland, and the penial complex; such variations are useful in taxonomic studies. For example, species of the same planorbid genus can be differentiated on the basis of the relative length and width of the vergic sac and the preputium. Species of different planorbid genera can be differentiated on the basis of whether the penis is simple and filiform (*Biomphalaria*), coiled and inverted, called ultrapenis (*Bulinus*); or has a stylet at its tip (*Gyraulus*).

c. Histology and Ultrastructure

A knowledge of the histology, histochemistry, and structure of gastropods is fundamental for an appreciation of the host-parasite relationships. Larval trematodes, beginning with the miracidium or the ingested embryonated egg, are in intimate contact with the snail's various tissues. The attachment and penetration of the miracidium, its

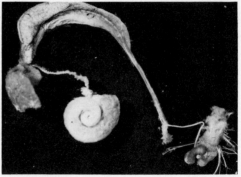

A   B

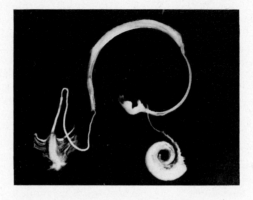

C

FIGURE 1-48.   Reproductive organs (genitalia) of
some pulmonate planorbid snails. (A) *Helisoma trivolvis* (Michigan), (B) *Helisoma corpulentum* (Minnesota), (C) *Biomphalaria alexandrina* (Egypt).

subsequent metamorphosis, the migration of the rediae or daughter sporocysts to distal
(posterior) organs and tissues, the further development to the cercarial stage, and the
exit of the cercariae from the snail, are processes which are usually accomplished at
the expense of the various tissues. The compatibility or incompatibility of snails can
be better explained by an adequate knowledge of the detailed structure and function of the snail's tissues. A number of reports have appeared on this
subject[29,97,98,99,100,125,126] and more recent accounts, especially on the ultrastructure, will
be referred to below. Figure 1-49 is a sagittal section in *Biomphalaria glabrata* showing
the general anatomy and helps in orientation during histological studies.

1. The Skin of the Head-Foot

This is what the free-swimming miracidium of the Digenea first encounters before
penetration of the gastropod. The surface of the foot is covered with an epithelium
made of tall columnar cells with oval or elongated nuclei, rich in chromatin. The cytoplasm is granular and is lightly basophilic. The cells are provided with long cilia on
the exposed surface of the organ. The epithelium rests on a basement membrane (Figure 1-50).

FIGURE 1-49.   Sagittal section through the snail *Biomphalaria glabrata*, to show general anatomical features and relationships between various organs. The stomach and part of the albumen gland were ruptured in the process of taking out sand granules before sectioning.

Ultrastructural studies confirmed light microscope observations on the structure of the skin of gastropod molluscs. However, it was elucidated that not all epithelial cells are ciliated, but that tracts of ciliated epithelial cells are interspersed between the normal columnar cells (Figure 1-51). The cilia are approximately 5 to 6 μm in length. The epithelial cell layer is penetrated at intervals by ducts from mucous glands lying deeper within the tissues. Wilson et al.,[183] working with *Lymnaea truncatula*, found that the ground cytoplasm of the columnar epithelial cells is very dense with numerous mitochondria and vesicles. The nucleus is elongate, occupying ½ to ⅔ of the length of the cell. The brush border of the epithelial cells, which shows by the light microscope, is revealed by the electron microscope to be microvilli 0.1 μm wide and 1.0 μm long on the surface of each epithelial cell. The outer surface of the snail is covered with a layer of mucous about 1.0 μm thick, and it may be that the microvilli serve to stabilize this layer.

Histochemical tests conducted on the snail surface indicated large amounts of protein in the cytoplasm of epithelial cells, because they give a strong positive bromophenol blue reaction. The cells also appear to contain much extractable RNA, indicating that they are sites of active synthesis. The combination of protein and RNA is presumably responsible for the high density of the epithelial cell cytoplasm revealed by the electron microscope. The carbohydrate present in epithelial cells can mostly be extracted with amylase, suggesting that it is predominantly glycogen.

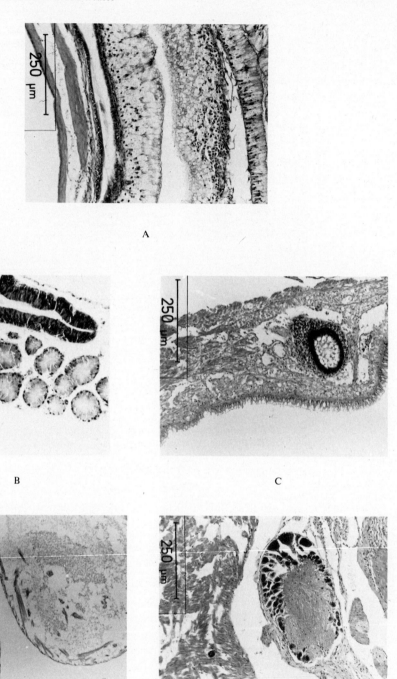

FIGURE 1-50.   Sections in the snail *Biomphalaria glabrata*, showing the microstructure of some organs and tissues, revealed by light microscopy. (A) Columellar muscle, esophagus, oviduct, and midamental gland, (B) Sperm duct cut longitudinally and the prostate tubules cut transversely, (C) Eye, showing lens in the center, and retina with its pigment layer, (D) Part of the auricle and ventricle of the heart, (E) Nerve ganglion surrounded by the epineurium, and a commissure is shown connecting it to other ganglia.

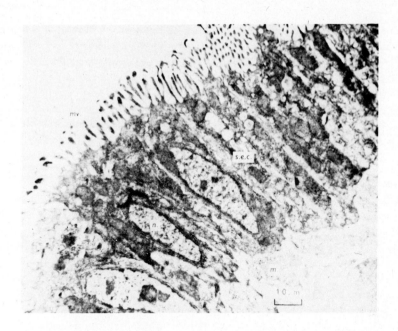

FIGURE 1-51.   The structure of the surface epithelium of the snail *Lymnaea trun-catula*. Section through snail epithelium, showing the columnar cells with microvil-lar surface. The cells rest on a basal lamella overlying muscle and connective tissue. (After Wilson, R., Pullin, R., and Denison, J., An investigation of the mechanism of infection by digenetic trematodes: the penetration of the miracidium of *Fasciola hepatica* into its snail host *Lymnaea truncatula, Parasitology*, 63, 491, 1971. With permission of Cambridge University Press.)

Underneath the exposed epithelium of the head-foot organ are numerous mucus cells. These are more abundant in the foot, and they tend to be in groups, with muscle fibers running through and between these groups and subgroups. Land snails (Stylom-matophora) have secretory ducts opening at the surface. In the basommatophorans which have been studied, however, each mucus cell comprises a secreting unit and secretes mucus through a gradually tapering process. The muscle fibers of the foot apparently serve to force the secreted mucus through the intercellular spaces of the epithelial cells. These gland cells were found to be strongly PAS and alcian blue posi-tive. This material is not amylase extractable, and it can reasonably be assumed that it is neutral and acid polysaccharide. The brush border region of the epithelium gives positive reactions for both carbohydrate and protein, perhaps indicating that material covering the snail surface has a dual origin in gland cells and in surface epithelium.

The muscular elements underneath the epithelium covering the head are not as ex-tensive as those in the foot.

2. The Mantle

The mantle covering of the pallial region has a surface consisting of columnar to cuboidal epithelial cells with parabasal, round nuclei. The cells of this epithelium have pigment in the cytoplasm (except in albino snails). The epithelium is continuous with that of the *tunica propria*, which covers the visceral region of the body. Here, however, the epithelial cells are flat or squamous, or partly cuboidal. There is a basement mem-brane and a supporting connective tissue beneath the mantle covering of the two re-gions. The supporting layer has many collagen-like fibers and smooth muscle cells.

Electron microscopy has revealed more details of the structure of the mantle.[119] The apical cell membrane of the epithelial cells has many irregular and occasionally branched microvilli protruding into the fluid-filled space between the mantle and the shell, the extra-pallial space. The epithelial cells are interconnected at the apical side of the lateral cell membranes by "*zonulae adherentes*". The highly lobulated nuclei of the epithelial cells occupy nearly the entire height of these cells. Large deposits of glycogen are usually present in the cytoplasm of the epithelial cells. Pigment granules also often occur in the epithelial cells, and these granules are positive for acid phosphatase; this enzyme is also present in the Golgi bodies and vesicles (primary lysosomes) scattered in the cell. Collagen-like fibers are densely packed and constitute the connective tissue layer lining the basal lamina of the epithelium. Muscle fibers at the base of the connective tissue layer are overlapping and stout smooth muscle fibers.

3. Respiratory Organs

The wall of the lung in pulmonate snails lies between the mantle epithelial surface and a respiratory epithelium lining the lung cavity. There is a basement lamina for each epithelium, but the space between the epithelia contains delicate smooth muscle fibers, connective tissue, some veins, and a blood sinus system which is in intimate contact with the epithelia and with the veins. The epithelium lining the lung cavity consists mainly of cuboidal cells.

4. The Kidney

The lumen of the kidney is lined by an epithelium consisting of low columnar, but mainly cuboidal, cells, beneath which there is a basement membrane, supported with a sheet of fibroblasts and a few smooth muscle fibers. The epithelium has a characteristic wavy appearance (Figure 1-52), evidently to increase the epithelial surface. The wall of the kidney is lined on the outside surface of the animal by the mantle. The kidney wall is formed of connective tissue, some fibroblasts, veins, and a rich supply of sinuses which are in intimate contact with the veins and with the renal epithelium.

Ultrastructural studies showed the cells of the kidney of *Lymnaea stagnalis* to contain large (5 to 20 μm) excretion granules, which are constricted off together with part of the cytoplasm.[17] In degenerating nephrocytes great numbers of lipid granules, probably arising from mitochondria, are found. Deposits of glycogen are present in the nephrocytes as well as in the cells of the ureter, suggesting the kidney to be a glycogen-storing organ. The presence of glycogen is accompanied by that of an elaborate agranular endoplasmic reticulum.

5. Alimentary Tract

The structure of the alimentary tract is of significance because some digenean species whose embryonated eggs do not hatch are ingested by the snail host; presumably the eggs hatch within. In general, the microstructure of the alimentary tract is the same throughout, with variations as to size and shape. The wall of the tract is enveloped on the outside by a sheath of connective tissue, where the organs lie free in the body cavity. This is followed by a layer of circular, and then a layer of longitudinal, smooth muscle fibers. The lumen of the tract is lined with a simple columnar epithelium, resting on a basement lamina.

The buccal mass consists mainly of strong musculature, to assist in the process of rasping of food; the mass contains the radular ribbon, surrounded by a radular sac, and a radular carrier (Figure 1-52). The main muscular support of the mass consists of spindle-shaped fibers in longitudinal sections. The radular ribbon, carrying the teeth, is supported by the odontophoral cartilage (or radular carrier), which acts as a cushion sharing in the constant movement of the ribbon, together with the radular protractor and retractor muscles. In the esophagus the ciliated simple epithelium lining the lumen is arranged in several longitudinal folds. Mucus cells or goblet cells are usually not found among the epithelial cells but do occur beneath the basement lamina in the muscle layers. The cells secrete PAS-positive materials through the intercellular

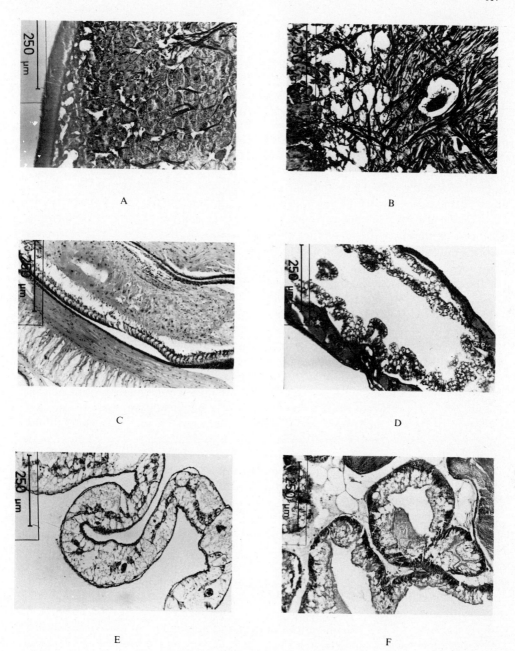

FIGURE 1-52.   Sections in the snail *Biomphalaria glabrata*, showing the microstructure of some organs and tissues, revealed by light microscopy. (A) Simple epithelium of the foot and underlying mucous glands, connective tissue, and muscle fibers, (B) Dense muscle fibers in the core of the foot and vascular connective tissue, (C) Buccal mass, showing radular sac, radula, and colostyle, (D) Kidney, note undulated cuboidal epithelium lining the lumen, (E) Undulated rectal ridge, (F) Tubules of the digestive gland and intertubular connective tissue.

spaces of the epithelial lining. In the stomach the gizzard, when present in some species, has the thickest muscular wall of the entire digestive tract. Its epithelium, or the layers underneath it, does not contain mucous-secreting goblet cells. Sand grains are usually present in the lumen, and it appears that this portion of the alimentary tract serves only as a mechanical grinder of food materials. The portion of the intestine following the stomach evidently displays an active secretory function.

6. Digestive Gland

Also known in the literature as "liver" or "hepatopancreas", it is a compound tubular gland, consisting of a main hepatic duct with one short dorsal branch and numerous secretory lobules. The main duct opens into the prointestine at its junction with the pylorus. The tubules of the digestive gland (Figure 1-52) are covered with a thin, connective tissue sheath which is continuous with that of the hepatic duct. There is a loose vascular connective tissue between the tubules continuous with that of the ovotestis. Numerous blood spaces, with varying numbers of vesicular cells, and pigment cells, are present in the interstices of the connective tissue. These vesicular connective tissue cells play an important metabolic part in the storage of glycogen. Several authors now agree that the epithelium of the tubules consists of two types of cells, digestive and secretory, with a few mucous cells scattered among them. The secretory cells (known also as lime cells, calcium cells, etc.) are usually pyramidal or rhomboidal in shape with their bases lying on the connective tissue sheath (Figure 1-53). The digestive cells are columnar with basal nuclei. It is believed by many authors that in healthy tissue they contain granular deposits of glycogen, lipid globules, and food vacuoles, but the cytoplasm near the base in the area of the nucleus is largely free of stored material.

The digestive cells constitute the principal glandular epithelial cells and show considerable polymorphism. This polymorphism is due to the several functions which the cell is believed to perform; thus the same cell passes through a cycle of several phases: absorption, digestion, and fragmentation of the tips of the cell into the lumen of the tubule. During the absorptive phase the cells possess a striated free border which is lost when the absorption phase is complete. During the succeeding digestive phase the cells are tall and have swollen tips. This is followed by fragmentation, when the swollen distal ends of the cells are constricted off into the lumen of the digestive gland tubule as nonnucleated spheres and carried to the stomach containing a residue of undigested waste.

The digestive cells of *Biomphalaria pfeifferi* were studied by light and electron microscopy and by histochemical tests.[119] Four different, arbitrarily chosen age stages are distinguished.

Stage 1 — The cell is in the form of a thin columnar cell with homogeneous basophilic cytoplasm. The outer nuclear membrane is studded with ribosomes. The cells bear a conspicuous brush border coated by PAS positive diastase-resistant material. At the ultrastructural level the brush border consists of densely packed, straight microvilli. The Golgi complex, which is located in the supranuclear region, has a very active appearance. Small mitochondria are distributed throughout the cell, except in the homogeneous apical region.

Stage 2 (Figure 1-54) — Small vesicles appear and accumulate in the cytoplasm just below the homogeneous apical region. The vesicles start fusing to form larger vacuoles. Since these vacuoles contain acid phosphatase, as indicated by light and by electron microscopy, they are secondary lysosomes. During this stage the cell becomes broader due to the increase in number and size of the vacuoles. Moreover, the density of the ribosomes, and hence the basophilia of the cytoplasm, decreases.

Stage 3 (Figure 1-54) — Coarse material which has a yellow or yellowish brown color appear in the oldest vacuoles, which lie just above the nucleus. Fusion of these vacuoles gives rise to one large vacuole which occupies the whole width of the cell. Histochemical tests show the yellowish granules in the vacuoles to contain acid phosphatase, in addition to lipofuscin, properties which are typical of residual bodies. The formation of endocytotic vacuoles in the apical region of the cell continues, and the cell becomes wider towards the top, often projecting into the lumen of the tubule. The

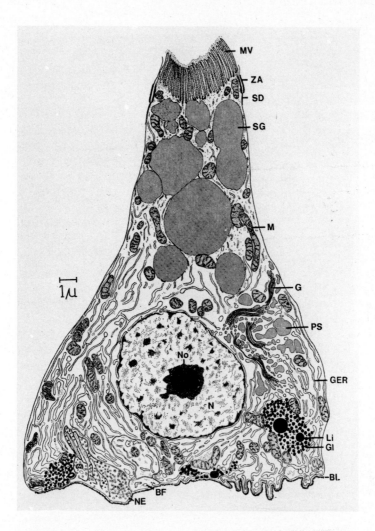

FIGURE 1-53. Secretory cell of the epithelium of a digestive gland tubule of *Biomphalaria pfeifferi*. Drawing from electron micrographs. BF, basal infolding; BL, basal lamina; G, Golgi apparatus; GER, granular endoplasmic reticulum; Gl, glycogen; Li, lipid; M, mitochondrion; MV, microvilli; N, nucleus; NE, neuroepithelial junction; No, nucleolus; PS, prosecretion granule; SD, septate desmosome; SG, secretion granule; and ZA, zonula adherens. (After Meuleman, E. A., *Neth. J. Zool.*, 20, 355, 1972.)

digestive cells vary in width between 3 and 20 and in height between 15 and 62μ; the differences depend on the number and size of contained vacuoles.

Stage 4 — The cell is shorter than in previous stages. The residual body is discharged, without rupture of the membrane, into the lumen of digestive gland, together with the apical cytoplasm.

7. Amoebocytes

These are nucleated amoeboid corpuscles corresponding to leucocytes of higher animals. They are found in the circulatory system, connective tissues, and several other organs. In addition to being amoeboid with lobose pseudopodia, they may be oval or round. Although the size varies in various gastropods and in the same individual, they measure about 12 by 9μ. The nucleus is usually eccentric in position, and is round, oval, or lentiform. The cytoplasm is lightly basophilic and granular.

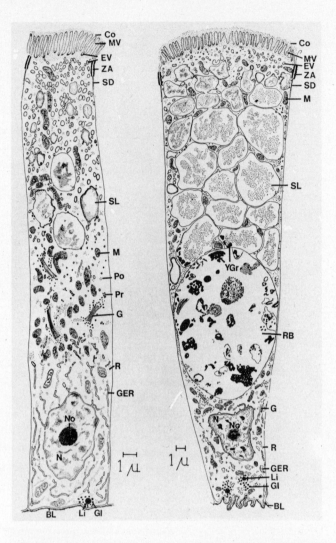

FIGURE 1-54.   Digestive cell of the digestive gland of the snail
Biomphalaria pfeifferi. (A) Stage 2, and (B) Stage 3. The stages reflect
the various functions performed by this cell (see text). BL, basal lamina; Co coat; EV, endocytosis vesicles; G, Golgi apparatus; GER,
granular endoplasmic reticulum; Gl, glycogen; Li, lipid; M, mitochondrion; MV, microvilli; N, nucleus; No, nucleolus; Po, polysome;
Pr, primary lysosomes; R, ribosomes; RB, residual body; SD, septate
desmosome; SL, secondary lysosome; YGr, yellow granule; and ZA,
zonula adherens. (After Meuleman, E. A., *Neth. J. Zool.,* 20, 355,
1972.)

Based on ultrastructure and enzyme histochemistry studies on the blood of *Lymnaea
stagnalis,* it was shown that the blood contains one type of cell, the amoebocyte.[152]
Ultrastructural observations confirmed that the shape of the cell and its nucleus varies
greatly; the cell possesses pseudopodia. The nucleus contains one inconspicuous nucleolus (Figure 1-55). The presence of many large and small vacuoles in the peripheral
cytoplasm indicates that the amoebocytes are actively involved in endocytosis. They
contain an extensive Golgi apparatus. The cytoplasm contains many membrane-
bound, lysosome-like structures and granular endoplasmic reticulum which is not very
prominent. The cells have round to elongate mitochondria. By histochemical tests the

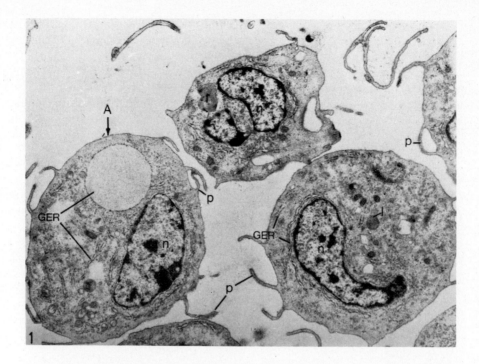

FIGURE 1-55. Amoebocytes of *Lymnaea stagnalis*. Low power electron micrograph (magnification × 6000.) of amoebocytes directly caught in the fixative (OsO_4/glut). In amoebocyte "A" part of the granular endoplasmic reticulum (GER) is dilated and contains fine granular material. p, pseudopodia; n, nucleus; and l, lysosome-like structure. (After Sminia, T., *Z. Zellforsch.*, 130, 501, 1972. With permission.)

cells show a very strong peroxidase activity, however, the acid phosphatase activity is weak; it increases after phagocytosis. Although most studies such as the above on pulmonate snails have shown that only one type of blood cells exists, the amoebocyte (hemocyte), Cheng and Auld[24] observed two morphologically distinguishable types in *Biomphalaria glabrata*: granulocytes and hyalinocytes.

8. Connective Tissue

The connective tissue contains various cellular and fibrous elements. The cellular components are fibroblasts, pigment cells, vesicular cells, and mucous cells, while the fibrous components are collagenous-like fibers and delicate fibrils on the muscle fibers. There are two different types of connective tissue in gastropods: the loose "vascular" connective tissue and the dense connective tissue. The vascular connective tissue is characterized by an open network of slender fibroblasts, appearing in section as numerous oval or irregularly round perforations. These latter, also known as circulation spaces or blood spaces, hold the hemolymph of the gastropod, which stains a homogeneous pink with eosin Y. The dense connective tissue contains numerous fibroblasts, pigment cells, collagenous-like fibers, and amoebocytes.

Ultrastructure studies of the connective tissue of *Lymnaea stagnalis* showed it to consist of eight different cell types: pore cells, granular cells, vesicular connective tissue cells, amoebocytes, fibroblasts, undifferentiated cells, pigment cells, and muscle cells.[152] Pore cells are characterized by numerous invaginations of the cell membrane bridged by cytoplasmic tongues. Probably pore cells produce hemocyanin. Granular cells contain numerous cysteine-rich glycoprotein granules. The contents of these granules are released by exocytosis. On the basis of this study it was suggested that these cells are involved in the production of blood proteins. The empty-looking vesicular connective tissue cells appeared to contain large amounts of glycogen, suggesting that these cells have nutritive functions.

Relative to pore cells, since, in the characteristic plasma membrane, invaginations of the cell granules are present which can be identified as hemocyanin molecules, it has been supposed that the cells produce this hemocyanin and that the crystalline material inside the GER consists of this pigment. More recent studies confirmed this assumption.[153] Inside the cisterns of the GER large numbers of granules are present which have the characteristic shape and dimensions of hemocyanin molecules.

9. The Genitalia: Ovotestis

This is the reproductive organ in pulmonate gastropods, both the Basommatophora and the Stylommatophora. Both ova and sperm are produced in the compartments of the ovotestis, known as acini. Each acinus is enclosed in a wall, "Ancel's layer", which is made of thin connective tissue two to three cells thick, with abundant collagenous-like fibers. This layer is usually followed by a germinal epithelium. The female germinal cells (oocytes and mature ova) are located at the apices of the acini, whereas the male germinal cells are found along the sides of the acinal wall. Spermatogonia by spermatogenesis result in the formation of spermatocytes, spermatids, and then spermatozoa. Spermatozoa appear in the center of the lumen of the acinus in the form of large bundles, and each spermatozoa has a slightly flattened cephalic portion and a long flagellum.

Female Tract — The female tract is mostly glandular, consisting of the oviduct (Figure 1-50), muciparous gland, uterus, oothecal gland, and vagina.

Male Tract — The sperm duct and the prostate (Figure 1-50) are glandular, while the vas deferens and the penial complex are muscular, with a few secretory cells.

Histochemical and electron microscope observations indicated that in the female and male part (in the tract as well as in the glands) of the reproductive system of *Biomphalaria glabrata*, glycogen ciliated cells alternate with secretory cells. The secretory cells of the albumen gland produce galactogen and proteins. In the rest of the female part, nine different secretory cell types can be distinguished on the basis of their location, the ultrastructure of the cells, and the histochemical nature of the products, which are sulphated and nonsulphated acid-mucopolysaccharides, glyco- or muco-proteins and neutral polysaccharides. In the sperm duct and prostate gland seven different secretory cell types are found. Each cell type has its special location and ultrastructure. These cells produce mainly lipoproteins.[35]

10. Nerve Ganglia

The center of a ganglion contains bundles of neurofibrils. Ganglion cells are located on the periphery of the ganglion, except where a nerve root or a commissure leaves the ganglion. The ganglion (Figure 1-50) is covered with a thin sheath, the perineurium, outside of which there is a relatively thick layer, the epineurium, containing arteries and blood spaces.

A detailed description of the central nervous system of *Lymnaea stagnalis* and *Biomphalaria glabrata*, as well as the location of neurosecretory cells, was given by Lever et al.[85] (Figure 1-56). Neurosecretory cells of *B. glabrata* are in both cerebral and parietal ganglia and on the visceral ganglion. In each cerebral ganglion a large group of these cells is found near the so-called medio-dorsal body, which lies partly upon the intercerebral commissure. Furthermore, some special cells of this type are located in the lateral lobe of this ganglion. The product of the medio-dorsal neurosecretory cells is transported to the median lip nerves. In the left parietal ganglion one dorsal cell and a group of rostral cells show neurosecretory characteristics. The right parietal ganglion

123

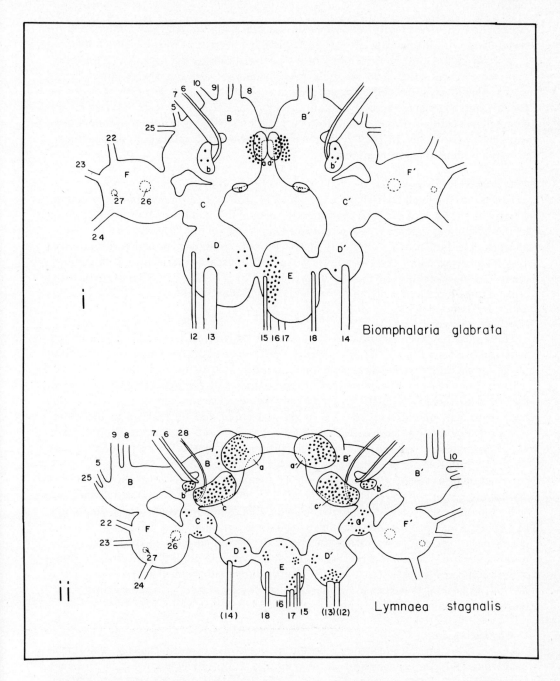

FIGURE 1-56. The location of neurosecretory cells in the central ganglia of *Biomphalaria glabrata* and of *Lymnaea stagnalis*. B, B': cerebral ganglia; C, C': pleural ganglia; D, D': parietal ganglia; E: visceral ganglion; F, F': pedal ganglia; a, a': medio-dorsal bodies; 5: cerebro-buccal connective; 6: n. opticus; 7: n. tentacularis; 8: n. frontolabialis superior; 9: n. labialis medius; 10: n. penis; 12: n. pallialis sinister externus; 13: n. pallialis sinister internus; 14: n. pallialis dexter; 15: n. genitalis; 16: n. analis; 17: n. intestinalis; 18: n. cutaneous pallialis; 22: n. pedalis superior; 23: n. pedalis medius; 24: n. pedalis inferior; 25: commissura subcerebralis; 26: commissura interpedalis anterior; 27'' commissura interpedalis posterior; and 28: n. nuchalis. (Redrawn after Lever, J., de Vries, C. M., and Jager, J. C., *Malacol. Int. J. Malacol.*, 2, 219, 1965.)

has only one neurosecretory cell, while the visceral ganglion contains a long band of such cells curving around the left parieto-visceral connective.

B. glabrata does not have as many neurosecretory cells as *L. stagnalis.* In the latter snail the cerebral ganglia do not only have a medio-dorsal but also a latero-dorsal group of neurosecretory cells. The cells are more numerous in these and in the parietal ganglia, and the pleural ganglia contain three distinct groups of such cells, which are completely lacking in *B. glabrata.* However, there is one exception to this comparison between the two snails. In *B. glabrata* the number of neurosecretory cells in the visceral ganglion is higher than in *L. stagnalis.*

11. Osphradium (Olfactory Organ)

This is an elongated pear-shaped sac which is located at the junction of the mantle collar with the neck of the snail, between the median line and the pneumostome siphon of the basommatophorans. It has not been reported for the terrestrial stylommatophorans. The lumen of this organ is lined with tall, columnar epithelial cells covered with long, dense cilia. There is a layer of smooth muscle fibers following the epithelium. The lower portion of the sac is surrounded by a peripheral ganglion, which receives a thick branch nerve from the left visceral ganglion.

12. Statocyst (Balancing Organ)

This organ has been reported in both aquatic and terrestrial snails. It is a paired spherical sac found in the latero-posterior corner of each pedal ganglion at the root of the commissure to the pleural ganglion. It is innervated by these ganglia.

By light microscopy the epithelium of the statocysts of *Lymnaea stagnalis* is shown to consist of three cells: high sense cells, low sense cells, and supporting cells.[52] The high sense cells measure up to 30 μ in the tangential and up to 40 μ in the radial direction. For the low sense cells these figures are 60 μ and 15 μ, respectively (Figure 1-57). Specimens of *L. stagnalis* from which both statocysts have been removed lose the ability for geotactic orientation upon a slope, in air as well as in water.

Ultrastructural studies on the statocysts of *Lymnaea stagnalis*[52] also showed the epithelium to consist of three cell types: high sense cells, low sense cells, and supporting cells. The high sense cells are bipolar, having cilia, some extremely small microvilli-like bulges which project into the lumen, and an axon originating at the cell basis. All ciliary basal feet point away from the center of the cell surface. In the center of the proximal part of the static nerve, which is composed of two bundles of thin and thick axons, respectively, a short, tube-like diverticulum of the lumen of the statocyst is present. In normal as well as in regenerating statocysts, statoliths are formed intracellularly in supporting cells and in undifferentiated cells, respectively.

13. Pigments in Snails

Pigments are present in various cells and tissues of gastropods. The snail *Cerithidea californica* Haldeman stores both carotenoid and chlorophyllic pigments. B-carotene, lutein, carotenoid acids, zeaxanthin, and chlorophyll derivatives have been identified. All available evidence, such as that obtained from spectral properties, color reactions, chromatographic behavior of pigments, and observation of ingested snail tissues in the gut of rediae, suggests that larval trematodes absorb the carotenoid and chlorophyll pigments from the snail host. The larvae appear to store the B-carotene in an unchanged condition.

d. Ecology of Gastropods

The habitat of gastropods does not only ensure the survival and propagation of these molluscs, but needs also to be favorable for the survival of the free-living larval dige-

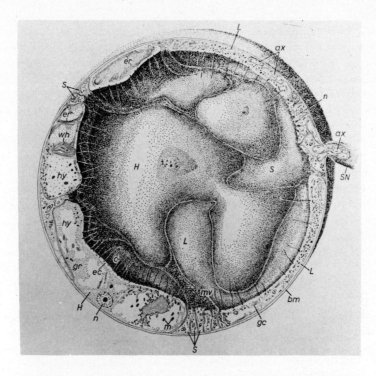

FIGURE 1-57.   Semischematic drawing of the lower half of a right-statocyst
of *Lymnaea stagnalis*. H, high sense cell; L, low sense cell; S, supporting
cells; SN, static nerve; ax, axon; bm, basement membrane; c, cilia; ec, ecto-
plasm; er, endoplasmic reticulum; gc, Golgi complex; gr. granular plasm;
hy, hyaline plasm; m, mitochondria; mv, microvilli; n, nucleus; and wh,
whorl. (After Geuze, 1968.)

neans and the transmission of the infection. A combined effect of certain physical,
chemical and biological factors determines the suitability of the habitat for molluscan
life and the completion of the life cycle of the digenetic trematode. A large number of
molluscs are marine, some are terrestrial, while others are fresh water. The marine
gastropods are prosobranchs, whereas the fresh-water forms are both prosobranchs
and pulmonates. The majority of gastropods which live on land are pulmonates of the
order Stylommatophora, and usually the term land molluscs, or terrestrial molluscs,
refers to such snails and slugs. However, particularly in the tropics, certain genera of
the prosobranch superfamilies Neritacea, Archaeotaenioglossa, Rissoacea, and Litto-
rinacea live on land, and are referred to as land operculates. In the discussion to follow
the ecology of marine gastropods will be treated first, followed by estuarine gastropods, then the terrestrial and fresh-water forms.

1. Marine Gastropods

Marine gastropods may be rock clingers in the sublittoral part of the rocky shore or
colonize the beach. Although habitat factors influencing these snails and their distribution vary from one species to the other, they are able to withstand variations in
temperature, light, humidity, salinity, periods of restricted feeding, and wave action.
Such forms include the limpets, the periwinkles, the top-shells (trochids), the dog-
whelk, and others. All the latter gastropods exhibit zonation which can be correlated
not only with position on the shore relative to tidal levels, but is causally related to it.
Naturally other factors have to be considered in this respect, such as climate, degree
of exposure to wave action, and other effects in the gastropod's niche. Whether living

in the lower or the upper parts of the shore near high tide level, they are completely dependent on the sea during early life. All littoral species discharge eggs or young to the sea, or the fixed spawn is immersed periodically. In the case of the common periwinkle, *Littorina littorea*, spawning occurs chiefly at night and on the flood tide when the females liberate numbers of egg capsules each containing three eggs. From these emerge the small motile larvae which, after a short life in the plankton, settle on the shore as young periwinkles.

Many of these forms have to adapt to resist periods of exposure with consequent desiccation and the effect of high temperature. The temperature of their body during exposure usually remains about 1° lower than that of the air, however, it was reported that some limpets and top shells differ as to their body temperature. Up to 20°C the body temperature is similar to that of neighboring inanimate objects, but between 20°C and 30°C the animal remains a little cooler than its surroundings. The body temperature may fall with the rising tide and the snails then become active.

Studies on the effect of temperature, apart from desiccation, on certain British littoral molluscs during the summer, have been reviewed by Fretter and Graham.[47] The gastropods *Patella vulgata, P. intermedia, P. aspera, Gibbula cineraria, G. umbilicalis, Monodonta lineata, Littorina littorea, L. saxatilis, L. neritoides, L. littoralis*, and *Nucella lapillus*, although they all showed some signs of distress and active movement usually ceased at 30°C, differed as to their thermal death point; the latter depended on their vertical distribution on the shore. *G. cineraria*, which occurs at very low levels, has the lowest thermal death point at 36.2°C, and irritability is no longer recorded at 34 to 35°C (heat coma). *L. neritoides* is the most tolerant of heat, as death occurs at 46.3°C, and heat coma at 38°C. This littorinid has the most southerly distribution, extending into the Black Sea and North Africa. Although *Nucella lapillus* has a rather similar vertical range to *Littorina littorea*, it is more restricted to shade near its upper limit, and correlated with this is its much lower lethal temperature of 40°C as compared with 46°C for *L. littorea*. The above findings might indicate that the intertidal gastropods could temporarily tolerate temperatures well in excess of those experienced on the shore, and the possibility of heat death under natural conditions is rather remote. Also relative to the thermal death point of certain periwinkles (*L. littorea* and *L. saxatilis* from different levels on the shore), conflicting reports exist based on investigations in Canada (Nova Scotia) and England. While the Canadian workers found differences in death temperature corresponding to different levels on the shore, the British workers, with specimens from Cardigan Bay, did not find significant differences. Some littorinids seem to be very tolerant to the lower temperature extreme. Among *L. littorea, L. saxatilis* and *L. littoralis* in Northern Russia, *L. saxatilis* was found to be the most tolerant of low temperatures, surviving more than 27 hr in air at −9.4°C, but the other species could also withstand the effect of low temperatures. This is further emphasized by the fact that *Littorina littorea* and a few other molluscs remain intertidal during the winter, but become inactive and cease to feed at about 8°C. Gastropods such as littorinids, limpets, topshells, and others must be resistant to freezing temperatures, judging from the fact that they have permanently colonized shores in cold climates. Many molluscs, however, migrate downshore or into the relatively warmer sublittoral during the autumn.

The activity of the gastropods which inhabit the shore varies with the tide, which seems to impose a rhythm on behavior. As the tide retreats the prosobranchs on rocks become inert before the surface dries, and it has been observed that before the tide retreats many have returned to a protective crevice. However those in rock pools will continue to move about and feed at low water even in bright sunshine. But for the majority, especially the herbivores and scavengers, there is a rhythm of locomotory activity and feeding. Some believe that this may be endogenous rhythm rather than a direct induction by environmental factors. Nevertheless, some workers described a fortnightly rhythm of activity in *Littorina saxatilis*, coinciding with the spring tides. Such a rhythmical activity is usually reflected in oxygen consumption. A persistent diurnal and tidal rhythm of oxygen consumption was described for *Littorina littorea* and *Urosalpinx cinerea*. When *Littorina irrorata* is exposed after a long period of sub-mergence, oxygen consumption increases, but it is dependent on the temperature. The consumption is maximal for *L. irrorata* at 35°C, and decreases rapidly at lower tem-peratures. The above indicates that shore prosobranchs can make use of atmospheric air in addition to their respiration through their gills. Indeed they are air breathers too. They have become adapted, in varying degrees, to utilizing atmospheric oxygen. In many such as the prosobranch littorinids and shore pulmonates (for example mem-bers of the Siphonaridae, Otinidae, and Amphibolidae), the roof of the mantle cavity is thin, vascularized, and acts as an additional respiratory surface. These have become readapted for aquatic breathing also by the development of a secondary gill. But all can breathe under water. Littorinids, as well as other shore snails, when the tide with-draws, close down the aperture with the operculum partially or completely, after taking in a bubble of air which is released as soon as the animals become active again on reimmersion. Many shore molluscs must rely partly or solely on cutaneous respiration and many small prosobranchs which inhabit intertidal pools have no gills and only small mantle cavities (for example, members of the Omalogyridae, Rissoellidae, and Pyramidellidae).

Summarizing the ecology of intertidal molluscs, particularly those on smooth rocky shores, Newell[123] indicated that the heat-light-desiccation complex, whose intensity will vary with times of exposure, directly or indirectly determines the upper limit of distri-bution of plants and animals, but that its effects can be masked or overriden by irreg-ularities and other purely local features, or by the adoption by the organisms of special habits and habitats. Thus the majority inhabit crevices or live under boulders instead of merely crawling or clinging on the rock faces, and thus escape the full rigors of desiccation and temperature changes, or of dislodgment by wave action. In spite of this, all must in some measure be adapted to withstand desiccation, to breathe atmos-pheric oxygen, to maintain their position, or actively to regain it. In other words, all are to some extent amphibious, for only the most tightly closing bivalve can completely isolate itself from aerial conditions.

2. Estuarine Gastropods

The quiet, brackish waters of estuaries, salt marshes, and saline pools throughout the world are favorite habitats for gastropods. Representatives of certain families (Po-tamidae) are restricted to the tropical and subtropical regions of all continents, whereas some other families (Littorinidae) have subtropical as well as temperate zone species. In the U.S., examples of the estuarine gastropods are: *Cerithidea scalariformis*, found from Georgia to Texas, being very abundant in the Florida keys; *Cerithidea califor-nica*, found from Bardinas Bay to San Diego in California; *Batillaria minima* on the muddy coasts of Florida; and *Nassarius obsoletus* on the mud flats of the Atlantic Coast. In addition, *Littorina irrorata* is abundant in the Gulf of Mexico and extends north to New Jersey and New York. In Louisiana, *L. irrorata* is a common snail in the brackish-water coastal marshes. The snails are often seen, during low tide, attached in large numbers to the sedges and other vegetation. Some are also found on marine vegetation along the relatively open waters of the Gulf of Mexico.

In the U.K. some prosobranchs such as *Hydrobia ulvae* and *H. ventrosa* have be-come adapted to brackish waters of the estuaries. The association of *H. ulvae* with the sea lettuce *Ulva latuca* accounts for its specific name, though Fretter and Graham[47] believe that this link is by no means so definite as was once thought, and the snail browses over the surface of other weeds such as *Enteromorpha*, and occurs in large numbers on apparently bare mud, or muddy sand or gravel. Small numbers of *H. ulvae* may be found in the lochs, where the typical individual has a rough eroded shell; it was suggested by some parasitologists that such individuals with eroded shells were parasitized. *H. ulvae* in some places occurs in shallow pools, where, after rain, the salinity of the pools may fall to 13%, and in hot weather, when a maximum temperature of 30°C has been recorded, it may rise as high as 34.7%. *H. ulvae* flourishes in these extremes of salinity and temperature. Probably there is a combination of features that regulates the distribution of these small hydrobiids, not salinity alone, and it has been suggested that in the British Isles, and probably elsewhere, distinct biological races of the prosobranchs exist which differ in their salinity tolerance. In Denmark oxygen consumption and temperature were related to the habitat of *H. ulvae* and few other molluscs. Oxygen consumption rises abruptly with temperature, and for *H. ulvae* it is 35 mg/kg/hr at 2°C and 490 mg/kg/hr at 20°C. Another brackish-water hydrobiid, *Truncatella subcylindrica*, lives in muddy habitats at the level of high tide, where it may be only occasionally wet by sea water, and in the British Isles it is commonly associated with the plants *Suaeda maritima* and *Halimione portulacoides*. *Assiminea grayana*, a member of the family Assimineidae, occurs at levels which are under water only at spring tides, where it may be found in large numbers on grass and sedge stems and in the mud at their base.

In addition to the above prosobranchs a number of marine forms penetrate the lower reaches of estuaries, e.g., *Patella vulgata*, *Gibbula umbilicalis*, *Littorina saxatilis*, *L. littorea*, *Buccinum undatum*, and *Nassarius reticulatus*. These animals presumably retain isotonicity and their tissues can tolerate transitory low salt concentration, but their lack of osmotic independence means that they do not have the ability to enter freshwater bodies.

The effect of solutions of different saline content on some brackish-water snails in Japan was tested in connection with their susceptibility to infection with lung flukes. It was found that the optimum salinity for *Assiminea parasitologica* is 0.25%, that of *A. latericea miyazakii* is 0.4%, and that for *A. japonica* is 0.6%. It was also found that the optimum salinity for the miracidium of *Paragonimus ohirai* is 25%, and that the only compatible assimineid of the three species studied is *A. parasitologica*.

3. Terrestrial Gastropods

Compared to snails living in other habitats, relatively few physical and chemical factors limit the occurrence of terrestrial snails. Among these factors are temperature, humidity, availability of calcium, pH of the soil, and of course availability of food. The preferred habitats of terrestrial pulmonate snails and slugs are sheltered areas, under decayed leaves covered with fungi and algae. These not only provide them with food but also create a microhabitat with a low and stable temperature and often high humidity. For this reason they may enter crevices and hide under stones in heavily shaded areas, e.g., *Cionella lubrica*, the snail host of *Dicrocoelium dendriticum* in the U.S. They may also hide under various objects, such as damp paper, discarded cardboard, flower pots, garbage cans, and similar objects. Forested river valleys often harbor large populations of land snails and slugs, especially valleys having outcrops of limestone rock. These are favorite habitats for *Anguispira alternata*, the snail host of some brachylaemid trematodes. Often, isolated woodlands in the midst of cultivated areas afford favorable snail habitats. Most snails are associated with distinctive kinds of environment. Some are restricted to the heavy, virgin forests in temperate climates; others are limited to the more open woodlands which have been cut over. Many species are found in regions forested with oak, maple, willow, and other deciduous trees; only a few are associated with coniferous trees, and this is the case in both Europe and North America. This is apparently because the majority of land snails are found in association with a soil pH on the alkaline side and a soil which is rich in calcium. Slugs are, however, tolerant of soils with low calcium content. It was demonstrated experimentally that there is a correlation between the thickness and weight of the shell and the amount of calcium in the food provided to the terrestrial snail. Several species are found in abundance on and near limestone cliffs. Tropical rain forests with lush vegetation, high humidity, and a moderate stable temperature harbor several species of land snails (both pulmonates and land operculates) and a wide variety of slugs. Such are the habitats in most of Central America, South America, and several islands of the Caribbean. In such areas, limestone cliffs close to the sea flourish with several land species such as *Pleurodonte orbiculata* and *Subulina octona*. Like other land snails and slugs, most of these species become very active after rain. They come out of their hiding places and move around looking for food.

Other important factors in the terrestrial mollusc's environment, in addition to calcium and the soil pH, are temperature and humidity. Land snails vary as to their tolerance of extremes of temperature. Some species tolerate, and are actually active at, temperatures close to 0°C and can hibernate at even lower temperatures. Many species can tolerate a temperature as high as 42°C, as long as the environmental humidity is high so that the mollusc avoids excessive water loss through the surface. It has been demonstrated that land snails and slugs are capable of temperature regulation and acclimation. Temperature regulation is accomplished through loss of some water through the surface, which thus maintains their body temperature below that of the environment. While fresh-water snails usually have a body temperature similar to the water around them, land snails, with some exceptions, have a body temperature which differs from that of the air. The capacity to regulate their body temperature in the presence of higher air temperature around them enables terrestrial snails to live in hot climates even at low or moderate humidity. Various species of snails, however, show differences as to the potential for water regulation, and thus some can adapt to certain places better than others. This no doubt causes differences in geographical distribution of certain species on a world-wide basis, as well as differences in distribution in the same area.

Certain species of land snails are more or less restricted to the vicinity of water bodies, or the bottom lands or flood plains of rivers and streams. Some, such as species of the genus *Succinea*, live among cattails and other plants bordering ponds and streams, or on any flotsam in the waterbody.

4. Fresh-Water Gastropods

There are a large variety of fresh-water habitats and the molluscs which inhabit them, in general, show considerable tolerance for extremes of physical, chemical, and biotic-trophic conditions. There are differences in the habitat requirements of the various groups of fresh-water prosobranchs and pulmonates. However, among representatives of these two main divisions, differences in habitat requirements are, in some cases, observable at the species level. In others, although optimum conditions are similar for all related species, extremes are tolerated better by some species than by others.

Bodies of fresh water available for gastropods are rivers, natural lakes, streams, swamps, ponds, flood-plains of rivers, man-made lakes, irrigation canals, drains, borrow-pits, and others. Gastropods vary as to their habitat preference and in their tolerance for such habitat characteristics as water velocity, water quality, associate fauna, and fluctuation in water level, and their preference for certain types of substrata, aquatic weeds, depth of water at which they live, and others. A few examples will be given below.

All members of the prosobranch family Pleuroceridae in North America require clean water. *Pleurocera acuta*, and other species belonging to the genus *Pleurocera*, prefer to inhabit relatively large bodies of water, as do most pleurocerids except *Goniobasis* spp. Pleurocerids are usually found in shallow water, only a few inches deep, though they may also be found at depths up to 3 ft. *Pleurocera acuta* snails occupy quiet and sheltered areas. Generally they can be regarded as bottom dwellers since they like to burrow under the sand most of the time. They may also burrow under layers of decaying leaves and other organic material.[33] *Goniobasis livescens*, on the other hand, live in extremely varied habitats, from natural springs to swift flowing rivers and open lakes. They are often found clinging or crawling on the sides of rocks and stones although in lake situations they, like *P. acuta*, also burrow under the sand bottom. The stones frequented by *Goniobasis* are often densely covered with algae and diatoms. In the Orient the family Pleuroceridae is represented by a single genus, namely, *Semisulcospira*. Species of this genus are intermediate hosts for several digenetic trematodes and live in a variety of habitats. *Semisulcospira libertina* may be found on rocks in swiftly running brooks, or in sluggish ditches.

The amphibious hydrobiid *Oncomelania nosophora* in Japan is the intermediate host of *Schistosoma japonicum* and is found chiefly on the soil above the edge of the water in irrigation ditches, in the Kofu area. In the Chikugo River area the snails are found along the water's edge in irrigation ditches, creeks, and uncultivated marshes, as well as on the soil in the river bed. In the Ukishima and Numazo areas they are found in uncultivated marshy areas and along the rice field irrigation ditches. In Japan and elsewhere in the Orient, it seems that the original habitats of *Oncomelania* spp. are the uncultivated low-lying marshes or undisturbed wet grass areas. *O. nosophora* and related species depend on the vegetation and soil for nourishment and for laying their eggs. These are laid singly or in short chains on the soil, on water-logged wood or porous stones, and are covered with an agglutinated sand jacket to give them a camouflage to match nearby lumps of mud and sand. A certain number of *O. hupensis* in China will be found just below the surface of the water at all seasons, but in the spring when the water rises and covers the snails, they prefer to remain at the water's edge or an inch above, where they lay their eggs. They are not to be found on the muddy bottom of the canals, where other snails are found such as *Semisulcospira* sp., *Alocinma longicornis*, and *Parafossarulus manchouricus*. However, *Assiminea* are occasionally found together with *O. hupensis*. In Leyte, the Philippines, *O. quadrasi* has a spotty, though widespread distribution. In general, it is limited to the many networks of winding, small creeks which feed into the larger streams and rivers. Its habitat is permanent, slow-running water which is affected by periodic flooding, and hence it is most frequently found on flat or slightly sloping land. The creeks are often modified into small, grassy, marsh areas, or into larger swamp areas with grass and trees.

Another Oriental hydrobiid, *Parafossarulus manchouricus* is the intermediate host of *Clonorchis sinensis*. It is aquatic and is common in the lakes, ditches, and canals of China, Japan, and Taiwan, where it is found on the muddy bottom. In Japan, *P. manchouricus* is found in the downstream area of the Tone River, where it occurs in sluggish water rich in aquatic vegetation; in Tokushima Prefecture, the snail is found in low swampy areas which are permanent and where the water is sluggish and the substratum is alluvial and rich in organic matter. Moreover, the snails are found in association with certain aquatic plants and snails, for example, *Semisulcospira bensoni*, *Lymnaea japonica*, and *Cipangopaludina malleata* are found together with *Parafossarulus manchouricus*. In Thailand, *Bithynia goniomphala* has a similar habitat and is found in swampy areas and in canals with a muddy bottom.

There are also several other hydrobiids which are intermediate hosts for digenetic trematodes. *Pomatiopsis lapidaria* is the host for *Paragonimus kellicotti* in North America. It is an amphibious snail which lives on banks of fresh-water bodies, including streams, marshy areas, and flood-plains of rivers; in these locations the snails are found in humid spots between leaves and rarely crawl on the upper surface of the leaves. *Aroapyrgus costaricensis* and *A. colombiensis* are small aquatic hydrobiids which serve as hosts for *Paragonimus mexicanus* in Costa Rica, and *P. caliensis* in Colombia, respectively. Their favorite habitats are swift, narrow, and shallow mountain streams. The snails are readily collected in quiet pools, where they are attached to dead and decaying leaves and twigs and in the case of *A. costaricensis* they are also found attached to the submerged roots of trees, *Anacardium excelsum*, which are abundant along the banks. Unlike *A. colombiensis, A. costaricensis* can be found in streams at various elevations, from sea level to 4620 ft.[112]

Members of the family Assimineidae are found in large amphibious colonies along the grassy banks of marshy estuaries and along large rivers far inland. Thus in some places they are found together with *Oncomelania*.[1]

Thiara granifera, one of the snail hosts of *Paragonimus westermani*, lives in fast-flowing but not torrential fresh-water streams with firm, muddy, or sandy bottom, and in some cases lives attached to concrete lining of canals. The snail is very common in the Orient and in the Pacific islands, but has also been introduced into several countries including the U.S. and Puerto Rico.

Members of the pulmonate fresh-water family Planorbidae serve as hosts for a variety of trematodes, among which are some representatives of the schistosomes, paramphistomes, plagiorchids, and echinostomes. They inhabit small as well as large bodies of water. They have a preference for shallow waters, either still or only slightly flowing, in streams, marshes, borrow-pits, flooded areas, irrigation canals, small rivers, and lakes. In these habitats the water has a moderate organic content, a muddy but firm substratum, moderate light penetration, little turbidity, submergent or emergent aquatic weeds, and abundant microflora. Other characteristics of the planorbid habitat are discussed in Volume I, Chapter 2. A large number of the habitats of planorbid snails are small temporary water bodies which dry up seasonally in several endemic or enzootic areas of snail-transmitted infections. In these situations the snails have shown the ability to withstand desiccation for long periods of time, especially if the drought is gradual and the snails protect themselves from direct sunlight under weeds and debris, or if they fall into the crevices of the dry mud bottom. The planorbids exhibit species and strain differences in their ability to withstand such adverse conditions until favorable conditions, usually brief, return.

Among the pulmonate fresh-water family Lymnaeidae there are representatives which serve as hosts for several trematodes such as some fasciolids, a few paramphistomes, certain echinostomes, strigeids, plagiorchids, and others. Lymnaeids are either aquatic or amphibious in habit. The amphibious forms are represented by *Lymnaea truncatula*, found in Europe, adjacent parts of Asia south nearly to the Persian Gulf and Indian Ocean, and in North Africa and the highlands of Kenya. Forms similar in habitat requirements are the North American *Lymnaea humilis, Fossaria parva, Fossaria cubensis*, and *Fossaria bulimoides*, and the Australian *Lymnaea tomentosa*. The typical habitat of these amphibious snails is a low-lying, poorly drained area of the moist muddy banks of ponds or slow-flowing streams. The aquatic species of *Lymnaea* are of world-wide distribution, and they favor shallow, but clean, slow-flowing water which is rich in dissolved oxygen and aquatic weeds. They are found in large and small lakes, as well as in rivers, streams, ponds, reservoirs, and seepage from these water bodies, which may have a gentle flow of water that is rich in microflora.

a. Ecological Parameters of Fresh-Water Habitats

The above description of the various habitat preferences of fresh-water pulmonates and operculates is a resumé of what we know about the ecology of fresh-water habitats. This does not mean that in each habitat described above the particular snail species should be present. Sometimes it is difficult to explain the absence of snails from habitats which are seemingly favorable. The combined effects of several parameters rather than an extreme of any one factor may account for this phenomenon. Such factors are usually categorized under physical, chemical, and biological, important among which are: amount of sunlight able to penetrate the water, amount of food available, strength of the current, nature of the substratum, ionic composition of the water, extent of growth of aquatic weeds, and presence or absence of parasites and predators. Although it is realized that several of these factors are interdependent, a convenient way of considering them will be to discuss them singly.

Size, volume and depth — These factors are only operative in the case of large bodies of fresh water, such as lakes, as they are relevant to whether these bodies of water are rich in nutrients (eutrophic), or poor (oligotrophic). Smaller bodies of water, such as the majority of habitats of fresh-water gastropods (including ponds, pools, swamps, canals, and ditches) are usually shallow enough and are eutrophic, provided that other factors are favorable.

Water Current — Movement of water due to currents and wave action help in aeration of the habitat. However, when the movement exceeds a certain level it may be an ecologically limiting factor to certain gastropod species. In many habitats, however, there are always the sheltered areas where the gastropod can live protected and produce large populations. Snails vary as to their tolerance of strong currents, the deciding factor being apparently their ability to cling to a surface. Large prosobranchs are more tolerant than smaller prosobranchs or pulmonates, and among pulmonates those with a large shell aperture seem to cling to the surface better than those with a smaller shell aperture. Thus members of the Pleuroceridae and the Thiaridae withstand swift, but not torrential, currents, better than the Hydrobiidae; among pulmonates the bulinids tolerate current velocity more than the biomphalarids.

Temperature — Temperature in a body of fresh water depends on altitude and latitude. Temperature changes usually affect the habitat by altering the rate of photosynthesis and bacterial decomposition. In addition to the annual temperature cycle, there are in some places diurnal temperature fluctuations. In the majority of habitats temperature does not seem to be a limiting factor for snail distribution but only influences their rate of reproduction. Snails are tolerant of a wide range of temperatures which prevail normally in their habitats throughout the world. They can live in a temperature slightly above 0°C, and up to 42°C.

In tropical areas there are slight seasonal variations in water temperature, and the snails breed throughout the year. In some tropical habitats, however, very high temperatures may be a limiting factor. Observations in East Africa seem to confirm that high temperatures are a major barrier preventing *Biomphalaria pfeifferi* from colonizing otherwise suitable natural habitats on the coastal plain of East Africa; but under certain circumstances, in that area or in other similar tropical areas of Africa, artificial habitats created by irrigation schemes could probably support populations of this species.

Permanence and Stability — Changes in water level affect the balance between producers (vegetation in a wide sense), consumers (mainly animals), and reducers (mainly bacteria), such a balance being essential for all the organisms living in the water. Small- or medium-size habitats are especially affected by fluctuation in water level. As far as snails are concerned, they thrive and build large populations in a permanent habitat, but changes in water level disturb their activities. If the change is extreme to the extent of little or no water being left, it leads to serious consequences. Several gastropod species living in non-permanent habitats have developed two adaptations: (1) they make use of the favorable periods, though very short in some cases, and reproduce rapidly to increase their population, and (2) they have become adapted to withstand the adverse conditions of drought, by estivation. Many individuals of the population will die, but some will survive, especially if they are protected, and there are records of survival of several months of drought for these species.

The filling up of the habitat with water may be affected by local rains or by rains further upstream. Several population studies have been conducted, especially on snails of medical and veterinary importance, in Africa, Asia, North and South America, and Europe, which demonstrated clearly that there is a definite correlation between climatic conditions, especially the rainfall cycle, and the population density and age structure of the snails and their infection rate with trematodes. My observations in western and southern Sudan and in the Senegal River Basin confirm findings in other parts of Africa. However, it should be noted that within these areas where there is a rainfall cycle there are some permanent habitats, such as irrigation schemes or large lakes, which are not subject to such drastic changes in water level.

In Brazil it has also been found that major seasonal fluctuations in snail population densities and cercarial infection rates are consequent upon rainfall and temperature, and the ecological changes brought about by them. In the northeast area of Brazil, the average annual rainfall for the coastal zone of the state of Pernambuco over a 13 year period was 1700 mm (67 in.); over 80% of this rain falls in the period from March through August as follows: 156, 253, 374, 293, 215, and 161 mm monthly. During the period from September through February the average rainfall varies from 26 to 66 mm/month. During the dry season many pools and small streams dry up gradually. Heavy rains usually begin to fall in March or April, flooding low areas and filling the streams. Water recedes gradually after the rainy season and by December most of the temporary bodies of water are dry.

It should be noted that although uninfected snails are capable of surviving periods of drought by estivation, infected snails do not normally survive as well. This has been found to be the case in infection with various trematodes. For *Schistosoma haematobium* in Africa it was revealed that fluctuations in snail population densities are accompanied with corresponding fluctuations in infection rates of this schistosome. Infection rates rise as population densities increase and as their age structures change. Although usually there is a build-up of the snail populations of several species after the rains, rainfall may affect snail populations of other species adversely by producing conditions inimical to breeding, probably through flooding.

In some other areas where the rainfall is insignificant, such as in the central and northern Sudan, other factors contribute to the seasonal changes in snail population densities. Such factors have been determined to be the seasonal presence of silt during the flood season,[105] seasonal changes in temperature, and deterioration in conditions of the permanent habitat.

Chemical Factors — For the effect of some chemical factors on a fresh-water planorbid snail, see Table 1-2. It has been shown that the chemical composition of the water is important in determining the presence or absence of several species of aquatic snails. Even closely related species show variations as to their tolerance of the ionic composition of the water. It has been demonstrated that *Biomphalaria glabrata*, and probably other species of *Biomphalaria*, have a higher tolerance of Cl as NaCl than *Bulinus* spp. have. It was determined experimentally that the maximum tolerated concentration of chlorides as NaCl is 2123 ppm for *Bulinus truncatus*, and 3641 for *Biomphalaria glabrata*. These experimental studies are supported by field observations in places where species of *Biomphalaria* and *Bulinus* occur. In the Nile Delta in Egypt one encounters populations of *Biomphalaria alexandrina* close to the Mediterranean Coast, but those of *Bulinus truncatus* occur further inland. *Bulinus truncatus* is absent in areas very close to the Persian Gulf Coast in Iraq and Iran, whereas they occur further inland in the border area of both countries. Water in irrigation canals and two large reservoirs in the Sudan contained 5 to 30 ppm of chlorides (as Cl), and 145 to 330 ppm of total solids dried at 180°C.[102] Evidently, these amounts are much below the maximum tolerated content by all the species living in these water bodies, such as the pulmonates *Bulinus (Bulinus) truncatus, Bulinus (Bulinus) forskalii, Bulinus (Physopsis) globosus, Biomphalaria ruppellii, Biomphalaria sudanica,* and *Lymnaea natalensis,* and the prosobranchs *Thiara tuberculata, Cleopatra bulimoides, Cleopatra cyclostomoides, Pila ovata, Lanistes carinatus, Gabbia senaariensis, Bellamya unicolor,* and others.

In Japan the hydrobiid snail *Oncomelania nosophora* dies when immersed in sea water for about 48 hr, and in a 3% saline solution it dies in 5 days. Thus populations of this Japanese snail situated close to the shore are seriously affected by the tide.

Bulinus and *Biomphalaria* tolerate Mg less well than Na or even K. On the other hand, the lethal concentration of K is equal to that of Mg in the case of *Biomphalaria,* and much less in the case of *Bulinus. Bulinus* has a greater resistance than *Biomphalaria* to the nitrates and equal resistance to the nitrites, in terms of quantity. As to other aspects of water chemical quality, field studies in Rhodesia indicated that the concentration of dissolved calcium bicarbonate had a demonstrable effect on the distribution of some species, notably the planorbid *Biomphalaria pfeifferi.* This species was found to be restricted to waters with calcium concentrations of more than 5 mg/ℓ as Ca^{++}, bicarbonate concentrations of over 20 mg/ℓ HCO_3^- as $CaCO_3$.[179] Williams[180] then, in a series of laboratory experiments using *Biomphalaria pfeifferi* and *Bulinus (Physopsis) globosus,* demonstrated that the maximum rate of population growth occurred in test waters with what he termed "medium" calcium concentrations of 5 to 40 mg/ℓ as Ca^{++} and bicarbonate concentrations of 20 to 200 mg/ℓ as $CaCO_3$. In irrigation canals and reservoirs in the Sudan, this author[102] had determined that well-established and large populations of *Bulinus* spp., *Biomphalaria* spp., *Lymnaea natalensis,* and others occurred in waters that had a total hardness ($CaCO_3$ in ppm) of 100 to 230.

The hydrogen ion concentration of the majority of fresh-water habitats does not seem to limit the distribution of gastropods. They can tolerate a wide range of 4.5 to 10. However, their shells tend to be thinner and more fragile in waters on the acidic side of this range. Dissolved oxygen is utilized by both fresh-water prosobranchs and pulmonates. Among the pulmonates, planorbids can survive in waters with less dissolved oxygen than lymnaeids, on account of the presence of hemoglobin in the hemolymph of the planorbids, and of hemocyanin in the lymnaeids. When in a given habitat, usually of small size, oxygen is lost in excess as a result of the respiration of the living organisms and especially by the decomposition of organic matter and dead bodies, the gastropod population is seriously affected.

Biological Factors — Parasites, predators, aquatic weeds, and microflora constitute the important associates in the gastropod's habitat. In addition to bacteria and fungi which affect gastropods, larval digenetic trematodes affect the well-being of the gastropod. A number of nematodes and leeches also parasitize gastropods and several animals prey on them. Among the latter are insects, crabs and crayfishes, other snails, fishes, amphibians, birds, and mammals. However, there is a biological balance in natural habitats which ensures the survival of a good number of the population. The beneficial presence of aquatic weeds and microflora has been indicated above. Gastropods occur, though in smaller numbers, in the absence of aquatic weeds, as long as their food (mainly microflora) is plentiful.

As indicated in the above discussion of various environmental parameters, some attempts have been made to evaluate the effect of single factors. However, it is felt that more data are still needed for an exact assessment to be made of the importance of individual environmental factors in controlling the occurrence and size of snail populations under natural conditions.

e. Physiology of Gastropods

Among molluscs, in general, there are some physiological processes which are fundamentally similar to those in other invertebrate groups, and the basic principles behind others are similar even to those in higher vertebrate groups. However, there are some processes which are distinctive to this group of animals. It is not intended here to cover details of the physiological processes. There are some publications which deal specifically with detailed aspects of this subject in the Mollusca, and the reader is referred to these publications.[55,122,133,178] The physiology of the gastropods only will be dealt with here rather than all molluscs, since gastropods comprise the majority of the intermediate hosts of the digeneans.

1. Digestion

Prosobranch gastropods use the radula in scraping their food, while many use their gills in ciliary feeding; the food consists of fine particles which enter the mantle cavity with the water current. Some prosobranchs live most of the time half buried in the sand or mud bottom and are selective deposit feeders. Special adaptations are to be found in prosobranchs, such as the crystalline style contained in a style sac in the stomach. The style, which is a hyaline rod composed of mucoproteins, releases amylolytic enzymes which help in the digestion of cellulose. Such a style sac is present only in herbivorous prosobranchs and those which are ciliary feeders, but is absent among carnivorous species. In the latter the stomach itself is reduced in size and receives digestive enzymes from the digestive gland. On the other hand the radula and buccal mass become more developed.

The majority of the pulmonates are herbivores. The fresh-water basommatophorans feed on fine particles and periphyton, the scummy layer which accumulates on the surface of aquatic weeds and other objects in the water; it is a favorite food for snails, and consists of several species of microflora and microfauna. Many species rarely feed on aquatic weeds in nature, but they do in the laboratory. Other species, such as *Marisa cornuarietis*, are known to be voracious feeders on aquatic vegetation. In addition, food particles in the water or floating on the surface layer are readily ingested. The Ancylidae (limpets) scrape the bottom for food, as they spend their life submerged. A variety of objects constitute the food for land snails and slugs, including leaves, berries, and decaying vegetable matter. Land snails and slugs do not possess the gizzard and the rest of the stomach of the fresh-water forms, but there is instead a well-developed long crop.

Digestion in gastropods is both extracellular and intracellular, the former taking place in the lumen of the digestive tract and the latter taking place by phagocytosis and pinocytosis in the epithelial cells of the digestive gland and in amoebocytes. Recently it became known that intracellular digestion also occurs by the uptake of particulate matter by epidermal cells of the fresh-water snail *Lymnaea stagnalis*.[194] It has been known for some time that one of the functions of the epidermis is the uptake of solutes from the medium. The cells of the digestive gland epithelia secrete protease, amylase, and lipase. Extracellular digestion is extensive in the long crop of land snails, through the action of enzymes. In addition, bacterial populations which are abundant in the crop and intestine break down the cellulose, the main constituent of the food of these herbivores. In carnivorous land snails, which feed on other smaller snails, on slugs and earthworms, enzymes are secreted by the esophagus and protease by the salivary glands. Cellulose, which is an important structural polysaccharide, is digested in prosobranchs by amylolytic enzymes released by the style, associated with the stomach. In no invertebrate is the crop or midgut as acid as the vertebrate stomach. However, in most molluscs the stomach pH is 5.3 to 5.8. The intestine has a pH of about 8.

Carbohydrates are digested to simple sugars, which are stored in the form of glycogen in various tissues, among which are cells of the digestive gland, connective tissues, and muscles. Some enzymes have been reported, such as amylases, and *Helix* has enzymes for the digestion of Beta glucosides. Fats are fully hydrolyzed to alcohol and fatty acids. The digestive gland secretes a lipase which is active at a pH of 6 to 7. Proteins are digested to their component amino acids which may then be absorbed and built into specific new proteins. Protein digestion is essentially a series of hydrolyses of peptide linkages. In some carnivorous gastropods certain enzymes for protein digestion are secreted. The digestive gland has carboxypeptidase, while the digestive tube and esophageal glands secrete dipeptidase and aminopeptidase.

A number of inorganic ions are essential for the snail. Inorganic ions enter in the composition of organic compounds, for example amino acids, and hence in proteins. Sulfur is important in mucopolysaccharides. Small traces of some metals are essential in nutrition and in enzymatic functions. All ions in molluscs are in general low except Ca. Magnesium is usually present in lower concentration than calcium, but in *Littorina* it is of higher concentration. Fresh-water snails can extract and concentrate ions from the external medium through their epidermal covering, but in addition certain ions are acquired from food. Gastropods, as well as other molluscs, regulate ion absorption and storage, but much less so than vertebrates do. The kidney in molluscs helps in ionic regulation. This is done through reabsorption of ions from filtrate, so that the fluid flushed out from the kidney of some fresh-water pulmonates is water.

The uptake of certain elements by fresh-water snails has been studied using radioactive isotopes. Some of the isotopes used are ^{32}P, ^{131}I, ^{64}Cu, ^{14}C, etc. Absorption of the element can be calculated quantitatively by observing the decrease in radioactivity of the liquid with time, after correcting for natural decay. Autoradiographs of snail sections reveal the sites of concentration of the element. Many of the elements used tend to be deposited in the digestive gland, but are also present in other organs. Cu, as ^{64}Cu, had a toxic action on the protoplasm where the element diffused into almost all the tissues of the snail, *Biomphalaria glabrata.* Iron 59 is the only isotope found to be transferred to the egg in appreciable quantities.

Actually snails can thus serve as biological indicators of low level radioactivity. Moreover, there is the possibility of the use of radioactive tagging in field studies on the ecology of gastropods. This is an improvement over experiments which were carried out in the past on the vagility of some species of snails where they were marked with paint to check on their movement later on.

2. Respiration

Prosobranchs possess one or more gills, and pulmonates possess a highly vascularized air-breathing lung where large quantities of blood come into close proximity with air. However, respiration throughout the surface of the animal is generally present to a certain extent in all aquatic molluscs, even in the presence of specific organs for respiration. Many shore gastropods rely partly or entirely on cutaneous respiration, and many small prosobranchs which inhabit intertidal pools have no gills and only small mantle cavities. In some pulmonates, the surface of the mantle, and to some extent the foot, because of their large area account for over 50% of gaseous exchange.

The respiratory pigment of gastropods is usually hemocyanin, a copper-containing compound carried in the plasma. In hemocyanin the copper is probably bound directly to the protein of the molecule, and hemocyanin usually constitutes more than 90% of the total blood protein. Hemocyanins are usually colorless when deoxygenated but have a strong absorption in the ultraviolet. They become slightly blue, however, when oxygenated. Oxygen combines reversibly with hemocyanin in the proportion of 1 oxygen molecule to 2 copper atoms. When the hemocyanin of *Helix pomatia* is analyzed for amino acids, it shows twice as much cystine and tryptophan, and more arginine, than in human hemoglobin. The hemocyanin of *H. pomatia* has a sedimentation constant of 98.9, a molecular weight of 6,680,000, and an isoelectric point of 5.0. The blood of most gastropods is saturated at low oxygen pressures, a factor which enables many gastropods to inhabit places with poor aeration.

Members of the family Planorbidae have in their hemolymph hemoglobin as a respiratory pigment which is saturated at oxygen pressures lower than those for hemocyanin. Planorbids are thus more suited than others to live under especially poor oxygen conditions. It should be noted that although planorbids have hemoglobin in their blood, they, as well as all other molluscs (except cephalopods and Monoplacophora) have hemoglobin in the tissues, where it occurs in buccal muscles, stomach muscles, and the nervous system. Like hemoglobin in other animals, the molluscan hemoglobin consists of an iron-porphyrin (heme), coupled to a protein, globin. The hemoglobin of some planorbids has a sedimentation constant of 33.7, a molecular weight of 1,539,000, and an isoelectric point of 4.7.

Metabolic rates differ considerably from one group of gastropods to the other, and variations are found for the same species, as a result of both intrinsic and extrinsic factors. In land snails and some aquatic snails oxygen consumption increases after hibernation. For all species it increases after activity and after feeding. In general, there is a relationship between temperature and oxygen consumption; the latter increases as the temperature rises. It was shown by von Brand et al.[174] that such a relationship exists up to 37°C in the case of *Biomphalaria glabrata*. At 41°C, however, the animals were definitely damaged. Their respiratory rate decreased and did not come back to the original level during the recovery period.

Another factor which has been investigated in relation to oxygen consumption is starvation. The influence of starvation on the oxygen consumption of *Biomphalaria glabrata*, *Helisoma duryi*, *Physa gyrina*, and *Physa* sp. was found to be pronounced and quite similar in the four species. A progressive lowering in metabolic level occurred until the snails finally died.

The rate of oxygen consumption of the eggs of some pulmonates was investigated and the findings were comparable to results obtained with higher animals.[130] The rate of oxygen consumption of the developing egg of *Biomphalaria glabrata* increases from the time of oviposition to hatching (9 days), following a curve strikingly similar to that found in egg-laying vertebrates.

3. Carbohydrate Metabolism

There is a broad range of blood sugar levels among the gastropods, and variations exist among closely related species. In general the levels are much lower than in vertebrate animals. As an example, *Helix pomatia* has a normal average sugar level of 15 mg%, and another terrestrial snail, *Achatina fulica* has an average level of 55 mg%. It has been demonstrated that the blood sugar level is an indication of the carbohydrate intake in the diet, and thus the level increases if the snail is fed on a diet rich in carbohydrate. The activity of the snail, its feeding, and its winter hibernation have their effect on the level of blood sugar. Hibernation applies to the inactive period of both aquatic and terrestrial snails. However, low blood sugars in the gastropod are accompanied with high levels of stored glycogen. This can be shown by determining the glycogen content of the total dry weight of the animal. It has been shown that stored glycogen level exhibits seasonal variation in temperate climates. The variations are evident in the same individual during the same season and seem to depend on reproductive changes occurring in the snail.

Among the land and fresh-water gastropods (both pulmonates and prosobranchs) of temperate climates, the albumen gland exhibits a seasonal pattern of polysaccharide synthesis. In autumn, glycogen is synthesized and stored in the gland as well as in other tissues. This food reserve is slowly catabolized during winter hibernation. In spring the galactogen is synthesized and accumulates specifically in the albumen gland. Thus although there is a decrease in glycogen at egg laying period, there is an increase in the total polysaccharide content of the snail due to the increase of galactogen, especially in the eggs. The galactose homopolymer is presumably a food reserve for the developing embryo, as it is the sole carbohydrate component of the embryonic perivitelline fluid and is catabolized during development. Thus hiberation is preceded by increased glycogen synthesis, and egg laying is preceded by an increased galactogen synthesis.

In the adult sexually mature snail galactogen is synthesized in the albumen gland from a uridine diphosphate galactose precursor. The polysaccharide is synthesized and accumulates in globules within secretory cells in response to abundant food, moisture, and warmth. This occurs whether in nature (in the spring) or in the laboratory. Goudsmit[55] noted that galactogen and glycogen are found in the same snail, and sometimes in the same cell. Uridine diphosphate glucose and uridine diphosphate galactose are precursors in glycogen and galactogen synthesis, respectively. These two sugar nucleotides are interconvertible via two routes: epimerase and nucleotidyl transferase, and such enzymes are present in cell-free extracts of albumen gland. The biochemical factors which may trigger the synthesis of galactogen in response to spring-like conditions, and the hormonal or enzymatic regulator molecules that stimulate glycogen synthesis and accumulation in preparation for winter hibernation are unknown. Relative to the accumulation of galactogen in the albumen gland after dormancy, a recent study by Rudolph,[144] using histochemical and biochemical analyses of this gland in the terrestrial snail *Triodopsis multilineata*, showed that the carbohydrate content rises from a dormant value of 20.5% to a high of 46.6% of the dry weight of the albumen gland on the 24th day postdormancy, with the main increase occurring between the 6th and 12th day postdormancy.

Some information is available on the carbohydrate metabolism of snails under anaerobic conditions. Anaerobic conditions for aquatic snails are produced when the snails are accidentally buried in the mud, when they escape the action of some molluscicides by burrowing into the mud, or when their habitat dries up and they go through a period of estivation. It was found in one study that under complete anaerobic conditions members of the Lymnaeidae and Physidae are considerably less resistant to lack of oxygen than the Planorbidae or several other prosobranchs used in the study.[172] The anaerobic metabolic level of the nonresistant species, as expressed by the rates of CO_2 production and carbohydrate consumption, was in general somewhat higher than that of resistant species, but a certain overlapping occurred. The resistant species did not accumulate lactic acid within their tissues during an anaerobic period, while the nonresistant species did so to a marked degree. Other findings were that in most species the anaerobic carbohydrate consumption was only slightly higher than the aerobic rate, which might be explained on the basis of possible occurrence of aerobic fermentations in these species. As to lactic acid production it was found that this acid was quantitatively a major end product of the anaerobic carbohydrate consumption only in *Lym-*

naea stagnalis and *Lymnaea natalensis*; in all other species unidentified end products must have prevailed. In another study by the same team of workers[118] it was shown that the species not resistant to anaerobiasis are killed primarily because of the accumulation of lactic acid, while the resistant species are more tolerant to the lack of oxygen due to the fact that they accumulate in their tissues the less toxic fatty acids rather than lactic acid. The acids formed by *Biomphalaria glabrata* and *Helisoma duryi* (some of the resistant species) were identified by chromatographic means and crystallographic data as propionic and acetic acids.

Carbohydrate metabolism has been investigated especially with reference to aquatic snails estivating as a result of the desiccation of their habitat. Since desiccating snails are also going through a period of starvation, the two factors were compared, using starved animals in water.[173] It was found that during starvation in water and during desiccation, polysaccharide and lipid stores become depleted. Lactic acid disappears completely from the tissues during desiccation and volatile acids diminish. Desiccation proper is probably responsible for the decrease in oxygen consumption but at high humidity starvation is a contributing factor. It was found that snails desiccating at high humidity have a purely aerobic metabolism, and that the relationship between the oxygen required for oxidation of polysaccharides and lipids and the total oxygen consumed indicates that protein may be the main substrate during prolonged periods of starvation in water or of desiccation.

4. Nitrogen Metabolism

Investigations on the nitrogenous constituents of various molluscs revealed that amino acids, either free or combined to form macromolecules among which the proteins are predominant, constitute a major part of the dry matter of the molluscan body. The pattern of the amino acid composition of the proteins and nonproteins of molluscan tissues is similar to that in other animals; however, no common pattern is found when one considers the free amino acid pools of molluscan tissues. The determination of the free amino acids has been done by the microbiological assay method or by automatic analyses by column chromatography. Among the proteins, analyses have been made on hemoglobins, hemocyanins, the conchiolins of the prismatic and the nacreous layers of the shell, and their amino acid composition. In addition, the molluscan body also contains ammonia, amines, uric acid, urea, and purines. Taurine is present only in marine forms, but not in fresh-water or terrestrial gastropods.

Certain enzymes are present in molluscs for the metabolism of nitrogenous compounds. For example in *Lymnaea stagnalis* there is L-amino acid oxidase in the digestive gland and muscles, and arginine decarboxyoxidase. L-Amino acid oxidase acts on arginine by oxidative deamination with the formation of α-keto-8-guanidinovaleric acid and γ-guanidinobutyric acid. Another two mechanisms in *Lymnaea* are involved in the metabolism of arginine: oxidative decarboxylation with formation of guanidobutyramide by the enzyme arginine decarboxyoxidase, and by the action of another enzyme arginase with the formation of ornithine and urea. Arginase has been detected not only in *Lymnaea* but in several other fresh-water and terrestrial gastropods, both prosobranchs (*Viviparus* app.), and pulmonates (*Planorbarius corneus*, *Helix* spp., *Achatina fulica*, and some slugs.[133] Urease has also been detected in some gastropods such as *Helix*. Fresh-water snails have arginase concentrations comparable to those in mammals, and in land snails such as *Helix* arginase concentrations may be eight times greater than in mammalian liver. However, *Helix* does not form urea by the ornithine cycle. The arginase concentration in snails parallels the uric acid content of the kidney.

There are also certain enzymes for purine metabolism; and although they differ from those of protein metabolism some of the end products are the same. A mollusc may excrete some uric acid from purines, regardless of whether its protein is degraded to ammonia, urea, or uric acid. Purines may be excreted as such, e.g., adenine or guanine, or they may be deaminated to xanthine or hypoxanthine, which is converted to uric acid by xanthine oxidase. If uricase is present, allantoin is then formed; allantoinase converts this to allantoic acid; allantoicase converts this to urea; urea may be broken down to ammonia by urease. Many gastropods degrade their purine to ammonia. The bivalve *Mytilus* has urease, xanthine oxidase, uricase, allantoinase, and allantoicase in its digestive gland. However, urease is lacking from the bivalve *Anodonta*, xanthine oxidase from the amphibious prosobranch snail *Pila*, whereas the aquatic pulmonate *Planorbis* has xanthine oxidase and uricase, and the terrestrial pulmonate *Helix* has only uricase.[133]

Nitrogen excretion is a labile character. It has changed during the evolution of gastropods according to their habitat, and thus gastropods represent the best example of phyletic change in form of nitrogenous products. Some gastropods have changed in transition from sea to fresh water to land, and with this there has been an accompanying transition toward uric acid excretion.[133] The uric acid in the kidney of marine prosobranchs is very low (2 to 4 mg/gm dry weight). Terrestrial snails, on the other hand, contain much uric acid (31 to 1000 mg/gm), whether pulmonates or operculates. Thus among gastropods the form of nitrogenous waste produced by protein catabolism is related to the availability of water. Extreme need for water retention in terrestrial gastropods has resulted in the evolution of a form of nitrogenous excretion which requires little water, and thus such gastropods excrete almost insoluble uric acid, resembling in this way snakes, lizards, and birds. Land gastropods are thus referred to as uricotelic, and the process itself is uricotely. White crystalline deposits accumulate in the kidney and are excreted as spherules at rather long intervals. Aquatic gastropods, on the other hand, excrete ammonia or urea, and the fluid which is flushed out of the kidney of *Lymnaea*, for example, is mainly water, serving to carry away excretory products after all inorganic ions have been retained.

5. Endocrinology

There are occasional statements in the literature pointing to the possible existence of hormones among molluscs. The authors, themselves, seem to be not quite sure of the presence and the effect of such hormones, and often the studies have never been followed through. In such cases the hormonal influence has only been inferred. The presence of a gonadal hormone controlling the development of the accessory reproductive organs in *Biomphalaria glabrata* has been inferred, based on observations of juvenile and adult uninfected snails, and others infected with the fluke *Ribeiroia ondatrae*.

However, some convincing results have emanated from investigations on the presence and functions of neurosecretory cells located on the cerebral ganglia and in the ganglia of the visceral ring. Several investigations have been conducted on the endocrinology of fresh-water snails at the Department of Biology, the Free University, Amsterdam, Holland, and on terrestrial snails and slugs at Bangor in the U.K. and in California. So far, morphological, histochemical, and physiological studies have been conducted on neurosecretion among gastropods, and studies have only begun on relating neurosecretion with infection of the gastropod with trematodes.

In *Lymnaea stagnalis* the position and number of neurosecretory cells (dorsal bodies) are fairly constant. The release of the neurosecretory material occurs in extensive neurohemal areas under the perineurium of the nerves, the intercerebral commissure, and the connectives, and in the connective tissue sheath around the central nervous system. Other proven endocrine organs in gastropods are the optic glands and the optic tentacles. A neuroendocrine role has been attributed to the dorsal bodies of fresh-water pulmonate snails by Lever.[85a] In all species a pair of mediodorsal bodies is found.

These are located at the dorsal side of the origins or the intercerebral commissure. In some species, e.g., *L. stagnalis*, a second pair of bodies is present at the laterodorsal side of the cerebral ganglia. The structural differences between the bodies are slight. The amount of cytoplasm shows great seasonal variations; in spring the dorsal body cells are rich in cytoplasm. Whether the dorsal bodies serve as neurohemal structures and/or as endocrine glands has been debated. Experimental work, however, has shown that their endocrine character is most probable.[68] Although some of the cells terminate in the neurohemal areas of the central nervous system, others do not have such terminations and are able to discharge their neurosecretion directly into the hemolymph and thus will have a hormonal influence on several physiological processes. The presence of the dorsal bodies has also been reported for terrestrial snails (*Helix*) and slugs (*Agriolimax*), and for prosobranchs.

The optic tentacles of the stylommatophoran pulmonates contain the eyes and the optic nerves connecting them with the cerebral ganglia. In addition, closely opposed to the eye is a large tentacular ganglion from which digitating extensions protrude into the dermomuscular layer. Between these nerve branches, glandular collar cells are found. Subepidermal cells and lateral-processed cells have been described along the stalk of the tentacles.[145]

Some evidence, although based so far on few studies, has shown that in gastropods reproduction is controlled by the endocrine structures which have been described for several species. In hermaphroditic gastropods the differentiation of male sex cells is dependent on the presence of a factor produced by the cerebral ganglia, whereas the female cells show autodifferentiation. There are also some indications that the optic tentacles in stylommatophorans play a part in regulating reproduction, but the mechanism by which this is accomplished is not very clear. Neurosecretions from the dorsal cells in *Lymnaea stagnalis* were shown to affect oviposition. After cauterization of these cells there is usually a decrease, though temporary, in the oviposition capacity of the snail.

Other effects have also been attributed to the neurosecretion of the dorsal bodies. One of these is the control of osmoregularity processes, involving water and ion regulation. It is known that fresh-water gastropods, such as *L. stagnalis*, maintain a high osmotic gradient between blood and environment. This gradient causes an inward flow of water and an outward diffusion of solutes. The processes for maintaining a steady state generally involve the production of large volumes of a hypotonic urine and an active uptake of ions. Experimental studies have indicated that deionized water will activate water elimination and the ion-uptake mechanisms, whereas a 0.1 M NaCl solution leads to a suppression of both processes.

In some gastropods (*Lymnaea* and *Aplysea*) extirpation of one or more ganglia of the central nervous system resulted in the swelling of the body, whereas reimplantation or injection of homogenates of these ganglia led to a decrease of the body weight. The production of diuretic factors is assumed to occur in these ganglia. It was also reported that changes occur in the electrical activity of some special neurosecretory cells after osmotic stimulation of the osphradium. Additional indications of neuroendocrine involvement in osmoregulation were shown, based on histological investigations dealing with osmotically induced changes in the amount of stainable material present in neuronal cell bodies.

A recent study by electron microscopy has thrown some light on the neuroendocrine involvement in osmoregulation in *Lymnaea stagnalis*[16] At the ultrastructural level two types of neurosecretory cells seem to be involved. In snails exposed to deionized water, a condition stimulating water elimination, and retention and/or uptake of ions, the release of secretory material from the axons of these types was observed. In animals exposed to hypertonic saline, known to reduce water elimination and ion uptake, the release activity declines and an enhanced accumulation of secretory granules occurs. The axons of one type not only end in the main neurohemal zones around the central nervous system, but also form a network around the ureter similar to that found around water- and ion-transporting epithelia in insects. It seems, therefore, that one type of neurosecretory cells activates water elimination, while the other type stimulates ion-uptake mechanisms.

Relative to the neurosecretory cells of molluscs it has been observed that there is a diurnal rhythm in their activity. This is also known to occur in the case of other invertebrates, and also in vertebrate animals. In the snail *Lymnaea stagnalis*, the process of neurosecretion in the caudo-dorsal cells, which produce an ovulation hormone, shows such a diurnal rhythm. The synthesis, transport, and release of the neurosecretory material is high during the evening and the early night and low during the rest of the day, while storage of material mainly occurs during the daytime. It has been established that environmental factors such as temperature, light, and noise may be involved, but light seems to be the most important. It has been shown that in the case of the marine opisthobranch *Aplysia californica*, the neurosecretory R15 neuron shows a spontaneous diurnal rhythm of electrical activity and that light/dark cycles control this rhythm.

Whether steroids are present in molluscs has always been an intriguing question. Some investigators have tried to demonstrate the presence or synthesis of steroids comparable to those which play a role in the regulation of reproduction in vertebrates. It was shown that 11-ketotestosterone and dehydroepiandrosterone are present in the slug *Agriolimax californicus*. These compounds seem to influence the release of the active principle of the optic tentacles in this species. Joose[68] is of the opinion that it is difficult to interpret such results because the morphology of the tentacles has revealed contradictory findings relative to the presence of endocrine structures.

f. Effect of Larval Digenea on the Gastropod Host

Most of the early information on the effect of larval digenea on the gastropod host has been included in several publications.[98,108,113,187,188] Although a slight pressure is exerted on neighboring tissues by the settling and metamorphosis of the miracidium, the effects of parasitism are not noticeable until a few weeks later when the migration and, later, the maturation of daughter sporocysts or rediae takes place. Congestion of the blood sinuses in the visceral mass is one of the first noticeable changes.

By far the most affected organ of the gastropod is the digestive gland and this is more pronounced in flukes having a redial stage in their life cycle than in those which do not. It is expected that a redia, because of the fact that it possesses a simple gut, consisting of a mouth, pharynx, and an intestine, causes more mechanical damage than the sporocyst, which lacks these organs. The uptake of nutrients by the sporocyst takes place through the tegument, but in the redia there is an uptake through the tegument in addition to actual feeding and destruction of the tissues. Infection with either larva causes toxic changes in the digestive cells of the epithelium; the distal walls of these cells often break down; the lateral walls also break down. Such changes result either in the formation of a syncytium, or the cells become cuboidal or sometimes squamous. Mechanical pressure of the larvae on the host tissues also takes place, in addition to the toxic effects, resulting in degenerative changes. Other histopathological alterations in the digestive gland of molluscs as a result of parasitism by larval trematodes have been reported, and these include: displacement of the tubules and loss of their branched structure, degeneration of the tubules' epithelium, increase in the number of cytoplasmic vacuoles, and reduction in size of the digestive gland. Some changes are also caused by starvation on account of drainage of nutrients by the parasites.

Some pathological alterations are in some cases due to lysis of the digestive gland epithelium and neighboring tissues. Naphthylamidase activity was localized in the tegument of rediae of certain species.

Ultrastructural changes of the secretory cells of the epithelium of the digestive gland tubules of *Biomphalaria pfeifferi* infected with *Schistosoma mansoni* for more than 12 weeks revealed cristolysis and reduction in size and in number of mitochondria, and a slight atrophy of the Golgi apparatus.[119] Many of the secretion granules in the cell apex become irregular in outline and contain patches of electron-dense material. Vacuoles containing myelin figures and electron-dense granular material are found in some of the secretory and also in a few of the digestive cells. Starvation can also cause similar pathology.[119]

In addition to the epithelium the connective tissue between the digestive gland tubules of infected snails is also affected, as revealed by light and electron-microscopy. The connective tissue matrix becomes more electron-dense and the collagen-like fibers more prominent.

It should be noted that even in a normal host-parasite relationship a snail tissue reaction at an advanced stage is formed around the sporocysts in the digestive gland. Pan[126] showed that sporocysts of *Schistosoma mansoni* maturing in the digestive gland of *Biomphalaria glabrata* provoke a reaction intermediate between the focal and generalized types of tissue response. Extensive infiltration of hypertrophic amoebocytes around daughter sporocysts in the digestive gland usually appears only after the generalized tissue reaction has commenced. In this laboratory, we have observed similar tissue reactions, and in some specimens the reaction may lead to the disintegration of the parasite. Furthermore, we demonstrated that cercariae trapped and dying in the loose vascular connective tissue, and in other tissues in the head-foot, mantle, kidney, and rectal ridge, incited a marked, generalized proliferative tissue reaction around them (Figure 1-58).

Infection with larval Digenea in the snail host also affects the biochemistry of the host and the physiological processes which take place in this invertebrate animal. It has been demonstrated that, with a few exceptions, there is a decrease in level of the host glycogen and in blood proteins, and fluctuations in the lipid content. These findings indicate that the host food reserves are utilized by the developing larvae.

The depletion of glycogen reserves is noticeable in the digestive gland epithelium, but more so in the connective tissue between the digestive gland tubules and that enveloping this gland and the ovotestis.

In addition to the effect of a low reserve of glycogen in the nutritional status of the snail, other effects have been reported. Depleted glycogen reserves make infected snails less tolerant of anaerobic conditions. Under field situations fresh-water snails are exposed to drought periods of their habitats during which they have to estivate. The effect of such periods is more pronounced in the case of infected snails than uninfected ones. Furthermore, estivating infected snails vary in the extent of their tolerance of drought conditions. It has been demonstrated in the case of *Biomphalaria glabrata* infected with *Schistosoma mansoni* in Brazil, and *Bulinus (Physopsis) ugandae* infected with *Schistosoma bovis* in the Sudan,[108] that the snails can only survive estivation if the schistosome is still at an early stage of development (mother sporocyst). Those with an advanced infection (daughter sporocysts and cercariae) die, apparently because of lowered nutrient reserve.

With reference to amino acids, it has been found that there is an overall reduction in the proteins and free amino acid contents of infected snails, especially methionine, which in some cases may be completely absent. The 17 amino acids, including methionine, which occur in the hemolymph of the snail (for example *Biomphalaria glabrata*)

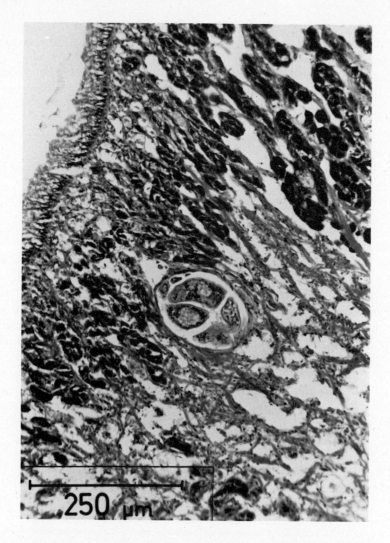

FIGURE 1-58.   Section in the foot of the snail *Biomphalaria glabrata*, infected
with *Schistosoma mansoni*, showing trapped cercariae, still in intact condition.

have been detected in the proteins of *Schistosoma mansoni* cercariae, and it has been
suggested that the parasite obtains at least part of its amino acid requirements from
the free amino acids of its host. That the host's hemoglobin, at least the heme-includ-
ing moiety, is taken up and utilized by the schistosome larvae has been demonstrated
by the use of ^{59}Fe as a tracer. Tissue protein concentrations are also altered as a result
of the infection.

With the depletion of both amino acids and glycogen as a result of parasitism, there
is an increase in the lipid content in the epithelium of the digestive gland cells, and
this has been demonstrated in the case of infections with various trematodes. That
such a rise in lipids and fatty acids may be the result of an increase in anaerobic metabolism of the carbohydrate reserves, on account of decreased oxygen, has been suggested by some investigators.

Relative to enzyme histochemistry and the changes that are associated with trematode infections, it has been demonstrated in the case of several larval Digenea, with a
few exceptions, that there are increases in the activities of acid and alkaline phospha-

tases of their respective molluscan hosts. This is probably associated with increased intracellular activities and with the exchange of polysaccharides between the host and the parasite. In the case of the snail *Oxytrema silicula* infected with rediae of the nanophyetid, *Nanophyetus salmincola* (see Volume II, Chapter 2), acid and alkaline phosphatases were found to be in greater quantities in snails infected with this parasite when compared to uninfected snails. Alkaline phosphatase was found to be associated with the rediae and developing cercariae, and this, together with similar findings in the case of other larval fluke infections, is thought to be a result of metabolism of the host's glycogen. It should be noted, however, that a decrease in alkaline phosphatase has been reported in other studies, such as those of a plagiorchioid cercarial infection in *Biomphalaria sudanica* and of *Fasciola hepatica* in *Lymnaea truncatula*. This, however, does not exclude phosphatase involvement in host carbohydrate breakdown.

Some physiological alterations have been reported in snails as a result of their infection with larval Digenea. Some of these are changes associated with growth, with fecundity, with life span, with heart rate and respiration, and with thermal tolerance.

Reports of effects of parasitism by several species of trematodes on the growth of their molluscan host have been contradictory. Enhanced growth has been reported as well as retardation in the growth of snails harboring the larval flukes. Reports of enhanced growth were mainly based on an observable increase in the size and thickne-s of the snail shell, and this is not necessarily accompanied by an increased growth of the soft parts. There is a general agreement that larval trematodes affect the fecundity of the snail, and the presence of a redial stage in the life cycle often results in castration of the snail. A comparison of the ovotestis region of the planorbid snail *Helisoma corpulentum* infected with either the echinostome *Petasiger chandleri* or the strigeid *Uvulifer ambloplitis* shows the difference between redial and sporocyst infections (Figure 1-59).[98] Rediae of *P. chandleri* occupy all the acini of the ovotestis and no snail ova or sperm are found in the acini, whereas the sporocysts of *U. ambloplitis* invade the digestive gland and the connective tissue of the ovotestis, but not the acini. Reports on the effects of sporocyst infections on fecundity have been contradictory. With both *Biomphalaria* and *Bulinus* snails infected with *S. mansoni* and *S. haematobium*, respectively, some authors observed a decline in egg laying by the snail followed by a complete suppression, whereas others reported a decline in egg laying, but egg laying was never inhibited. It should be noted that although the gonad may not be affected, especially when a redial stage is lacking, the albumen gland, on the other hand, is affected by parasitism, and it usually undergoes a reduction in size and in some cases becomes necrotic in appearance. Since the gland's main function is to provide the nutrient proteins required by the developing embryos, egg production will be inhibited even without any demonstrable effect on the gonad as a result of infection with digenean intramolluscan larval stages. Infection with rediae may, in some cases, inflict mechanical damage to the albumen gland also, as seen in Figure 1-59.

In spite of the parasitic stress on the snail caused by larval trematodes, it has been observed that infected individuals live for periods equal to, or slightly less than, uninfected ones. Here again, we find differing reports, and most of the laboratory and field data has been collected from observations on the medically important schistosome intermediate hosts. Under laboratory conditions, infected *Biomphalaria* spp. have been reported to have a median life expectancy of about 6 months and some field observations in Brazil support this contention, while others, in Egypt, claim that the life span of this infected snail is much shorter. Our experience, however, has been that under laboratory conditions, although a small percentage of the infected snails die, especially at the stage of cercarial emergence, most of the infected snails can survive well if maintained under favorable conditions.

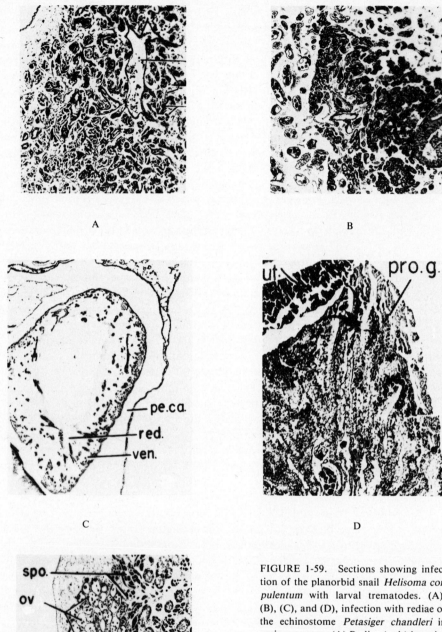

FIGURE 1-59.   Sections showing infection of the planorbid snail *Helisoma corpulentum* with larval trematodes. (A), (B), (C), and (D), infection with rediae of the echinostome *Petasiger chandleri* in various organs. (A) Rediae (red.) have invaded and destroyed all the acini of the ovotestis (ov.). (B) Albumin gland. (C) Ventricle (ven.). (D) Prostate gland (prog. g.). (E) Infection with sporocysts (spo.) of the strigeid *Uvulifer ambloplitis.* Note presence of the sporocysts only between the acini of the ovotestis. d.g., digestive gland; h.d., hermaphroditic duct; pe. ca., pericardial cavity; and ut., uterus. (After Malek, E. A., *Am. Midl. Nat.,* 48, 94, 1952.)

Other effects of parasitism on the snail are in the form of an increase in the heart rate and an increase in the oxygen uptake, as these aspects reflect the metabolic rate of the host. That infected snails have a lower maximum thermal tolerance limit and a lower hemolymph osmolarity as compared to uninfected snails has also been reported.

Infection with certain trematodes may also affect the behavior of the snail and its migratory cycle. It has been reported by several authors that infection with the heterophyid *Cryptocotyle lingua* affects the zonation and winter migration of the Atlantic coast (Europe and North America) common periwinkle, *Littorina littorea*. The snails apparently become immobilized by the infection and migrate more slowly, and to a lesser extent, than the noninfected snails.

g. Taxonomy of Gastropods and Gastropod Families

Before embarking on the subject of taxonomy of gastropods it is relevant to indicate the importance of clarifying the specific identity of a gastropod in relation to its infection with digenetic trematodes. Parasitologists work very closely with the snail intermediate hosts and the less confusion they encounter in the taxonomy of these snails, the better they appreciate the susceptibility or nonsusceptibility of the molluscs, as well as other aspects of the host-parasite relationships. Unfortunately in the case of molluscs the old taxonomic problem of what natural unit constitutes a population, or a subspecies, a species, or a group of species, has not been resolved for a great number of molluscs. One of the problems has been a misunderstanding of the dynamic species concept. Species vary in space, as well as time, and therefore the species should not be regarded as a clearly defined unit, but should be recognized as a complex of variable populations. This is of significance in the understanding of host-parasite relationships. Infraspecific variations take place among populations of snails which become isolated due, for example, to the drying up of large habitats. This would also affect the parasite, and divergence runs parallel in both snails and trematode. The most usual evidence for infraspecific divergence is in the differential susceptibility to infection found between certain local races of snails and races of parasites other than those for which they normally serve as hosts.

It was suggested by Wright[188] that speciation in the digenetic trematodes is most often based on parallel evolution of the flukes with their molluscan rather than their definitive hosts. Discontinuous distribution of the molluscs provides the necessary geographical isolation between centers in which the parasite cycle is completed, and these centers are the gene pools on which local selection pressures act to produce differing genotypes. Thus the divergence of the trematode is intimately associated with that of the snails. In addition to changes brought about in susceptibility, variation in local forms of snails may be apparent in their resistance to desiccation and in their breeding habits.

1. Criteria Used in Snail Taxonomy
a. Morphology

The taxonomy of the species involved in the transmission of trematodiasis has, in general, been in a state of confusion for many years. Early descriptions of snails were based on certain shell features, in some cases on one shell only, and thus do not provide satisfactory bases for differentiation of species. The shell exhibits considerable individual variation and also varies with the age of the snail, the type of habitat, and, in the case of aquatic snails, with the quality of the water where it has been reared. Differences due to water quality have been demonstrated experimentally, and these observations are not new; the Reverend Cooke[29a] reported similar differences in 1896. Because of shell variations, many artificial species have been created, a problem which causes confusion to trematodologists trying to establish host-parasite relationships. Even with the relatively recent use of anatomical details in taxonomy of snails there are still numerous discrepancies in identification. These discrepancies are probably due to the "lumpers" and the "splitters" among malacologists. One malacologist might lump together a number of variations found in a group of specimens, considering them either as the normal range of variation found in most species, or as characters produced through different environments, and call the snails "ecotypes". Another taxonomist, on the other hand, because of these variations, might split the nomenclature of the specimens he found and consider them as either separate species or subspecies. Like several other animals, populations of snails, because of geographical isolation, exhibit genetic divergence. This is especially true with land snails occurring on certain islands, which provided excellent opportunities for the study of the effect of isolation on the genetic constitution of certain populations and the resulting lack of interbreeding among such populations. This isolation also exists among fresh-water forms, and results in the formation of infraspecific variations. Many live in small temporary pools, completely isolated from any other population. These and other similar habitats have caused the isolation to proceed long enough to produce genetic divergence. The barriers can be broken when the snails are passively dispersed. This has taken place through floods or change in river beds. The movement of river waters downstream helps to disperse snails, but the populations need not necessarily become established downstream, unless they encounter favorable conditions. Examples of this are the great rivers of the world. The Nile no doubt brings downstream from Central Africa and the Ethiopian highlands certain species of *Biomphalaria* and *Bulinus*, yet *Bulinus (Physopsis)* spp. are never found north of the Central Sudan, and the same is also true with *Biomphalaria sudanica*. In the U.S., snails such as *Helisoma corpulentum, Lymnaea stagnalis*, and certain other lymnaeids are abundant in the upper reaches of the Mississippi River, but such species are never found in the Central or the Southern part of the country. Dispersal of fresh-water snails may also take place through human agency, and dispersal by birds has been documented on several occasions. The pulmonate fresh-water basommatophorans are hermaphrodites and are capable of self-fertilization; thus if one individual is carried by humans or by birds it is capable of establishing a population in the new habitat. The marine dioecious gastropods on the other hand are widely distributed, their pelagic larvae are carried for long distances, and unlike the terrestrial or fresh-water pulmonates, have many opportunities for genetic interchange with potentially interbreeding populations.

It became apparent to many malacologists that the morphology of the shell and soft parts are in some cases of limited value as criteria for establishing a satisfactory systematic arrangement of taxa in the lower taxonomic categories. To give only one example: the African planorbid genus *Bulinus* is currently divided into three subgenera. Within the subgenera the various species are very similar morphologically, although biochemically and cytologically some of them are quite distinct. This is especially true of the species of the subgenus *Bulinus* s.s., which comprises at least two, and perhaps more, species groups. Species are generally placed into these species groups according to the shapes of their redular mesocones, or sometimes according to their chromosome numbers. However, these two sets of characteristics do not always show a correlation. Thus new approaches to the problem of molluscan taxonomy and systematics, in addition to morphology, have to be sought. In recent years the new approaches have involved cytological studies, biochemical studies with reference to paper chromatography and electrophoresis, and serological methods. These new approaches will be treated singly. As will be noticed, such new methods have dealt mainly with the medically and economically important molluscs.

b. Cytology

Cytological studies on several gastropods revealed that in the subclass Prosobranchiata the chromosome numbers range from n = 7 to n = 36. In the subclass Pulmonata, haploid chromosome numbers range from n = 5 to n = 44, excluding polyploidy, among the more than 400 species studied. In the order Basommatophora the range is from n = 15 to n = 19 (excluding polyploidy), while in the order Stylommatophora it ranges from n = 5 to n = 44. Although some molluscan cytologists claim that cytological information has been shown to be helpful in species discrimination in both the prosobranchs and the pulmonates, others believe that such a tool is only of value at the higher taxa. However, considerable research was conducted on the cytotaxonomy of the planorbid vectors of the schistosomes, and the data obtained was revealed to be useful in the systematics of the genus *Bulinus*, but not of *Biomphalaria*. The genus *Bulinus* is remarkable in that it contains a series of polyploid populations or species characterized by exact multiples of the basic chromosome number (n = 18) of the family Planorbidae to which they belong. Some species have the normal diploid chromosome number (n = 18, 2n = 36), whereas various other species are either tetraploid (n = 36, 2n = 72), hexaploid (n = 54, 2n = 108), or octoploid (n = 72, 2n = 244). All polyploids, as well as some of their close diploid relatives, occur in the subgenus *Bulinus* s.s. This bulinine subgenus comprises the more northerly *Bulinus (Bulinus) truncatus* species group, whose members are mostly tetraploid (but with one octoploid species), and are readily susceptible to infection with North African *Schistosoma haematobium*, and the more southerly *Bulinus (Bulinus) tropicus* species group, which is diploid and nearly always refractory to infection with *S. haematobium*. Such cytological methods are, therefore, helpful in differentiating species belonging to these two groups, especially in areas where their distribution overlaps. It was also demonstrated that polyploidy in members of the genus *Bulinus* is always associated with susceptibility to infection with *S. haematobium*. In the subgenus *Bulinus* there are very few morphological differences between the various species, regardless of their chromosome number. The differences that do exist are slight. Karyotypic differences other than chromosome numbers also seem to be minor. However, each of the chromosome number groups can be distinguished readily by micro-Ouchterlony double diffusion precipitin tests, as well as by electrophoretic separations of muscle proteins. The results have been included in several publications by Burch and associates.[19]

c. Serology

Studies in systematic serology have been used to elucidate relationships in some animal groups such as vertebrates and arthropods. In these studies the most frequently used antigens have been blood serum proteins, either complete or fractionated. These have proven to be superior to tissue extracts, especially when turbidometric measurements of suspended precipitates have to be made. Advances were made in that field when the Ouchterlony plate gel-diffusion technique was introduced, especially as this technique allows qualitative comparisons to be made, and does not require clear antigen solutions. In addition to the Ouchterlony diffusion plates, other methods such as ring-tests and acrylamide and agar gel immunoelectrophoresis have been used for precipitating antigen — "antibody" systems as immunological techniques in molluscan systematics. The trend in systematic serology as applied to molluscs has increased, as evidenced by a number of studies on members of the gastropod families Lymnaeidae, Planorbidae, Pleuroceridae, Hydrobiidae, and also among the Bivalvia.

Preparation of antisera is performed by homogenizing certain organs of the snail, or the whole snail, and injecting the homogenate subcutaneously at five or six sites in the nuchal region of rabbits. Injecting rabbits twice with about a 3-week interval proved to be an effective method. The rabbits are bled about one week after the second injection in order to collect the antiserum. Antisera to particular organs and to snail egg proteins of certain species are then stored at 4°C as a reference bank for future taxonomical studies.

When both immunological and cytological methods were used in the study of members of the subgenus *Bulinus* a good correlation was found between chromosome numbers and immunological characteristics.[19] In no instance did a species with one particular haploid chromosome number, when tested with antiserum for a species with a different chromosome number, fail to give a reaction for nonhomologous antigens ("nonidentity"). On the other hand, populations of the same chromosome number, with only two possible exceptions, showed no "nonidentity" reaction.

In the case of the family Lymnaeidae, immunological studies indicated that distinct "species-groups" (i.e., taxonomic units usually referred to as "genera" or "subgenera") do indeed occur, and immunological data correlate well with previous morphological information. Moreover, *Radix relicta* from Yugoslavia showed by its "identity" reaction to be related to *Radix auricularia*. However, "nonidentity" reactions occurred when antigens from *R. relicta* were tested with antisera from species representing other lymnaeid species-groups (*Austropeplea, Bulimnea, Fossaria, Lymnaea, Pseudosuccinea, and Stagnicola*), from widely separated regions, such as northern and central Europe, Japan, India, South Africa, and the U.S.[18]

d. Electrophoresis

In the field of malacology electrophoretic techniques have been used for several purposes, among which are characterization of hemolymph proteins, biochemical pathology, developmental biology, analysis and purification of specific substances, and in taxonomy. For taxonomical purposes several electrophoretic techniques have been employed including cellulose acetate electrophoresis, starch gel electrophoresis, and polyacrylamide gel electrophoresis. The snails studied were biomphalarids, bulinids, oncomelanids, lymnaeids, and species of the Oriental genus *Semisulcospira*; whole snails, hemolymph, eggs, foot muscles, digestive gland, or crystalline style were used. The substances stained for in these studies included proteins, antigens, esterases, phosphatases, and other enzymes.

It was demonstrated in some of the above studies that planorbid egg proteins, when subjected to electrophoresis, are sensitive indicators of differences at the population level; the esterases of the digestive gland from mature snails maintained on a standard diet were also thought to be useful for comparative purposes. However, variations in electrophoretic patterns of esterases of the digestive gland occur, and these were evaluated using biomphalarid species from various parts of the neotropics and lymnaeid snails from North America.[114] The findings showed that there are intrapopulation and interpopulation differences in the same species. But when tissues from several individuals of the same population, or of different populations of the same species, are pooled, one can obtain a pattern which is to some extent distinctive of the species. Moreover, at the species level, and among the species and populations which we used, there were similarities suggesting close relationships between some populations of *Biomphalaria glabrata* and *B. tenagophila* on the one hand, and of certain populations of *B. peregrina* and of *B. obstructa* on the other hand (Figure 1-60). On the basis of the findings we suggested a correlation between certain patterns that indicate biochemical similarities or differences among the planorbid snail populations, and the susceptibility of the species or the population to infection with the schistosomes. In the case of the lymnaeid snails studied, no such correlation could be suggested, but the findings indicated that each of *Pseudosuccinea columella* from certain localities in Louisiana, *Fossaria humilis* from Michigan, and *Stagnicola palustris*, also from Michigan, showed consistent and distinctive patterns, which can be useful in their identification.

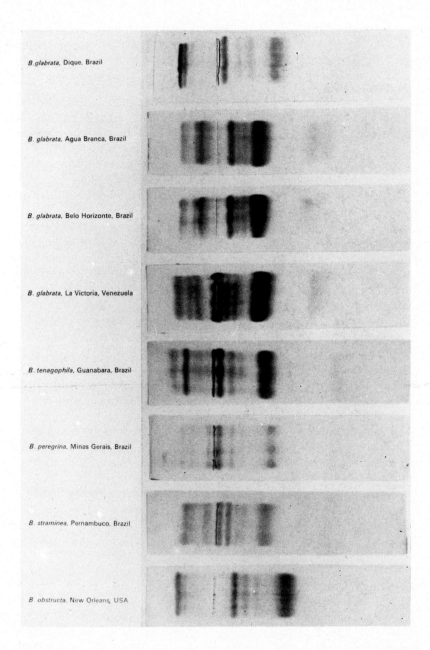

B.glabrata, Dique. Brazil

B. glabrata, Agua Branca, Brazil

B. glabrata, Belo Horizonte, Brazil

B. glabrata, La Victoria, Venezuela

B. tenagophila, Guanabara, Brazil

B. peregrina, Minas Gerais, Brazil

B. straminea, Pernambuco, Brazil

B. obstructa, New Orleans, USA

FIGURE 1-60.   Electrophoretic patterns of the digestive gland esterases of some *Biomphalaria* species, field collected from different areas of the Americas. (After Malek, E. A. and File, S. K., *Bull. W.H.O.,* 45, 819, 1971.)

It is to be gathered from the above discussion of the various methods used in gastropod taxonomy and systematics that each one of these old or new methods has its pitfalls, and thus a multidisciplinary or integrated approach has been recommended.

h. Gastropod Families

It is not intended in this book to give a detailed account of molluscan systematics. The major groups were treated in the introductory part of this subsection. Some of the gastropod families, which comprise members of significance in the transmission of trematodiases (Table 1-4) and certain nematode infections, will be listed below, together with their diagnostic morphological features and important representatives.

1. The Pulmonate Families
a. Family Planorbidae

Shell usually discoidal, sinistral (ultradextral), may be globose or physoid. Animal sinistral, pulmonary, and genital apertures on left side. Tentacles long, filiform, and cylindrical, with eyes situated at their inner bases. Structure of penial complex variable, but always consists of a preputium and a vergic sac. Radula with bicuspid central tooth, large bi- or tricuspid lateral teeth and multicuspid marginal teeth. A highly vascular pseudobranch present on left side. Hemolymph with hemoglobin, hence reddish. Five subfamilies include:

Subfamily Planorbinae — Shell discoidal. Preputium without preputial gland; vergic sac without flagella. Ratio preputium to vergic sac variable, and of diagnostic value. Genus *Biomphalaria* Preston, 1910. Several species occur in Africa, southwestern Asia, southern U.S., Mexico, Central and South America, and islands of the Caribbean. Important synonyms are Tropicorbis Brown and Pilsbry, 1914; Australorbis Pilsbry, 1934; and Taphius H. and H. Adams, 1855.

Subfamily Bulininae — Shell sinistral usually globose or turreted (*Bulinus* Muller, 1781), or may be discoidal (*Indoplanorbis* Annandale and Prashad, 1920). In *Bulinus (Bulinus) truncatus* and other species and *Bulinus (Physopsis)* spp. shell is globose, whereas *Bulinus (Bulinus) forskalii* and related species the shell is with a long, slender spire. Members of the genus *Bulinus* are in Africa, the Middle East, southern Europe, and some Mediterranean islands. Members of the genus *Indoplanorbis* are in India, Thailand, the Malay Peninsula, and Sumatra.

Subfamily Segmentininae — Small planorbids with low, smooth, and glossy discoidal shells; aperture heart shaped. Prostate tubules arranged in a single row along prostate duct. One or two flegella at end of vergic sac. Genus *Segmentina* Fleming 1822, in Asia and Africa; genus *Hippeutis* Charpentier, 1837 in Asia.

Subfamily Helisomatinae — Shell discoidal, sinistral. Prostate fan shaped in cross section; prostate tubules compound, branching several times. Preputium with cup-shaped preputial gland. The genus *Planorbarius* Froriep, 1806 is the only representative in Europe and North Africa. The other members of this subfamily are American, for example the genus *Helisoma* Swainson, 1840.

b. Family Lymnaeidae

Shell dextral, ovately oblong; spire more or less attenuated and varies in height. Columella spiraled or twisted. Shell varies in thickness and in size of body whorl. Animal (soft parts) dextral. Tentacles flattened, triangular in cross section. Central tooth of radula unicuspid; lateral teeth bi- or tricuspid, marginal teeth serrated. Prostate bulblike or cylindrical. Ratio vergic sac to preputium variable. Pseudobranch absent. Cosmopolitan in distribution. Genera *Lymnaea* Lamarck, 1799; *Pseudosuccinea* Baker, 1908; *Fossaria* Westerlund, 1885; and Stagnicola Leach, 1830.

c. Family Physidae

Shell sinistral, spiral, shining, or dull, smooth or with microscopic transverse striae, spire may be elongated or very short. Animal sinistral; tentacles long, slender, and cylindrical. Foot narrow, pointed posteriorly. Inner margin of mantle plain or digitate extending over shell. Radula arranged in V-shaped rows; central tooth multicuspid; lateral and marginal teeth multicuspid. Vergic sac usually twice as long as preputium, but varies in length. Preputium with large gland on upper surface. Hemolymph coloreless. Pseudobranch absent. Cosmopolitan in distribution. Genera *Physa* Draparnaud, 1801; and Aplexa Fleming, 1820.

d. Family Ancylidae

Shell patelliform (caplike or limpet shaped). Tentacles short, blunt, cylindrical with eyes on their inner base. Radula with uni- or bicuspid central tooth, bicuspid or comb-like lateral teeth. Verge in some species resembles ultrapenis of the planorbid *Bulinus.* Pseudobranch present. Cosmopolitan in distribution. Genera *Ferrissia* Walker, 1903; and *Gundlachia* Pfeiffer, 1849.

2. The Prosobranch or Operculate Families

The following families comprise fresh-water or amphibious forms.

a. Family Hydrobiidae

Shell small, dextral, conical, or subconical. Operculum concentric or paucispiral, corneus or calcarious. Animal with a long snout; tentacles long thin, round. Mantle edge smooth. Gills platelike consisting of 20 to 50 lamellae. Verge, simple, bifid or trifid or with papillae or protrusions; verge on dorsal surface of neck arising on median line or slightly to the right. Central tooth of radula multicuspid with basal denticles; lateral teeth hatchet-shaped and multicuspid; marginal teeth slender and multicuspid.

Subfamily Hydrobiinae — Operculum thin, paucispiral. Verge simple, bifid or trifid. Eggs laid singly; some members are ovoviviparous. Genus *Oncomelania* Gredler, 1881, with five species (considered as subspecies of *O. hupensis* by some authors). They are amphibious snails and are restricted to the Orient. Genus *Pomatiopsis* Tryon, 1862, is North American, in the Mississippi River drainage. Anatomically similar to *Oncomelania*, but with a few minor differences. It is similar to *Oncomelania*, being amphibious. Genus *Littoridina* Souleyet, 1852. The verge is large and includes five or six small digitate processes. Members are in South and Central America and southern U.S. Genus *Aroapyrgus* Baker, 1930. Verge simple, snail ovoviviparous and aquatic. In South and Central America. Genus *Lithoglyphopsis* Thiele, 1928. Aquatic snails in Southeast Asia.

Subfamily Buliminae (= Bithyninae) — Shell ovate, comparatively large (10 mm); peristome generally thickened. Operculum thick, calcareous, usually concentric. Verge exserted, curved, fleshy, bifid. Central tooth of radula with numerous basal denticles. Oriental genera *Parafossarulus* Annandale 1924, and *Bulimus* Scopoli, 1777. The latter genus is also referred to in the literature as *Bithynia* Leach, 1818; *Bithinia* Gray, 1821; and *Bythinia* (a misspelling for *Bithynia*). In addition to the Orient, *Bulimus* is also represented in Europe and was introduced in the Great Lakes area of the U.S.

b. Family Synceridae (= Assimineidae)

Small amphibious snails; outer lip of aperture not thickened. Animal with a single pair of short eye stalks instead of tentacles. Genus *Syncera* Gray, 1821.

c. Family Thiaridae (= Melaniidae)

Shell high, turreted sometimes reaching 1 to 2 in. Sculpture conspicuous, consisting of spiral and axial intersections forming ridges, knobs or tubercles. Operculum corneous multispiral or paucispiral. Mantle edge smooth or with digitiform processes. Snout long. Central tooth of radula with five to seven denticles on the anterior edge; no basal denticles. One lateral tooth and two marginal teeth with few or many cusps. The majority ovoviviparous, a few oviparous; many females parthenogenetic. Cosmopolitan in distribution. Easily introduced in new regions. Genera *Thiara* (Bolten) Roding, 1798, and *Brotia* Adams, 1866. *Thiara* is in the literature under *Tarebia* H. and H. Adams, 1854, or *Melanoides* Olivier, 1804.

Table 1-4

MEDICALLY AND ECONOMICALLY IMPORTANT TREMATODES AND THEIR MOLLUSCAN INTERMEDIATE HOSTS[a]

Trematode			Molluscan intermediate host		
Species	Family	Definitive host	Species	Family	Geographical distribution
Schistosomes Mammalian schistosomes					
Schistosoma mansoni	Schistosomatidae	Human	*Biomphalaria glabrata*	Planorbidae	Puerto Rico, Dominican Republic, St. Kitts, Guadeloupe, Martinique, St. Lucia, Surinam, Venezuela
Schistosoma mansoni	Schistosomatidae	Human, opossum, several species of rodents	*B. glabrata*	Planorbidae	Brazil
			B. tenagophila	Planorbidae	Brazil
			B. straminea	Planorbidae	Brazil
			B. philippiana	Planorbidae	Ecuador[b] Chile[b]
			B. chilensis	Planorbidae	
			B. albicans	Planorbidae	Puerto Rico[b]
			B. riisei	Planorbidae	Puerto Rico[b]
		Human	*B. pfeifferi gaudi*	Planorbidae	Gambia, Portuguese-Guinea, Sierra Leone, Liberia, Guinea, Mauritania, Senegal, the Sudan, Ivory Coast, Ghana, Nigeria
			B. sudanica	Planorbidae	Cameroon, Zaire, Kenya, Uganda, Zaire, Tanzania, Malawi, Sudan
			B. rüppellii	Planorbidae	Zaire, Sudan, Eritrea, Ethiopia, Kenya, Uganda, Tanzania
			B. pfeifferi	Planorbidae	Uganda, Tanzania, Rhodesia, South Africa, Mozambique, Madagascar

Parasite	Final host	Snail intermediate host	Snail family	Distribution
	Human, gerbils, shrews	*B. choanomphala, B. smithi, B. stanleyi*	Planorbidae	Uganda
	Human	*B. alexandrina*	Planorbidae	Egypt
	Human, rodents, shrews	*B. alexandrina*	Planorbidae	Saudi Arabia, Yemen
		B. rüppellii	Planorbidae	Congo
		B. sudanica	Planorbidae	Congo
		B. bridouxiana	Planorbidae	Congo
	Human, dogface baboon	*B. rüppellii*	Planorbidae	Kenya
Schistosoma rodhaini Schistosomatidae	Rodents, dog, felines, human	*B. bridouxiana*	*Planorbidae*	Congo, Uganda
Schistosoma haematobium Schistosomatidae	Human	*B. sudanica*	Planorbidae	Morocco, Algeria, Tunisia, Libya, Egypt, Sudan, Turkey, Israel, Syria, Saudi Arabia, Yemen, Iraq, Iran
		Bulinus (Bulinus) truncatus		
		B. (B.) truncatus rohlfsi	Planorbidae	Cameroons, Ghana, Gambia, Mauritania
		B. (B.) guernei	Planorbidae	Senegal
		B. (B.) sengalensis	Planorbidae	Gambia, Senegal
		B. (Physopsis) jousseaumei	Planorbidae	Gambia, Portuguese Guinea, Mali, Senegal
		B. (Ph.) globosus	Planorbidae	Gambia, Portuguese Guinea, Sierra Leone, Liberia, Cameroons, Ghana, Nigeria, Angola, Sudan, Uganda, Tanzania, Rhodesia, South Africa, Mozambique
	Human, dogface baboon, monkeys	*B. (Ph.) globosus,*	Planorbidae	Kenya
		B. (Ph.) africanus		
		B. (Ph.) globosus,		
		B. (Ph.) nasutus,		
		B. (B.) coulboisi		
	Human	*Ferrissia tenuis*	Ancylidae	India
		Planorbarius metidjensis (= P. dufourii)	Planorbidae	Portugal

Table 1-4 (continued)

MEDICALLY AND ECONOMICALLY IMPORTANT TREMATODES AND THEIR MOLLUSCAN INTERMEDIATE HOSTS[a]

Trematode			Molluscan intermediate host		
Species	Family	Definitive host	Species	Family	Geographical distribution
Schistosoma intercalatum	Schistosomatidae	Human	*B. (Ph.) globosus, B. (Ph.) africanus*	Planorbidae	Zaire
Schistosoma mattheei	Schistosomatidae	Human	*B. (B.) forskalii*	Planorbidae	Gabon
		Cattle, sheep, goat, impala, zebra, human	*B. (Ph.) africanus B. (Ph.) globosus*	Planorbidae	South Africa, Transvaal, Rhodesia
Schistosoma margrebowiei	Schistosomatidae	Antelopes	*Bulinus tropicus Bulinus depressus*	Planorbidae	South Africa
Schistosoma, bovis	Schistosomatidae	Sheep, goat, cattle	*Bulinus (B.) truncatus*	Planorbidae	Southern Europe, southwestern Asia
Schistosoma bovis	Schistosomatidae	Sheep, goat, cattle, equines	*B. (B.) truncatus B. (Ph.) nasutus, B. (B.) forskalii, B. (B.) senegalensis*	Planorbidae	Africa
Schistosoma bovis	Schistosomatidae	Sheep, goat, cattle, human, pig, equines, camel	*B. (B.) truncatus, B. (Ph.) ugandae*	Planorbidae	Sudan
Schistosoma spindale	Schistosomatidae	Cattle, buffaloe	*Indoplanorbis exustus*	Planorbidae	India, Malaysia, Sumatra
Schistosoma leiperi	Schistosomatidae	Cattle, sheep, goat, antelopes buffalo	*B. (Ph.) africanus B. (Ph.) globosus*	Planorbidae	Zambia, South Africa
Schistosoma indicum	Schistosomatidae	Horse, goat, sheep, dog, camel	*Indoplanorbis exustus*	Planorbidae	India
Schistosoma incognitum	Schistosomatidae	Pig, Dog	*Lymnaea luteola*	Lymnaeidae	India
Schistosoma nasale	Schistosomatidae	Cattle	*Indoplanorbis exustus*	Planorbidae	India
			Lymnaea luteola	Lymnaeidae	India

157

Schistosoma japonicum	Schistosomatidae	Human, dog, cat, rat, mice, cattle, water buffalo, pig, horse, sheep, goat, badger	*L. acuminata* *Oncomelania nosophora, O. hupensis, O. quadrasi*	Lymnaeidae Hydrobidae	Japan China Philippines
		Dog, water buffalo, goat, pig	*O. formosana*	Hydrobidae	Formosa (Taiwan)
Schistosoma mekongi	Schistosomatidae	Human, dog	*Lithoglyphopsis aperta*	Hydrobidae	Laos, Cambodia
Bivitellobilharzia loxodontae	Schistosomatidae	Elephant	*Lymnaea sp.*	Lymnaeidae	Zaire
Schistosomatium douthitti	Schistosomatidae	Muskrat, deer mouse, meadow mouse, and other small rodents	*Stagnicola palustris* *Lymnaea stagnalis*	Lymnaeidae	Northern U.S., Canada, and Alaska
Heterobilharzia americana	Schistosomatidae	Raccoon, bobcat, dog, nutria, rabbit	*Fossaria cubensis* *Pseudosuccinea columella*	Lymnaeidae Lymnaeidae	Louisiana, Florida, Texas, North Carolina
Orientobilharzia turkestanicum	Schistosomatidae	Sheep, goat, water buffalo, cattle, equines, camel	*Lymnaea tenera euphratica* *Lymnaea rubiginosa* *L. gedrosiana*	Lymnaeidae	Iraq, Iran, U.S.S.R.
O. datti	Schistosomatidae	Water buffalo, cattle	*Lymnaea luteola*	Lymnaeidae	India
Avian schistosomes					
Ornithobilharzia canaliculata	Schistosomatidae	Royal tern	*Batillaria minima*	Potamidae (=Cerithiidae)	Florida
Austrobilharzia variglandis	Schistosomatidae	Ruddy tern stone	*Littorina pintado*	Littorinidae	Hawaii
Austrobilharzia variglandis	Schistosomatidae	Lesser scaup duck	*Nassarius obsoletus*	Nassariidae	Rhode Island, New Jersey
Austrobilharzia variglandis	Schistosomatidae	Red-breasted merganser	*Nassarius obsoletus*	Nassariidae	Connecticut

Table 1-4 (continued)

MEDICALLY AND ECONOMICALLY IMPORTANT TREMATODES AND THEIR MOLLUSCAN INTERMEDIATE HOSTS[a]

Trematode		Definitive host	Molluscan intermediate host		
Species	Family		Species	Family	Geographical distribution
Austrobilharzia variglandis	Schistosomatidae		Littorina planaxis	Littorinidae	California
Austrobilharzia penneri	Schistosomatidae	Chick, pigeon	Cerithidea scalariformis	Cerithiidae	Florida
Gigantobilharzia huttoni	Schistosomatidae	White pelican	Haminoae guadeloupensis	Akeridae	Florida
G. sturniae	Schistosomatidae	Large starling, sparrow, wagtail	Segmentina hemisphaerula	Planorbidae	Japan
G. gyrauli	Schistosomatidae	Blackbird	Gyraulus parvus	Planorbidae	Wisconsin
G. huronensis	Schistosomatidae	Goldfinch, cardinal	Physa gyrina	Physidae	Michigan
Trichobilharzia ocellata	Schistosomatidae	Ducks, teals	Lymnaea stagnalis Stagnicola palustris	Lymnaeidae	Europe, Canada, Michigan, Wisconsin
T. physellae	Schistosomatidae	Ducks	Physa anatina	Physidae	Louisiana
T. physellae	Schistosomatidae	Ducks	Physa parkeri	Physidae	Canada, Michigan, Wisconsin
T. stagnicolae	Schistosomatidae	Canary (experimental)	Stagnicola emarginata	Lymnaeidae	Canada, Michigan, Wisconsin
T. yokogawai	Schistosomatidae	Ducks	Lymnaea swinhoei	Lymnaeidae	Formosa
Trichobilharzia sp.	Schistosomatidae	Ducks, cormorant	Lymnaea natalensis	Lymnaeidae	Zaire
Liver flukes					
Fasciola hepatica	Fasciolidae	Sheep, cattle, other herbivores, human	Lymnaea truncatula	Lymnaeidae	Europe, North Africa, Egypt, Eastern and Central Asia, Asia Minor
			Fossaria bulimoides, Pseudosuccinea columella, Fossaria modicella	Lymnaeidae	South, Southwest, West, North Central, U.S.

Parasite	Family	Snail intermediate host	Definitive hosts	Snail family	Geographic location
		F. cubensis		Lymnaeidae	Puerto Rico, Louisiana
		P. columella		Lymnaeidae	Colombia
		L. viatrix		Lymnaeidae	Argentina
		L. rubiginosa		Lymnaeidae	Malayasia
		L. Philippensis		Lymnaeidae	Philippines
		L. swinhoei			
		L. pervia		Lymnaeidae	Japan
		L. japonicum		Lymnaeidae	
		L. tomentosa		Lymnaeidae	Australia
Fasciola gigantica	Fasciolidae	*Lymnaea natalensis*	Sheep, cattle, other herbivores, human	Lymnaeidae	Africa
		L. auricularia rufescens		Lymnaeidae	Pakistan, India
		L. acuminata		Lymnaeidae	India
		Fossaria ollula, Pseudosuccinea columella[a]		Lymnaeidae	Hawaii
		L. rubiginosa		Lymnaeidae	Malaya
Fascioloides magna	Fasciolidae	*Fossaria parva, Stagnicola palustris, S. caperata*	Cattle, deer, elk, moose, sheep	Lymnaeidae	North America
Fascioloides magna	Fasciolidae	*Lymnaea truncatula*	Deer, elk, cattle, sheep	Lymnaeidae	Czechoslovakia
Parafasciolopsis fasciolaemorpha	Fasciolidae	*Coretus corneus*	Moose, elk, deer, wild goat	Planorbidae	Poland, Russia
Clonorchis sinensis	Opisthorchiidae	*Parafossarulus manchouricus*	Human, dog, cat	Hydrobiidae	China, Japan, Formosa, Korea
		Bithynia (= Bulimus) fuchsianus		Hydrobiidae	China
Opisthorchis felineus	Opisthorchiidae	*Bithynia (= Bulimus) leachi*	Cat, dog, human	Hydrobiidae	Central and Eastern Europe and Siberia
Dicrocoelium dendriicum	Dicrocoeliidae	*Helicella candidula*	Sheep, cattle, deer, rabbit, human	Helicellidae	Europe, Western Asia
		Zebrina detrita		Enidae	
		Cionella lubrica	Sheep, cattle	Cionellidae	United States

Table 1-4 (continued)

MEDICALLY AND ECONOMICALLY IMPORTANT TREMATODES AND THEIR MOLLUSCAN INTERMEDIATE HOSTS[a]

Trematode			Molluscan intermediate host		
Species	Family	Definitive host	Species	Family	Geographical distribution
Intestinal flukes					
Fasciolopsis buski	Fasciolidae	Human, pig,	*Segmentina hemisphaerula, Hippeutis cantori, Segmentina trochoideus*	Planorbidae	China, Indochina, Formosa
					India (Assam)
					India (Assam)
Echinostoma ilocanum	Echinostomatidae	Rats, human	1st intermediate host:		
			Gyraulus convexiusculus	Planorbidae	Philippines
			Hippeutis umbilicalis	Planorbidae	Philippines
			Gyraulus prashadi	Planorbidae	India
			G. convexusculus	Planorbidae	Java
			2nd intermediate host:		
			Pila conica	Pilidae	Philippines
			Viviparus javanicus	Viviparidae	Java
Echinochasmus perfoliatus	Echinostomatidae	Dog, cat, human	*Parafossarulus manchouricus*	Hydrobiidae	Orient
Gastrodiscoides hominis	Paramphistomatidae	Pig, human	*Helicorbis coenosus*	Planorbidae	India (Assam)
Gastrodiscus aegyptiacus	Paramphistomatidae	Horse, donkey, mule	*Bulinus (Bulinus) forskalii*	Planorbidae	Africa
Heterophyes heterophyes	Heterophyidae	Cat, dog, fox, human	*Cerithidea cingulata*	Potamidae (= Cerithiidae)	Japan
Heterophyes heterophyes	Heterophyidae	Cat, dog, fox, human	*Pirenella conica*	Potamidae (= Cerithiidae)	Egypt

Parasite	Family	Host	Snail	Snail Family	Location
Metagonimus yokogawai	Heterophyidae	Cat, dog, human	*Semisulcospira libertina*	Pleuroceridae	Orient
		Cat, dog, human	*Thiara granifera*	Thiaridae	Orient
	Heterophyidae		*Semisulcospira cancellta*	Pleuroceridae	Russia
			S. laevigata		
Prohemistomum vivax	Cyathocotylidae	Cat, dog, kite	*Cleopatra bulimoides*	Thiaridae	Egypt
Haplorchis taichui and H. pumilio	Heterophyidae	Birds, mammals (experimental)	*Thiara granifera*	Thiaridae	Orient
			Semisulcospira libertina	Pleuroceridae	
Haplorchis taichui	Heterophyidae	Cat, human	*Thiara tuberculata*	Thiaridae	Thailand
Stellantchasmus formosanum	Heterophyidae	Birds	*Thiara granifera*	Thiaridae	Orient
			Thiara tuberculata	Thiaridae	Orient
			Semisulcospira libertina	Pleuroceridae	Orient
Stellantchasmus falcatus	Heterophyidae	Cat, humans	*Thiara tuberculata*	Thiaridae	Thailand
Metagonimoides oregonensis	Heterophyidae	Raccoon, mink	*Oxytrema (Goniobasis) silicula*	Pleuroceridae	U.S., Canada
Apophallus imperator	Heterophyidae	Cat, pigeon	*Amnicola limosa*	Hydrobiidae	U.S., Canada
Euryhelmis monorchis	Heterophyidae	Mink	*Pomatiopsis lapidaria*	Hydrobiidae	U.S.
Cryptocotyle lingua	Heterophyidae	Gulls, dog	*Littorina littorea*	Littorinidae	Northern Europe, U.S.
Rumen Flukes					
Paramphistomum microbothrium	Paramphistomatidae	Cattle, sheep, other herbivores	*Bulinus (Bulinus) truncatus*	Planorbidae	Various countries in Africa
			B. (Physopsis) ugandae		
			B. (Ph.) globosus		
			B. (Bulinus) tropicus		
			B. (B.) truncatus		
Paramphistomum microbothrioides	Paramphistomatidae	Cattle, sheep, other herbivores	*Fossaria modicella*	Planorbidae	Iran, Israel, Sardinia
			Fossaria bulimoides	Lymnaeidae	U.S., Puerto Rico

Table 1-4 (continued)
MEDICALLY AND ECONOMICALLY IMPORTANT TREMATODES AND THEIR MOLLUSCAN INTERMEDIATE HOSTS[a]

Trematode			Molluscan intermediate host		
Species	Family	Definitive host	Species	Family	Geographical distribution
Paramphistomum cervi	Paramphistomatidae	Cattle, sheep, other herbivores	*Planorbis planorbis Anisus vortex Armiger crista*	Planorbidae	Europe
Paramphistomum ichikawai	Paramphistomatidae	Cattle	*Segnitilia alphena*	Planorbidae	Australia
Cotylophoron cotylophorum	Paramphistomatidae	Cattle	*Indoplanorbis exustus*	Planorbidae	India
Calicophoron calicophorum	Paramphistomatidae	Cattle	*Pygmanisus polorius*	Planorbidae	Australia
Ceylonocotyle streptocoelium	Paramphistomatidae	Cattle, sheep	*Glyptanisus gilberti*	Planorbidae	Australia
Lung flukes					
Paragonimus westermani	Paragonimidae	Human, tiger, lion, cat, dog, badger	*Semisulcospira libertina*	Pleuroceridae	Japan, China, Korea, Formosa
			Thiara granifera, T. toucheana, T. tuberculata	Thiaridae	China, Formosa
			Semisulcospira amurensis	Pleuroceridae	Korea
			S. bensoni	Pleuroceridae	Japan
			Brotia asperata	Thiaridae	Philippines
			Assiminea lutea	Assimineidae	China
	Paragonimidae	Rats	*Assiminea parasitologica*	Assimineidae	Japan
			Angustassiminea nitida	Assimineidae	Japan

Paragonimus ohirai	Paragonimidae	Rats	*Assiminea parasitologica*	Assimineidae	Japan
			A. yoshidayukioi		Japan
			Angustassiminea nitida		Japan
Paragonimus miyazakii	Paragonimidae	Weasel, dog, cat, marten, human	*Bythinella nipponica*	Hydrobiidae	Japan
Paragonimus sadoensis	Paragonimidae	Weasel, cat, dog	*Tricula minima*	Hydrobiidae	Japan
Paragonimus skrjabini	Paragonimidae	Man	*Tricula sp.*	Hydrobiidae	China
Paragonimus africanus	Paragonimidae	Man, free-living drill	*Potadoma freethii*	Thiaridae	Cameroon, Africa
Paragonimus kellicotti	Paragonimidae	Mink, cat, pig, raccoon	*Pomatiopsis lapidaria*	Hydrobiidae	U.S.
Paragonimus caliensis	Paragonimidae	Opossums	*Aroapyrgus colombiensis*	Hydrobiidae	Colombia
Paragonimus mexicanus	Paragonimidae	Opossums, raccoon, cat, fox, man	*Aroapyrgus costaricensis*	Hydrobiidae	Costa Rica

a   Adapted from Malek and Cheng (1974), with additions.[113]
b   Experimental infection.

d. Family Pilidae (= Ampullaridae)

Shell very large, smooth, green or brown in color. Usually dextral except *Lanistes*. Spire short; body whorl large. Lip of aperture simple. Operculum large, calcareous, or corneous; concentric. Tentacles long, filiform. Cosmopolitan in distribution. Genera *Pila* Roding, 1798, in Africa and Asia; *Pomacea* Perry, 1811, in Central and South America, southern U.S. and Caribbean islands; *Lanistes* Montfort, 1810, in Africa; and *Marisa* Gray, 1824, in South America.

e. Family Pleuroceridae

Shell thick, solid; sutures shallow, whorls flat sided. Spire long tapering into a point. Operculum paucispiral. Animal oviparous or ovoviviparous. Representatives in North America and one genus in the Orient. Genera *Pleurocera* Rafinesque, 1818, and *Goniobasis* Lea, 1862, in the U.S.; *Semisulcospira* Boettger, 1886, in the Orient.

The following prosobranch families comprise marine or brackish water forms.

f. Family Nassariidae

Shell thick; aperture drawn into a short canal. Small carnivorous snails of worldwide distribution. Genus *Nassarius* Froriep, 1806.

g. Family Potamidae (= Cerithiidae)

Shell with heavy spiral striations; aperture with a small basal canal; columella tightly coiled. Tentacles thick at base but narrow abruptly at distal end. Cosmopolitan in tropics and subtropics in tidal zones and brackish estuaries. Genera *Pirenella* Gray, 1847; *Cerithidea* Swainson, 1840, with several species; and *Batillaria* Benson, 1842.

h. Family Littorinidae

Shell thick, imperforate, conical. Spire low, pointed. Sutures shallow. Operculum corneous. Members commonly known as "periwinkles." Genus *Littorina* Ferussac, 1821, with several species represented along the coasts of the U.S. and Europe.

III. CONTROL OF TREMATODE INFECTIONS

Efforts to combat trematode infections have been directed toward breaking the life cycle at one or more points and thus interrupting transmission. This means combat of the parasite at the adult stage, at the egg-miracidium stage, or at the cercarial or metacercarial stage. Control can thus be affected by treatment of the human host with one of a few chemotherapeutic agents, by preventing human excreta from reaching the water through the provision of latrines, by snail control through use of molluscicides and other means, and in the case of schistosomiasis through health education aimed at making the population aware of the dangers of the disease and how to avoid the infection. Similar health education methods aimed at preventing the public from eating raw or inadequately cooked crabs, crayfish, fish, and vegetation are the most important measures in the control of trematode infections other than the schistosomes. In some cases reservoir hosts of these flukes actually pollute the water containing the snail hosts more than humans do, and maintain a high level of endemicity. Furthermore, in practice, none of the above measures have been given adequate trial, except in the case of schistosomiasis in a few endemic areas. Some measures have also been suggested and are directed towards killing the trematode inside the snail, by biological means, but have not so far been attempted except in one or two experimental areas. The latter measures, in addition to chemical control of the snails, are discussed here.

A. Molluscicides

Advances have been made in recent decades in the field of molluscicides. The effectiveness of certain compounds has been demonstrated, when used alone or in combination with chemotherapy in some endemic areas of schistosomiasis. Control of snails by use of molluscicides has the advantage of controlling snail hosts of more than one of the snail-transmitted helminthiases, and this no doubt is considered in the cost/benefit analyses. Indeed this has been the case in some endemic areas of human schistosomiasis where the control operation also involved the snail hosts of bovine schistosomiasis, of fascioliasis, and of paramphistomiasis. There are some "available" and some "potential" molluscicides. The classification as to available and potential is based on effectiveness, completeness of laboratory evaluation, field screening, field trials, demonstration of transmission control, and commercial availability of the product. On this basis Bayluscide ® (niclosamide), Frescon (N-Tritylmorpholine), sodium pentachlorophenate (NAPCP), and copper sulfate may be considered as "available molluscicides." A number of other compounds have good potentials, and in addition to the fact that they have been subjected to laboratory evaluations, they have received limited field evaluation and field screening in a few endemic areas. Available and potential (candidate) molluscicides vary as to their chemical composition, properties, formulations, dispersion, and their suitability for use by certain equipment employed for their application. A number of publications, mainly by WHO, dealt in detail with properties, advantages and disadvantages of molluscicides.[113,139,140,184,185,186]

There is actually a need for a variety of molluscicides, and for the availability of different formulations of the same molluscicide, because of the great variation in snail species, their habits, and the nature of terrain and climate where the molluscicides are to be used. Moreover, there are variations in habitats of the snail hosts, as to physical, chemical and biological features.

Table 1-5, which is the most recent and complete table by WHO (1971), shows properties of the known products, and includes figures on their side effects. The following, however, is a summary of the properties of the available molluscicides, and those of a few candidate compounds.

1. Available Molluscicides

a. Frescon (N-Tritylmorpholine)

Originally known as WL-8008, it is produced by the Shell Company. It has been very thoroughly investigated by the company, and by several investigators in endemic areas. It is N-triphenylmethyl-morpholine, but the triphenylmethyl portion is designated "trityl", providing the name N-tritylmorpholine. It is available in the following formulations: emulsifiable concentrates, water-dispersable powder and granules, spreading oils, and baits. Analytical tests have been developed for field use. Mud, vegetation, and light do not affect its efficacy, but the latter is affected due to hydrolysis of the compound at pH values below 7.0. The emulsifiable concentrate is stable in storage and in the field.

Frescon has been found effective against aquatic planorbid hosts of schistosomiasis, and against amphibious and aquatic lymnaeid hosts of fascioliasis. Satisfactory results, using Frescon, have been obtained in Rhodesia, Tanzania, Egypt, and Brazil, against snail hosts of schistosomiasis.

b. Bayluscide ® (Niclosamide)

This compound is produced by the Farbenfabriken Bayer, Germany. Bayluscide ® is the ethanolamine salt of 5,2″-dichloro-4′-nitrosalicylanilide. It was commercialized in the form of its ethanolamine salt, and formulated as a 70% wettable powder (Bay-

Table 1-5
MOLLUSCICIDES AND THEIR PROPERTIES[186]

Common name	Niclosamide	N-trityl-morpholine	NaPCP	Copper sulfate	ZDZ	Yurimin	Copper oxide
Physical form of technical material	Crystalline solid	Crystalline solid	Crystalline solid	Crystalline solid	Amorphous solid	Crystalline solid	Amorphous solid
Active ingredient	Ethanolamine salt of 2'5-dichloro-4'-nitrosalicyl-anilide	N-trityl-morpholine	Sodium pentachlorophenate	Copper ion	Zinc di-methyl-di-thiocarbamate	3,5-dibromo-4-hydroxy-4'-nitroazobenzene	Copper ion
Solubility in water	230ppm (pH dependent)	—	33%	32%	65 ppm	Very slight	—
TOXICITY							
Snail LC_{90} (ppm × h)	3—8	0.5—4	20—100	20—100	25—60	4—5	7—100
Snail eggs LC_{90} (ppm × h)	2—4	240	3—30	50—100	50—100	—	50—100
Cercaria LC_{90} (ppm)	0.3	—	—	—	—	—	—
Fish LC_{90} (ppm)	0.05—0.3 (LC_{50})	2—4	—	—	—	0.16—0.83 (LC_{50})	—
Rats, acute oral, LD_{50} (mg/kg)	5,000	1,400	40—250	—	1,400	168 (mice)	2,000
Herbicidal activity	No	No	No	Yes	No	No	No
STABILITY (affected by)							
U.V. light	Yes	—	Yes	No	—	No	No
Mud turbidity	Yes	No	No	Yes	No	Yes	Yes
pH	Optimum 6—8	Yes	No	Yes	—	Yes	Yes
Algae, plants	No	No	No	Yes	No	—	Yes
Storage	No	No	No	No	No	—	Yes
HANDLING QUALITIES							
Safe	Yes	Yes	Varies	Yes	Yes	Yes	Yes
Simple	Yes	Yes	Yes	Yes	Yes	Yes	Yes
FORMULATIONS	70% W.P. 25% E.C.	16.5% E.C.	75% flakes 80% pellets 80% briquettes		Granules 50% Powder 90%	Granules 5%	Powder

	TPTA	TBTO	TBTA	TPLA	TBS	Endod	Nicotinanilid
FIELD DOSAGE							
Aquatic snails (ppm × h)	4—8	1—2	50—80	20+	100	—	60
Amphibious snails on moist soil (g/m²)	0.2	—	0.4—10	—	10	5	Not effective
Common name	TPTA	TBTO	TBTA	TPLA	TBS	Endod	Nicotinanilid
Physical form of technical material	Crystalline solid	Liquid	Crystalline solid	Crystalline solid	Crystalline solid	Powdered berries	Amorphous solid
Active ingredient	Triphenyltin acetate	Tri-n-butyl-tin oxide	Tri-butyl-tin acetate	Triphenyl lead acetate	3,4',5-tribromosalicyl		
Solubility in water	>500 ppm	10—30 ppm	80—150 ppm	1.5%	—	High	>100 ppm
Toxicity							
Snail LC_{90} (ppm × h)	1.9	0.9—1.3	7.2	3.6	2—4	240—480	4—6
Snail eggs LC_{90} (ppm × h)	0.6	0.55	0.55	—	0.2	10,000	1—2
Cercaria LC_{90} (ppm × h)	0.05	0.001	0.05	0.07		100	10—100
Fish LC_{90} (ppm × h)	2.4—4.8	1—3	2	12.3		1,200	
Rats, acute oral, LD_{50} (mg/kg)	<200	<250	<250			220	2g/kg (mice)
Herbicidal activity	Yes	No	Yes			No	No
Stability (affected by)							
U.V. light	Slight	No	No	Slight	Yes	No	No
Mud turbidity	Yes	Yes	Yes	Yes	No	No	
PH	No	No	No		No	No	
Algae, plants	Yes	Yes	Yes	Slight	No	No	Slight
Storage	No	No	No	Slight		No	No
Handling Qualities							
Safe	Yes	Yes	Yes	No	—	Yes	Yes
Simple	No	Yes	Yes	No	—	Yes	Yes
Formulations	Powder	6—10% slow release pellets 95% liquid	6—10% slow release pellets powder	Powder	Powder	Powder	Powder
Field Dosage							
Aquatic snails (ppm × h)	4—10	4—20	10—15	—	20—30	300—600	20+
Amphibious snails on moist soil	5—8g/m²	—	—	—	—	86 ppm (laboratory)	

luscide ® or Bayer 73), for which the International Organization for Standardization (IOS) has recommended the name niclosamide. A liquid emulsifiable concentrate (25% w/v active ingredient) was later developed and proved to be very effective against all stages of snails and their eggs at lower concentrations in the laboratory.[106] It is more suited for field application because the wettable powder formulation tends to clog nozzles of equipment used in applying it. The chemical acts quickly in low concentrations with about equal efficacy for 1- and 24-hr exposures. A pH range of 5 to 9, normally found in natural habitats, has no effect on niclosamide. It has been used with good results in Tanzania, South Africa, Iran (Volume II, Plate II), Egypt, Rhodesia, Puerto Rico, and the Philippines.

c. Sodium Pentachlorophenate (NAPCP)

This and related compounds of pentachlorophenol have been widely used for a variety of agricultural and industrial purposes, as wood preservatives, as termite deterrents, as weed killers, and as molluscicides. NAPCP is highly soluble and is formulated as flakes, pellets, briquettes, and another solid formulation has also been prepared which has a longer release time than the briquettes, extending to about 60 hr. NAPCP is stable but is somewhat irritating if handled or applied improperly. It is ovicidal and molluscicidal against fresh-water snails, but is less effective as an ovicide at molluscicidal concentrations against amphibious snails (*Oncomelania*). It is unaffected by the normal pH range of natural water; it is reduced by hard water; and is irreversibly absorbed by mud. Its effectiveness is reduced especially by sunlight, but also by high alkalinity and by adsorption on mud.

d. Copper Sulfate

Copper sulfate is a stable, easy-to-handle compound, usually applied as a solution in aquatic habitats only. It has been used as a molluscicide since 1920, but its use is now limited to a few countries. It is comparatively cheap, but 20 to 30 times the laboratory toxic level must be used in the field because its effectiveness is reduced in the presence of organic matter, certain types of dissolved solids, and a high pH. Its downstream carriage is very poor. It is now generally agreed that copper sulfate is too ineffective for major control efforts, and cannot compete with other available products.

2. Potential Molluscicides

This is a group of compounds with molluscicidal properties with a varied combination of merits. In addition to the fact that they have not been subjected but to very limited field evaluations, most of them are not available commercially in large quantities.

a. Yurimin (P-99)

Yurimin, a Japanese product, is 3,5-dibromo-4-hydroxy-4'-nitrobenzene. Its performance has been satisfactory against both aquatic and amphibious snails. It is stable, unaffected by sunlight or water hardness, but an acidic pH of 4.5 to 5.2 decreases efficacy. In Japan it was proven in laboratory and field tests to be several times more effective, when compared to NAPCP, against the amphibious snails *Oncomelania nosophora*.

b. Insoluble Copper Compounds
1. Copper Pentachlorophenate

This stable, relatively insoluble compound is obtained by mixing solutions of NAPCP and copper sulphate at a ratio of 2:1. It incorporates properties which far exceed those of the latter two compounds when applied separately. It is ovicidal and gives a residual effect in nonflowing water and on moist-soil habitats. It has proved successful in control operations for schistosomiasis in Venezuela and for fascioliasis in Australia.

2. Copper Carbonate

This stable product has been used as the molluscicide of choice for many years to control nonhuman schistosomes in lake habitats in the U.S. It was proven effective in Brazil and on moist-soil habitats in the Orient.

3. Cuprous Oxide

This compound is slow acting and has a good residual effect when formulated as the stabilized "chevreul salt."

c. Organo-Tin Compounds

Several organo-tin compounds had been known to be effective agricultural fungicides, and are major components of marine antifoulant paints, stabilizers for food packaging, and antihelminthics. Bis (tri-n-butyltin) oxide, tri-n-butyltin acetate, and tri-n-propylin oxide have also been proven to possess molluscicidal properties comparable to those of Bayluscide. Newly laid eggs and newly hatched snails are especially sensitive to the effect of these organotins. They are stable compounds and are not easily affected by the physicochemical features of the habitats. There are at present emulsifiable liquid concentrates and wettable powder formulations. Organotins are especially suitable for slow-release formulations, where they are incorporated in one of several elastomers, such as natural rubber, ethylene-propylene terpolymer, and styrene-butadine.

d. Mollutox ®

Mollutox ® is a registered trade name, and is produced by the Chemical and Insecticidal Company, Abou-Zabal, Egypt. It is a new formulation of the ethanolamine salt of 5,2′-dichloro 4′-nitrosalicylanilide, that is, Bayluscide® (niclosamide). Although new, the compound has already received laboratory as well as field evaluation. A colorimetric test in water samples has been developed.

3. Molluscicides of Plant Origin

Several plants and/or their fruits have shown molluscicidal properties, in such a way that it has been recommended that they are to be grown alongside canals and other water bodies; in addition, they can also be harvested and processed locally. Among these are trees and shrubs in Brazil, Puerto Rico, Africa, and Asia, including *Sapindus saponaria*, *Balanites aegyptica*, *Jatropha* sp., *Thea opeosa*, *Croton tiglim*, *Schima argenta*, *Phytolacca dodecandra*, and *Phytolacca* spp. The local name of *Phytolacca dodecandra* in Ethiopia is "Endod".

B. Biological Control

Biological control measures have been attempted and were directed against the snail hosts and the trematode larval stages.

A few pathogens and predators which attack the trematodes may also be useful as biological control agents. *Chaetogaster*, spp. are oligochaete annelids, live in close association with fresh-water snails, and have been demonstrated to prey on miracidia of *Fasciola hepatica*[73] and *Schistosoma mansoni*.[120] It is believed also that these oligochaetes prey on the cercarial stage in the water. Nonsusceptible snails in the same habitat with susceptible snails consume a large number of miracidia; the latter would be killed by strong tissue reactions. No doubt other organisms in the water prey on miracidia. What happens in nature about the regulatory capacity of these organisms to the miracidial populations of various trematodes is not known.

Guppies (*Lebistes reticulatus*) have been shown in laboratory and field tests to eat a large number of cercariae. Pellegrino and Demaria[129] noted that infection rates and worm burdens of white mice exposed to cercariae of *S. mansoni* in a natural habitat containing guppies were greatly reduced. Knight et al.[75] have quantified the cercariophagic activity of these fish by using cercariae of *S. mansoni* labeled with radioselenium.

The intramolluscan larval trematode stages also have their enemies which have been suggested as means of control of snail-transmitted diseases. The enemies include microsporidians and antagonistic larval stages of other trematodes. Microsporidians attack and destroy trematode larvae in some snails, and the literature contains several references on this subject.[21] However, these protozoan hyperparasites are difficult to identify and it is difficult to induce them to cause epidemics.

Observations on antagonistic effects exerted on intramolluscan larvae of medically important trematodes by larvae of other trematode species have been made and reported in several publications by investigators at the Hooper Foundation, San Francisco, Calif.[63,64] The small immature daughter rediae of the echinostome *Paryphostomum segregatum* are attracted to the sporocysts of *S. mansoni* and grow better when feeding on them than on snail tissues.[61] Indirect antagonism may not be due to physical attack on the sporocyst but due to toxic metabolites or to competition for nutrients. The predatism of an echinostome antagonist on schistosome sporocysts was recorded in a field trial.[63,90] The success of control through the use of antagonistic species may be possible only in small bodies of water and requires large scale life cycle maintenance of the antagonistic species to always make available a large number of eggs to be "seeded" in the habitat.

C. Environmental Control

Environmental control measures for attacking the trematode larval stages through proper disposal of sewage will be discussed under each chapter of this book. Similarly, there are environmental (engineering) measures for snail control and these are designed to upset the habitat of the snail and to prevent establishment of colonies. Some environmental (engineering) measures are known for various habitats. Often combined environmental and chemical measures are recommended.

REFERENCES

1. **Abbott, R. T.,** Handbook of medically important mollusks of the Orient and the Western Pacific, *Bull. Mus. Comp. Zool. Harv. Univ.,* 100 (3), 328, 1948.
2. **Allison, L. N.,** *Leucochloridiomorpha constantiae* (Mueller) (Brachylaemidae), its life cycle and taxonomic relationships among digenetic trematodes, *Trans. Am. Microsc. Soc.,* 62, 127, 1943.
3. **deAndrade, R. M.,** Hydrochemical data on the habitats of Planorbidae in Rio de Janeiro, Brazil, *Rev. Bras. Malariol.,* 6, 473, 1954.
4. **Antunes, C. M. F.,** Study of the effects of Gamma-radiation on eggs and miracidia of *Schistosoma mansoni, Rev. Inst. Med. Trop. S. Paulo,* 13, 383, 1971
5. **deAzevedo, J. Fraga, Cambournac, F., and Pinto, A. R.,** The bilharziasis of the Poutuguese territories of Africa, *C. R. 5, Congr. Int. Med. Trop. Paludisme, Istanbul,* 2, 311, 1954.
6. **Bair, R. D. and Etges, F.J.,** *Schistosoma mansoni:* factors affecting hatching of eggs, *Exp. Parasitol.,* 33, 155, 1973.
7. **Barbosa, F. S.,** Survival and cercaria production of Brazilian *Biomphalaria glabrata* and *B. straminea* infected with *Schistosoma mansoni, J. Parasitol.,* 61, 151, 1975.

8. **Barretto, A. C. and Barbosa, F. S.**, Vector qualities of the hosts of *Schistosoma mansoni* in Northeastern Brazil. IV. Elimination of cercariae of *Schistosoma mansoni* by *Australorbis glabratus* of different diameters, *Ann. Soc. Biol. Pernambuco*, 16, 13, 1959.

9. **Basch, P. F., DiConza, J. J., and Johnson, B. E.**, Strigeid trematodes (*Cotylurus lutzi*) cultured *in vitro*: production of normal eggs with continuance of life cycle, *J. Parasitol.*, 59, 319, 1973.

10. **Basch, P. F. and DiConza, J. J.**, The miracidium-sporocyst transition in *Schistosoma mansoni*: surface changes *in vitro* with ultrastructural correlation, *J. Parasitol.*, 60, 935, 1974.

11. **Beaver, P. C., Little, M. D., Tucker, C. F., and Reed, R. J.**, Mesocercaria in the skin of man in Louisiana, *Am. J. Trop. Med. Hyg.*, 26, 422, 1977

12. **Becker, W. and Lutz, W.**, Die Wirkung von Salzlösungen unterschiedlicher Ion enzusammensetzung auf die Schwimmaktivität der Cercarien von *Schistosoma mansoni*, *Ztschr. Parasitenkd.*, 50, 99, 1976.

13. **Benex, J. and Deschiens, R.**, Incidence des facteurs externes sur la longevité et la vitalité des miracidium de *Schistosoma mansoni*, *Bull. Soc. Pathol. Exot.*, 56, 987, 1963.

14. **Berntzen, A. K. and Macy, R. W.**, *In vitro* cultivation of the digenetic trematode *Sphaeridiotrema globulus* (Rudolphi) from the metacercarial stage to egg production, *J. Parasitol.*, 55, 136, 1969.

15. **Berrie, A. D.**, Snail problems in African schistosomiasis, in *Advances in Parasitology*, Vol. 8, Dawes, B., Ed., Academic Press, London, 1970, 43.

16. **Bonga, S. E. W.**, Neuroendocrine involvement in osmoregulation in a freshwater mollusc, *Lymnaea stagnalis*, *Gen. Comp. Endocrinol. Suppl.*, 3, 308, 1972.

17. **Bonga, S. E. W. and Boer, H. H.**, Ultrastructure of the renopericardial system in the pond snail *Lymnaea stagnalis (L.)*, *Z. Zellforsch. Mikrosk. Anat.*, 94, 513, 1969.

18. **Burch, J. B. and Hadžišče, S.**, Immunological relationships of two species of lymnaeidae (Mollusca: Basommatophora) from Macedonia, Yugoslavia, *Malacol. Rev.*, 7, 25, 1974.

19. **Burch, J. B. and Lindsay, G. K.**, An immunocytological study of *Bulinus* s.s. (Basommatophora: Planorbidae), *Malacol. Rev.*, 3, 1, 1970.

20. **Cain, G. D.**, Studies on hemogolbins in some digenetic trematodes, *J. Parasitol.*, 55, 301, 1969.

20a. **Campbell, D. L., Frappaolo, P. J. F., Stirewalt, M. A., and Dresden, M. H.**, *Schistosoma mansoni*: Partial characterization of enzyme(s) secreted from the preacetabular glands of cercariae, *Exp. Parasitol.*, 40, 33, 1976.

21. **Canning, E. U. and Basch, P. F.**, *Perezia helminthorum* sp. nov., a microsporidian hyperparasite of trematode larvae from Malaysian snails, *Parasitology*, 58, 341, 1968.

22. **Carriker, M. R. and Bilstad, N. M.**, Histology of the alimentary system of the snail *Lymnaea stagnalis appressa* Say, *Trans. Am. Microsc. Soc.*, 65, 250, 1946.

23. **Cheever, A. W. and Duvall, R. H.**, Single and repeated infections of grivet monkeys with *Schistosoma mansoni*: parasitological and pathological observations over a 31-month period, *Am. J. Trop. Med. Hyg.*, 23, 884, 1974.

24. **Cheng, T. C. and Auld, K. R.**, Hemocytes of the pulmonate gastropod *Biomphalaria glabrata*, *J. Invertebr. Pathol.*, 30, 119, 1977.

25. **Cheng, T. C. and Snyder, R. W.**, Studies on host-parasite relationships between larval trematodes and their hosts. IV. A histochemical determination of glucose and its role in the metabolism of molluscan host and parasite, *Trans. Am. Microsc. Soc.*, 82, 343, 1963.

26. **Chernin, E. and Bower, C.**, Experimental transmission of *Schistosoma mansoni* in brackish waters, *Parasitology*, 63, 31, 1971.

27. **Chu, K. Y. and Dawood, I. K.**, Cercarial production from *Biomphalaria alexandrina* infected with *Schistosoma mansoni*, *Bull. W. H. O.*, 42, 569, 1970.

28. **Coles, G. C.**, Fluke biochemistry — *Fasciola* and *Schistosoma*, *Helminth. Abst. Ser. A*, 44, 147, 1975.

29. **Coles, G. C. and Mann, H.**, Schistosomiasis and water works practice in Uganda, *East Afr. Med. J.*, 48, 40, 1971.

29a. **Cooke, A. H.**, Molluscs, in *The Cambridge Natural History*, Vol. 3, Harmer, S. F. and Shipley, A. E., Eds., Cambridge University Press, 1896, 1.

30. **Cort, W. W., Ameel, D. J., and van der Woude, A.**, Germinal development in the sporocysts and rediae of the digenetic trematodes, *Exp. Parasitol.*, 3, 185, 1954.

31. **Dawes, B.**, A study of the miracidium of *Fasciola hepatica* and an account of the mode of penetration of the sporocyst into *Limnaea truncatula*, *Libro Homenaje al Dr. Eduardo Caballero y Caballero*, 95, 1960.

32. **Dawes, B.**, *The Trematoda*, Cambridge University Press, Cambridge, 1968.

33. **Dazo, B. C.**, The morphology and natural history of *Pleurocera acuta* and *Goniobasis livescens* (Gastropoda: Cerithiacea: Pleuroceridae), *Malacol. Int. J. Malacol.*, 3, 1, 1965.

34. **de Brie, Jehan,** Le ben Berger, 1379 (original not discovered, but extracts published in Bon Berger by Paul Lacroix: Isid Liseux, Paris, 1879).

35. **De Jong-Brink, M.,** Histochemical and electron microscope observations on the reproductive tract of *Biomphalaria glabrata (Australorbis glabratus)*, intermediate host of *Schistosoma mansoni, Z. Zellforsch. Mikrosk. Anat.,* 102, 507, 1969.

36. **Deschiens, R.,** Incidence de la minéralisation de l'eau sur les mollusques vecteurs des bilharziasis, *Bull. Soc. Pathol. Exot.,* 47, 915, 1954.

37. **De Witt, W. B.,** Effects of temperature on penetration of mice by cercariae of *Schistosoma mansoni, Am. J. Trop. Med. Hyg.,* 14, 579, 1965.

38. **Dixon, K. E.,** The structure and composition of the cyst wall of the metacercaria of *Cloacitrema narrabeenensis* (Howell and Bearup, 1967) (Digenea: Philophthalmidae), *Int. J. Parasitol.,* 5, 113, 1975.

39. **Dorsey, C. H. and Stirewalt, M. A.,** *Schistosoma mansoni.* Fine structure of cercarial acetabular glands, *Exp. Parasitol.,* 30, 199, 1971.

40. **Dresden, M. H. and Edlin, E. M.,** *Schistosoma mansoni:* effect of some cations on the proteolytic enzymes of cercariae, *Exp. Parasitol.,* 35, 299, 1974.

41. **Dusanic, D. G.,** Histochemical observations of alkaline phosphatase in *Schistosoma mansoni, J. Infect. Dis.,* 105, 1, 1959.

42. **Erasmus, D. A.,** Studies on the morphology, biology and development of a strigeid cercaria (Cercaria x Baylis 1930), *Parasitology,* 48, 312, 1958.

43. **Erasmus, D. A.,** *The Biology of Trematodes,* Crane, Russak & Co., New York, 1972, 312.

44. **Faust, E. C. and Meleney, H. E.,** Studies on schistosomiasis japonica, *Am. J. Hyg. Monogr. Ser.,* 3, 399, 1924.

45. **Frank, G. H.,** Some factors affecting the fecundity of *Biomphalaria pfeifferi* (Kraus) in glass aquaria, *Bull. W. H. O.,* 29, 531, 1963.

46. **Freeman, R. S., et al.,** Fatal human infection with mesocercariae of the trematode *Alaria americana, Am. J. Trop. Med. Hyg.,* 25, 803, 1976.

47. **Fretter, V. and Graham, A.,** *British Prosobranch Molluscs,* Royal Society of London, London, 1962.

48. **Frick, L. P. and Hilleyer, G. V.,** The influence of pH and temperature on the cercaricidal activity of chlorine, *Mil. Med.,* 131(4), 372, 1966.

49. **Fried, B. and Butler, M. S.,** Histochemical and thin layer chromatographic analyses of neutral lipids in metacercarial and adult *Cotylurus* sp. (Trematoda: Strigeidae), *J. Parasitol.,* 63, 831, 1977.

50. **Fried, B. and Shapiro, I. L.,** Accumulation and excretion of neutral lipids in the metacercaria of *Leucochloridiomorpha constantiae, J. Parasitol.,* 61, 906, 1975.

51. **Gabucinus, H. F.,** Commentarius de lumbricis alvum occupantibus et eorum cura (Ludg. Batav.), 1549.

52. **Geuze, J. J.,** Observations on the function and the structure of the statocysts of *Lymnaea stagnalis* (L.), PhD. Thesis, Free University, Amsterdam, E. J. Brill, Leiden, 1968.

53. **Ginetsinkaya, T. A. and Dobrovolskii, A. A.,** A new method for observing sensillae of trematode larvae and the systematic importance of these structures, *Dokl. Akad. Nauk SSSR,* 151, 460, 1963.

54. **Goddard, C. K. and Martin, A. W.,** Carbohydrate metabolism, in *Physiology of Mollusca,* Vol. 2, Wilbur, K. M. and Yonge, C. M., Eds., Academic Press, New York, 1966, 275.

55. **Goudsmit, E. M.,** Carbohydrates and carbohydrate metabolism in Mollusca, in *Chemical Zoology,* Vol. 7, Florkin, M. and Scheer, B. T., Eds., Academic Press, New York, 1972, 219.

56. **Halton, D. W.,** Observation on the nutrition of digenetic trematodes, *Parasitology,* 57, 639, 1967.

57. **Harrison, A. D.,** The effects of calcium bicarbonate concentration on the oxygen consumption of the freshwater snail *Biomphalaria pfeifferi* (Pulmonata: Planorbidae), *Arch. Hydrobiol.,* 65, 63, 1968.

58. **Harrison, A. D., Nduku, W., and Hooper, A. S. C.,** The effect of a high magnesium-to-calcium ratio in the egg-laying rate of an aquatic planorbid snail, *Biomphalaria pfeifferi, Ann. Trop. Med. Parasitol.,* 60, 212, 1966.

59. **Harrison, A. D., Williams, N. V., and Greig, G.,** Studies on the effects of calcium bicarbonate concentrations on the biology of *Biomphalaria pfeifferi* (Kraus) (Gastropoda: Pulmonata). *Hydrobiologia,* 36, 317, 1970.

60. **Harry, H. W. and Cumbie, B. G.,** Stream gradient as a criterion of lotic habitats suitable for *Australorbis glabratus* (Say) in Puerto Rico, *Am. J. Trop. Med. Hyg.,* 5, 921, 1956.

61. **Heyneman, D. and Lim, H. K.,** Attraction of predaceous rediae of the echinostome, *Paryphostomum segregatum,* to sporocysts of *Schistosoma mansoni* and inhibition of development of rediae induced by the prey sporocysts, *J. Parasitol.,* 56(4), 1944, 1970.

62. **Heyneman, D. and Umathevy, T.,** Differentiation of trematode cercariae of the genus *Plagiorchis* by use of the patterns of their argentophilic cuticular structures, *Med. J. Malaya,* 20, 353, 1966.

63. **Heyneman, D. and Umathevy, T.**, A field experiment to test the possibility of using double infections of host snails as a possible biological control of schistosomiasis, *Med. J. Malaya*, 21, 373, 1967.

64. **Heyneman, D. and Umathevy, T.**, Interaction of trematodes by production within natural double infections in the host snail *Indoplanorbis exustus, Nature (London)*, 217, 283, 1968.

65. **Holliman, R. B., et al.**, Studies on centrifugation and hatching of *Schistosoma mansoni* eggs, *Amer. Midl. Nat.*, 87, 251, 1972.

66. **Ishii, Y. and Miyazaki, I.**, Comparative study of the eggshell of American *Paragonimus* through the scanning electron microscope, *Jpn. J. Parasitol.*, 19, 541, 1970.

67. **Jobin, W. R. and Ippen, A. T.**, Ecological design of irrigation canals for snail control, *Science*, 145, 1324, 1964.

68. **Joosse, J.**, Endocrinology of reproduction in molluscs, *Gen. Comp. Endocrinol. Suppl.*, 3, 591, 1972.

69. **Kassim, O. and Gilbertson, D. E.**, Hatching of *Schistosoma mansoni* eggs and observations on motility of miracidia, *J. Parasitol.*, 62, 715, 1976.

70. **Katsurada, F.**, The schistosomiasis japonica in Saga Prefecture, *Okayama Igakai Zasshi*, No. 175, Japanese text, 1904.

71. **Kemp, W., Damian, R., and Greene, N.**, *Schistosoma mansoni*: immunohistochemical localization of the CHR reaction in the cercarial glycocalyx, *Exp. Parasitol.*, 33, 27, 1973.

72. **Kerbert, C.**, Zur Trematoden-Kenntnis, *Zool. Anz.*, 1, 271, 1878.

73. **Khalil, L. F.**, On the capture and destruction of miracidia by *Chaetogaster limnaei* (Oligochaeta), *J. Helminthol.*, 35, 269, 1961.

74. **Kloetzel, K. and Correa, L. Dos R.**, Some quantitative data from Jabuticatubas (State of Minas Gerais, Brazil) by miracidia of *Schistosoma mansoni. Rev. Inst. Med. Trop. S. Paulo*, 10, 59, 1968.

75. **Knight, W. B., Ritchie, L. S., Liard, F., and Chiriboga, J.**, Cercariophagic activity of guppy fish (*Lesbistes reticulata*) using cercariae labeled with radioselenium (^{75}Sc), *Am. J. Trop. Med. Hyg.*, 19, 620, 1970.

76. **Koie, M., Christensen, N. Ø., and Nansen, P.**, Stereoscan studies of eggs, free-swimming and penetrating miracidia and early sporocysts of *Fasciola hepatica, Z. Parasitenkd.*, 51, 79, 1976.

77. **Koie, M. and Frandsen, F.**, Stereoscan observations of the miracidium and early sporocyst of *Schistosoma mansoni, Z. Parasitenkd.*, 50, 335, 1976.

78. **Kruidenier, F. J.**, The formation and function of mucoids in virgulate cercariae, including a study of the virgula organ, *Am. Midl. Nat.*, 46, 660, 1951.

79. **Krupa, P. L.**, Ultrastructural topography of a trematode eggshell, *Exp. Parasitol.*, 35, 244, 1974.

80. **Kuntz, R. E. and Chandler, A. C.**, Studies on Egyptian trematodes with special reference to the heterophyids of mammals. II. Embryonic development of *Heterophyes aequalis* Loos, *J. Parasitol.*, 42, 613, 1956.

81. **Kuntz, R. E., Tulloh, G. S., Davidson, D. L., and Huang, T. C.**, Scanning electron microscopy of the integumental surfaces of *Schistosoma haematobium, J. Parasitol.*, 62, 63, 1976.

82. **LaRue, G. R.**, The classification of digenetic Trematoda: a review and a new system, *Exp. Parasitol.*, 6, 306, 1957.

83. **Leuckart, R.**, Zur Entwicklungsgeschichte des Leberegels, *Zool. Anz.*, 99, 641, 1881.

84. **Leuckart, R.**, Zur Entwicklungsgeschichte des Leberegels (*Distomum hepaticum*), *Arch. Naturgesch.*, 48, 80, 1882.

85. **Lever, J., de Vries, C. M., and Jager, J. C.**, On the anatomy of the central nervous system and the location of neurosecretory cells in *Australorbis glabratus, Malacol. Int. J. Malacol.*, 2, 219, 1965.

85a. **Lever, J., Kok, M., Meuleman, E. A., and Joosse, J.**, On the location of Gomori-positive neurosecretory cells in the central ganglia of *Lymnaea stagnalis, Proc. K. Ned. Akad. Wet.*, Amsterdam, C64, 640, 1961.

86. **Lewert, R. M., Hopkins, D. R., and Mandlowitz, S.**, The role of calcium and magnesium ions in invasiveness of schistosome cercariae, *Am. J. Trop. Med. Hyg.*, 15, 314, 1966.

87. **Lewert, R. M. and Para, B. J.**, The physiological incorporation of carbon 14 in *Schistosoma mansoni* cercariae, *J. Infect. Dis.*, 116, 171, 1966.

88. **Lewert, R. M., Para, J., and Mehmet, A.**, Miracidial uptake of glucose in intact eggs of *Schistosoma mansoni, J. Parasitol.*, 56, 1250, 1970.

89. **Lie, K. L.**, Studies on Echinostomatidae (Trematoda) in Malaya. XIII. Integumentary papillae on six species of echinostome cercariae, *J. Parasitol.*, 52, 1041, 1966.

90. **Lie, K. J., Kwo, E. H., and Owyang, C. K.**, A field trial to test the possible control of *Schistosoma spindale* by means of interspecific trematode antagonism, *Southeast Asian J. Trop. Med. Public Health*, 1, 19, 1970.

91. **Lo Verde, P. T.**, Scanning electron micrscope observations on the miracidium of *Schistosoma, Int. J. Parasitol.*, 5, 95, 1975.

92. **Luttermoser, G. W.,** Studies on the chemotherapy of experimental schistosomiasis. III. Harvest of *Schistosoma mansoni* cercariae by forced nocturnal emergence from *Australorbis glabratus, J. Parasitol.,* 41, 201, 1955.

93. **Maldonado, J. F., Acosta Matienzo, J., and Herrera, F. V.,** Biological studies on the miracidium of *Schistosoma mansoni.* Part 3. The role of light and temperature in hatching, *P. R. J. Public Health Trop. Med.,* 25, 359, 1950.

94. **Maldonado, J. F., Acosta Matienzo, J., and Herrera, F. V.,** Biological studies on the miracidium of *Schistosoma mansoni.* Part 4. The role of pH in hatching and longevity, *P. R. J. Public Health Trop. Med.,* 26, 85, 1950.

95. **Maldonado, J. F., Acosta Matienzo, J., and Thillet, C. J.,** Biological studies on the miracidium of *Schistosoma mansoni.* Part 2. Behavior of the unhatched miracidium in undiluted stools under diverse environmental conditions, *P. R. J. Public Health Trop. Med.,* 25, 153, 1949.

96. **Malek, E. A.,** Susceptibility of the snail *Biomphalaria boissyi* to infection with certain strains of *Schistosoma mansoni, Am. J. Trop. Med.,* 30, 887, 1950.

97. **Malek, E. A.,** The preputial organ of snails in the genus *Helisoma* (Gastropoda: Pulmonata), *Am. Midl. Nat.,* 48, 94, 1952.

98. **Malek, E. A.,** Morphology, Bionomics and Host-Parasite Relations of Planorbidae (Mollusca: Pulmonata), Ph.D. thesis, University of Michigan, Ann Arbor, 1952.

99. **Malek, E.A.,** Morphological studies on the family Planorbidae (Mollusca: Pulmonata), I. Genital organs of *Helisoma trivolvis* (Say) (Subfamily Helisomatinae F. C. Baker, 1945), *Trans. Am. Microsc. Soc.,* 73, 103, 1954.

100. **Malek, E. A.,** Morphological studies on the family Planorbidae (Mollusca: Pulmonata). II. The genital organs of *Biomphalaria boissyi* (Subfamily Planorbinae H. A. Pilsbry, 1934). *Trans. Am. Microsc. Soc.,* 73, 285, 1954.

101. **Malek, E. A.,** Anatomy of *Biomphalaria boissyi* as related to its infection with *Schistosoma mansoni, Am. Midl. Nat.,* 54, 394, 1955.

102. **Malek, E.A.,** Factors conditioning the habitat of bilharziasis intermediate hosts of the family Planorbidae, *Bull. W. H. O.,* 18, 785, 1958.

103. **Malek, E. A.,** The ecology of schistosomiasis, in *Studies in Disease Ecology,* Vol. 2, May, J. M., Ed., Hafner Co., New York, 1961, 261.

104. **Malek, E. A.,** Report on Precontrol Studies of Bilharziasis in St. Lucia, mimeographed report, Pan American Health Organization/WHO, Washington, D.C., 1962.

105. **Malek, E. A.,** Bilharziasis control in pump schemes near Khartoum, Sudan and an evaluation of the efficacy of chemical and mechanical barriers, *Bull. W. H. O.,* 27, 41, 1962.

106. **Malek, E. A.,** Laboratory Evaluation of an Emulsifiable Concentrate Formulation of Niclosamide, unpublished report, World Health Organization, Geneva, WHO/Schisto/71.9, 1971.

107. **Malek, E. A.,** Some factors influencing the worm recovery rates from white mice infected with *Schistosoma mansoni, Proc. 3rd. Int. Congr. Parasit.,* Munich, 2, 804, 1974.

108. **Malek, E. A.,** Studies on the parasitism of mollusks as models for comparative pathology, *J. Invertebr. Pathol.,* 29, 1, 1977.

109. **Malek, E. A.,** Geographical distribution, hosts, and biology of *Schistosomatium douthitti* (Cort, 1914) Price, 1931, *Can. J. Zool.,* 55, 661, 1977.

110. **Malek, E. A.,** Realistic goals in the use of molluscicides in different endemic areas of schistosomiasis, *Proc. Int. Conf. Schisto. Cairo,* 359, 1978.

111. **Malek, E. A.,** unpublished data.

112. **Malek, E. A., Brenes, R., and Rojas, G.,** *Aroapyrgus costaricensis,* hydrobiid snail host of paragonimiasis in Costa Rica, *J. Parasitol.,* 61, 355, 1975.

113. **Malek, E. A. and Cheng, T. C.,** *Medical and Economic Malacology,* Academic Press, New York, 1974.

114. **Malek, E. A. and File, S. K.,** Electrophoretic studies on the digestive gland esterases of some biomphalarid and lymnaeid snails, *Bull. W. H. O.,* 45, 819, 1971.

115. **Mason, P. R.,** Stimulation of the activity of *Schistosoma mansoni* miracidia by snail-conditioned water, *Parasitology,* 75, 325, 1977.

116. **McConnell, J. F. P.,** Remarks on the anatomy and pathological relations of a new species of liver fluke, *Lancet,* 2, 271, 1875.

117. **McClelland, W. F. J.,** Production of *Schistosoma haematobium* and *Schistosoma mansoni* cercariae in Tanzania, *Exp. Parasitol.,* 20, 205, 1967.

118. **Mehlman, B. and von Brand, T.,** Further studies on the anaerobic metabolism of some freshwater snails, *Biol. Bull. (Woods Hole, Mass.),* 100, 199, 1951.

119. **Meuleman, E. A.,** Host-parasite interrelationships between the freshwater pulmonate *Biomphalaria pfeifferi* and the trematode *Schistosoma mansoni, Neth. J. Zool.,* 22, 355, 1972.

120. **Michelson, E. H.,** The protective action of *Chaetogaster limnaei* on snails exposed to *Schistosoma mansoni, J. Parasitol.,* 50, 441, 1964.

121. **Moore, M. N. and Halton, D. W.,** A histochemical study of the rediae and cercariae of *Fasciola hepatica, Z. Parasitenkd.,* 47, 45, 1975.

122. **Morton, J. E.,** *Molluscs,* Hutchinson University Library, London, 1958.

123. **Newell, G. E.,** Physiological aspects of the ecology of intertidal molluscs, in *Physiology of Mollusca,* Vol. 1, Wilbur, K. M. and Yonge, C. M., Eds., Academic Press, New York, 1964, 59.

124. **Olivier, L., von Brand, T., and Mehlman, B.,** The influence of lack of oxygen on *Schistosoma mansoni* cercariae and on infected *Australorbis glabratus, Exp. Parasitol.,* 2, 258, 1953.

125. **Pan, C. T.,** The general histology and topographic microanatomy of *Australorbis glabratus, Bull. Mus. Comp. Zool., Harv. Univ.,* 119(3), 299, 1958.

126. **Pan, C. T.,** Studies on the host parasite relationship between *Schistosoma mansoni* and the snail *Australorbis glabratus, Am. J. Trop. Med. Hyg.,* 14, 931, 1965.

127. **Pearson, J. C.,** A phylogeny of life-cycle patterns of the Digenea, in *Advances in Parasitology,* Vol. 10, Dawes, B., Ed., Academic Press, London, 1972, 153.

128. **Pearson, J. C.,** Observations on the morphology and life-cycle of *Neodiplostomum intermedium* (Trematoda: Diplostomatidae), *Parasitology,* 51, 133, 1961.

129. **Pellegrino, J. and Demaria, M.,** Results of exposing mice to natural pond water harboring a highly infected colony of *Australorbis glabratus, Am. J. Trop. Med. Hyg.,* 15, 333, 1966.

130. **Perlowagora-Szumlewicz, A. and von Brand, T.,** Studies on the oxygen consumption of *Australorbis glabratus* eggs, *J. Wash. Acad. Sci.,* 47, 11, 1957.

131. **Perlowagora-Szumlewicz, A. and von Brand, T.,** Physiology — Observations on the oxygen consumption of young *Australorbis glabratus, J. Wash. Acad. Sci.,* 48, 38, 1958.

132. **Pitchford, R. J., Meyling, A. H., Meyling, J., and Dutoit, J. F.,** Cercarial shedding patterns of various schistosome species under outdoor conditions in the Transvaal, *Ann. Trop. Med. Parasitol.,* 63, 359, 1969.

133. **Prosser, C. L. and Brown, F. A.,** *Comparative Animal Physiology,* 2nd ed., W. B. Saunders, Philadelphia, 1961.

134. **Purnell, R. E.,** Host-parasite relationships in schistosomiasis. 1. The effect of temperature on the infection of *Biomphalaria sudanica tanganyicensis* with *Schistosoma mansoni* miracidia and of laboratory mice with *Schistosoma mansoni* cercariae. *Ann. Trop. Med. Parasitol.,* 60, 90, 1966.

135. **Race, G. J., Martin, J. H., Moore, D. V., and Larsch, J. E.,** Scanning and transmission electron-microscopy of *Schistosoma mansoni* eggs, cercariae, and adults, *Am. J. Trop. Med. Hyg.,* 20, 914, 1971.

136. **Race, G. J., Michaels, R. M., Martin, J. H., Larsch, J. E., and Matthews, J. L.,** *Schistosoma mansoni* eggs; an electron microscope study of shell pores and microbarbs, *Proc. Soc. Exp. Biol. Med.,* 130, 990, 1969.

137. **Richard, J.,** La chetotaxie des cercaires de schistosomes, *C. R. Acad. Sci. Paris,* 266, 1865, 1968.

138. **Richard, J.,** La chetotaxie des cercaires. Valeur systematique et phyletique, *Mem. Mus. Natl. Hist. Nat. Ser. A Zool.,* 67, 179, 1971.

139. **Ritchie, L. S.,** Chemical control of snails, in *Epidemiology and Control of Schistosomiasis,* Ansari, N., Ed., S. Karger, Basel, 1973, 435.

140. **Ritchie, L. S. and Malek, E. A.,** Molluscicides: Status of their Evaluation, Formulation and Methods of Application, unpublished document, World Health Organization, Geneva, PD/MOL/1969, 1.

141. **Rowan, W. B.,** The mode of hatching of the egg of *Fasciola hepatica, Exp. Parasitol.,* 5, 118, 1956.

142. **Rowan, W. B.,** The mode of hatching of the egg of *Fasciola hepatica.* II. Colloidal nature of the viscous cushion, *Exp. Parasitol.,* 6, 131, 1957.

143. **Rowan, W. B.,** Schistosomiasis and the chlorination of sewage effluent, *Am. J. Trop. Med. Hyg.,* 13, 577, 1964.

144. **Rudolph, P. H.,** Accumulation of galactogen in the albumen gland of *Triodopsis multilineata* (Pulmonata: Stylommatophora) after dormancy, *Malacol. Rev.,* 8, 57, 1975.

145. **Runham, N. W. and Hunter, P. J.,** *Terrestrial Slugs,* Hutchinson and Co., London, 1970.

146. **Sakamoto, K. and Ishii, Y.,** Fine Structure of schistosome eggs as seen through the scanning electron microscope, *Am. J. Trop. Med. Hyg.,* 25, 841, 1976.

147. **Sambon, L. W.,** Remarks on *Schistosomum mansoni, J. Trop. Med. Hyg.,* 10, 303, 1907.

148. **Shattock, M. S., Fraser, R. J., and Garnett, P. A.,** Seasonal variations of cercarial output from *Biomphalaria pfeifferi* and *Bulinus (Physopsis) globosus* in a natural habitat in Southern Rhodesia, *Bull. W. H. O.,* 33, 276, 1965.

149. **Short, R. B. and Cartlett, M. L.,** Argentophilic "papillae" of *Schistosoma mansoni* cercariae, *J. Parasitol.,* 59, 1041, 1973.

150. **Short, R. B. and Gagné, H. T.,** Fine structure of a possible photoreceptor in cercariae of *Schistosoma mansoni, J. Parasitol.,* 61, 69, 1975.

151. **Short, R. B. and Kuntz, R. E.,** Patterns of argentophilic papillae of *Schistosoma rodhaini* and *S. mansoni* cercariae, *J. Parasitol.,* 62, 420, 1976.

152. **Sminia, T.**, Structure and function of blood and connective tissue cells of the fresh water pulmonate *Lymnaea stagnalis* studied by electron microscopy and enzyme histochemistry, *Z. Zellforsch. Mikrosk. Anat.*, 130, 497, 1972.

153. **Sminia, T. and Boer, H. H.**, Haemocyanin production in pore cells of the freshwater snail *Lymnaea stagnalis, Z. Zellforsch. Mikrosk. Anat.*, 145, 443, 1973.

154. **Standen, D. D.**, The effect of temperature, light, and salinity upon the hatching of the ova of *Schistosoma mansoni, Trans. R. Soc. Trop. Med. Hyg.*, 45, 225, 1951.

155. **Stein, P. C. and Lumsden, R. D.**, *Schistosoma mansoni*: topochemical features of cercariae, schistosomula, and adults, *Exp. Parasitol.*, 33, 499, 1973.

156. **Stirewalt, M. A.**, *Schistosoma mansoni*: cercaria to schistosomule, in *Advances in Parasitology*, Vol. 12, Dawes, B., Ed., Academic Press, London, 1974, 115.

157. **Stirewalt, M.**, Quantitative collection and proteolytic activity of preacetabular gland enzyme(s) of cercariae of *Schistosoma mansoni, Am. J. Trop. Med. Hyg.*, 27, 548, 1978.

158. **Stirewalt, M. A. and Fregeau, W. A.**, Effect of selected experimental conditions on penetration and maturation of *Schistosoma mansoni* in mice. 1. Environmental, *Exp. Parasitol.*, 17, 168, 1965.

159. **Stirewalt, M. A. and Kruidenier, F. J.**, Activity of the acetabular secretory apparatus of cercariae of *Schistosoma mansoni* under experimental conditions, *Exp. Parasitol.*, 11, 191, 1961.

160. **Stoll, N. E.**, This wormy world, *J. Parasitol.*, 33, 1, 1947.

161. **Sturrock, B. M. and Sturrock, R. F.**, Laboratory studies of the host-parasite relationship of *Schistosoma mansoni* and *Biomphalaria glabrata* from St. Lucia, West Indies, *Ann. Trop. Med. Parasitol.*, 64, 357, 1970.

162. **Sturrock, R. F.**, The influence of temperature on the biology of *Biomphalaria pfeifferi* (Krauss), an intermediate host of *Schistosoma mansoni, Ann. Trop. Med. Parasitol.*, 60, 100, 1966.

163. **Sugiura, S., Sasaki, T., Hosaka, V., and Ono, R.**, A study of several factors influencing hatching of *Schistosoma japonicum* eggs, *J. Parasitol.*, 40, 381, 1954.

164. **Thomas, A. P.**, Report of experiments on the development of the liver fluke *(Fasciola hepatica), J. R. Agric. Soc. Eng.*, 17, 1, 1881.

165. **Thomas, A. P.**, The rot in sheep, or the life history of the liver fluke, *Nature (London)*, 26, 606, 1882.

166. **Threadgold, L. T.**, The ultrastructure of the "cuticle" of *Fasciola hepatica, Exp. Cell Res.*, 30, 238, 1963.

166a. **Threadgold, L. T. and Arme, C.**, Electron microscope studies of *Fasciola hepatica.* XI, Autophagy and parenchymal cell function, *Exper. Parasitol.*, 35, 389, 1974.

167. **Ulmer, M. J.**, *Postharmostomum helicis* (Leidy, 1847) Robinson 1949, (Trematoda), its life history and a revision of the subfamily Brachylaeminae, *Trans. Am. Microsc. Soc.*, 70, 319, 1951.

168. **Ulmer, M. J.**, Studies on the nervous system of *Postharmostomum helicis* (Leidy, 1847) Robinson 1949 (Trematoda: Brachylaimatidae), *Trans. Am. Microsc. Soc.*, 72, 370, 1953.

169. **Upatham, E. S.**, Effects of some physico-chemical factors on the infection of *Biomphalaria glabrata* (Say) by miracidia of *Schistosoma mansoni* Sambon in St. Lucia, West Indies. *J. Helminthol.*, 46, 307, 1972.

170. **Upatham, E. S.**, The effect of water temperature on the penetration and development of St. Lucian *Schistosoma mansoni* miracidia in local *Biomphalaria glabrata. Southeast Asian J. Trop. Med. Public Health*, 4, 367, 1973.

171. **van Someren, V. D.**, The habitats and tolerance ranges of *Lymnaea (Radix) caillaudi*, the intermediate snail host of liver fluke in East Africa, *J. Anim. Ecol.*, 15, 170, 1946.

172. **von Brand, T., Baernstein, H. D., and Mehlman, B.**, Studies on the anaerobic metabolism and the aerobic carbohydrate consumption of some fresh water snails, *Biol. Bull. (Woods Hole, Mass.)*, 98, 266, 1950.

173. **von Brand, T., McMahon, P., and Nolan, M. O.**, Physiological observations on starvation and desiccation of the snail *Australorbis glabratus, Biol. Bull. (Woods Hole, Mass.)*, 113, 89, 1957.

174. **von Brand, T., Nolan, M. O., and Mann, E. R.**, Observations on the respiration of *Australorbis glabratus* and some other aquatic snails, *Biol. Bull. (Woods Hole, Mass.)*, 95, 199, 1948.

175. **Webbe, G.**, The effect of water velocities on the infection of animals exposed to *Schistosoma mansoni* cercariae, *Ann. Trop. Med. Parasitol.*, 60, 78, 1966.

176. **Webbe, G.**, The effect of water velocities on the infection of *Biomphalaria sudanica tanganyicensis* exposed to different numbers of *Schistosoma mansoni* miracidia, *Ann. Trop. Med. Parasitol.*, 60, 85, 1966.

177. **Webbe, G. and James C.**, Host-parasite relationships of *Bulinus globosus* and *B. truncatus* with strains of *Schistosoma haematobium, J. Helminthol.*, 46, 185, 1972.

178. **Wilbur, K. M. and Yonge, C. M.**, Eds., *Physiology of Mollusca*, Vol. 1 and 2, Academic Press, New York, 1964, 1966.

179. **Williams, N. V.**, Studies on aquatic pulmonate snails in central Africa. I. Field distribution in relation to water chemistry, *Malacol. Int. J. Malacol.,* 10, 153, 1970.
180. **Williams, N. V.**, Studies on aquatic pulmonate snails in Central Africa. II. Experimental investigation of field distribution patterns, *Malacol. Int. J. Malacol.,* 10, 165, 1970.
181. **Wilson, R. A.**, Hatching mechanism of the egg of *Fasciola hepatica* L., *Parasitology,* 58, 79, 1968.
182. **Wilson, R. A.**, Fine structure of the tegument of the miracidium of *Fasciola hepatica* L., *J. Parasitol.,* 55, 124, 1969.
183. **Wilson, R. A., Pullin, R., and Denison, J.**, An investigation of the mechanism of infection by digenetic trematodes: the penetration of the miracidium of *Fasciola hepatica* into its snail host *Lymnaea truncatula, Parasitology,* 63, 491, 1971.
184. World Health Organization, Molluscicide screening and evaluation, *Bull. W. H. O.,* 33, 567, 1965.
185. Snail control in the prevention of bilharziasis, Monograph Series No. 50, World Health Organization, Geneva, 1965.
186. World Health Organization, Report on "Meeting of Directors of Collaboratng Laboratories on Molluscicide Testing and Evaluation," Washington, D.C., 1970, unpublished document, World Health Organization, Geneva, WHO/Schisto/1971.6.
187. **Wright, C. A.**, The pathogenesis of helminths in the Mollusca, *Helminth. Abs.,* 35, 207, 1966.
188. **Wright, C. A.**, *Flukes and Snails,* Hafner Press, New York, 1973.
189. **Wright, C. A., Southgate, V. R., and Knowles, R. J.**, What is *Schistosoma intercalatum* Fisher, 1934?, *Trans. R. Soc. Trop. Med. Hyg.,* 66, 28, 1972.
190. **Wright, D. G. S.**, Responses of miracidia of *Schistosoma mansoni* to an equal energy spectrum of monochromatic light, *Can. J. Zool.,* 52, 857, 1974.
191. **Wu, K.**, Deux nouvelles plantes pouvant transmettre le *Fasciolopsis buski,* revue generale, *Ann. Parasitol. Hum. Comp.,* 15, 458, 1937.
192. **Wykoff, D. E.**, Studies on *Clonorchis sinensis.* IV. Production of eggs in experimentally infected rabbits, *J. Parasitol.,* 45, 91, 1959.
193. **Yamaguti, S.**, Systema Helminthum, in *The Digenetic Trematodes of Vertebrates,* Vol. 1 (Part 1), Interscience, New York, 1958.
194. **Zylstra, V.**, Uptake of particulate matter by the epidermis of the freshwater snail *Lymnaea stagnalis, Neth. J. Zool.,* 22, 299, 1972.
195. **Barbosa, F. S., Barbosa, I., and Arruda, F.**, *Schistosoma mansoni.* Natural infection of cattle in Brazil, *Science,* 138, 831, 1962.
196. **Cabrera, B. D.**, Schistosomiasis japonica in field rats in Leyte, Philippines, *Southeast Asian J. Trop. Med. Public Health,* 7, 50, 1976.
197. **Coelho, P. M. Z., et al.**, Crab-eating raccoon *Procyon cancrivorous nigripes* (Mivart 1885) (Carnivora: Procyonidae) naturally infected with *Schistosoma mansoni* in Minas Gerais State, Brazil, *J. Parasitol.,* 62, 748, 1976.
198. **De Paoli, A.**, *Schistosoma haematobium* in the chimpanzee — a natural infection, *Am. J. Trop. Med. Hyg.,* 14, 561, 1965.
199. **Fenwick, A.**, Baboons as reservoir hosts of *Schistosoma mansoni, Trans. R. Soc. Trop. Med. Hyg.,* 63, 557, 1969.
200. **Kuntz, R. E., Huang, T. C., and Moore, J. A.**, Patas monkey (*Erythrocebus patas*) naturally infected with *Schistosoma mansoni, J. Parasitol.,* 63, 166, 1977.
201. **Malek, E. A.**, The ecology of schistosomiasis, in *Studies in Disease Ecology,* Vol. 2, May, J. M., Ed., Hafner Co., New York, 1961, 261.
202. **Malek, E. A.**, Report on Precontrol Studies of Bilharziasis in St. Lucia, mimeographed report, Pan American Health Organization/WHO, Washington, D.C., 1963.
203. **Malek, E. A.**, unpublished data.
204. **Mansour, N. S.**, *Schistosoma mansoni* and *Schistosoma haematobium* found as a natural double infection in the Nile rat, *Arvicanthis n. niloticus,* from a human endemic area in Egypt, *J. Parasitol.,* 59, 424, 1973.
205. **Martins, A. V.**, Non-human vertebrate hosts of *Schistosoma haematobium* and *Schistosoma mansoni, Bull. W. H. O.,* 18, 931, 1958.
206. **Nelson, G. S., Teesdale, C., and Highton, R. B.**, The role of animals as reservoirs of bilharziasis in Africa, in *Ciba Foundation Symposium on Bilharzia,* J. & A. Churchill, London, 1962, 127.
207. **Pesigan, T. P., et al.**, Studies on *Schistosoma japonicum* infection in the Philippines. 1. General considerations and epidemiology, *Bull. W. H. O.,* 18, 345, 1958.
208. **Theron, A., Pointier, J. P., and Combes, C.**, Approche e cologique du probleme de la responsibilite de l'homme et du rat dans le fonctionnement d'un site de transmission a *Schistosoma mansoni* en Guadeloupe, *Ann. Parasitol. Hum. Comp. (Paris),* 53, 223, 1978.
209. **Wright, C. A., Southgate, V. R., and Knowles, R. J.**, What is *Schistosoma intercalatum* Fisher, 1934?, *Trans. R. Soc. Trop. Med. Hyg.,* 66, 28, 1972.

Chapter 2

SCHISTOSOMIASIS

I. INTRODUCTION

Schistosomiasis, or Bilharziasis or Bilharzia, as it is known in many endemic areas, is one of the most important public health problems in many countries in the tropics and subtropics. It is a snail-borne disease which is contracted by the cercarial stage of the parasites penetrating the skin of individuals who come in contact with infested waters. It is estimated that the disease affects the health of about 200 million people infected in 71 countries in Africa, the Middle East, South America, some islands of the Caribbean, and in the Orient. In the last three decades considerable information has been gathered on the distribution of the disease, its clinico-pathological manifestations, and its public health significance. These accomplishments have been made through various workers, whose efforts were stimulated and coordinated through the auspices of the United Nations World Health Organization and certain national governments, where the disease is endemic, and through investigations in developed countries which have interest in advancement of scientific knowledge and help to less developed countries. It became evident that the disease represents a social and economic burden of great magnitude, unfortunately in countries which can least afford to lose any of their productive capacity.

Much of the needed increase in food production, improvement of living standards, and introduction of industry in developing countries depends on the implementation of water resources development and hydroelectric schemes, as well as on the reclamation of land for agricultural purposes. There is evidence that such development projects have already aggravated preexisting low prevalences of schistosomiasis and have spread the disease into new areas. Any large-scale extension of irrigated territories will provide bigger and more permanent breeding conditions for the snail intermediate hosts, which are obligatory for the development and propagation of the schistosome parasites that produce the disease. In addition, development projects have caused resettlement of population, increased man-water contacts, and have attracted a considerable number of temporary laborers or permanent settlers who had been infected in their homeland. The planners of water-development projects should consider the disease repercussions and adopt control measures.

The public health significance of schistosomiasis has been documented in the last two decades. It has become evident that it is one of the major water-related diseases. Fortunately, it is now receiving regular attention in the preparatory phase of any large-scale water resource project financed by the United Nations Development Program (UNDP), the World Bank, or the U.S. Agency for International Development (USAID). This procedure has been established to prevent, as much as possible, aggravation of already existing disease, a grave situation which would cancel out some of the economic benefits to be derived from such projects.

Schistosoma mansoni and *S. japonicum* have always been regarded as the causative agents of severe disease. However, *S. haematobium*, although regarded as a pathologic agent in some countries like Egypt, has been denied much significance in other countries such as African countries south of the Sahara. But in recent years, based on community and group studies in Tanzania, Nigeria, and Senegal, it has been demonstrated that *S. haematobium* is responsible for severe lesions. Among these lesions are calcified bladder, deformity of the ureter and hydronephrosis, lesions wh.ch are common among adolescents and children, as well as in adults. Associated with schistosome infections are three disease syndromes: Katayama fever, which occurs in heavy initial infection, the acute syndrome starting when the infection becomes patent and deposition of the parasite's eggs in the tissues and their excretion in the urine or feces is taking place, and chronic schistosomiasis, when the host granulomatous response to the eggs has involved several organs of the body. The occurrence of significant degrees of pathology is related to the intensity of infection, but also to the intensity of host response. This is modified in the course of natural infection and by other factors, including malnutrition and alterations in the metabolic state of the host, and concomitant diseases. It is often difficult to determine the relative contribution of each of these diseases to the lowered health of the individual and community. Schistosomiasis is an insidious disease, inasmuch as it exerts alarming clinical manifestations in exceptional cases. Such cases are demonstrated only after heavy exposure of previously unexposed individuals. It is known that the lesions of the disease build up slowly and the damaging effect on the host is compounded over several years. The resistance of the infected individual is lowered, and consequently he becomes prone to other diseases.

In spite of the fact that schistosomiasis is one of the major medical and public health problems in many countries, it is not a reportable disease. Even in New York City, the City Health Department is able to tabulate annually only 5% of the estimated infected Puerto Rican population living in the city. It would be advantageous to include schistosomiasis in the list of reportable diseases in endemic areas. This will provide the countries concerned with a better basis to make surveillance and evaluation of its present and prospective status a simpler task. If any control efforts are being made in particular countries, then reporting of schistosomiasis will make it easy to evaluate progress or failure in the attack on the disease within a country.

Reliable data on morbidity and mortality are almost lacking. Schistosomiasis has a delayed onset of symptoms, and in many cases the symptoms do not attain serious proportions so the disease passes unnoticed. Many if not most cases observed medically result from incidental findings in patients seeking treatment of other conditions. The disease is rarely established as a direct cause of death, and official mortality rates, such as they are, no doubt represent only a fraction of the true rates.

Immunity to schistosomiasis has been an intriguing subject to investigators. That man is not susceptible to certain animal schistosomes and that certain individuals are less susceptible than others to the human species are evidences of innate immunity. Moreover, studies on lower primates and other animals have demonstrated a certain degree of acquired immunity to superinfection. The tendency for the prevalence curve to fall after its early peak offers indirect evidence for acquisition of immunity to superinfection in man. The fact that man lives at all where in endemic areas he is exposed continually to the cercariae in the water surely indicates an acquired resistance to reinfection. It is thus the belief of many workers in schistosomiasis that immunity is a major factor controlling the prevalence and intensity of the disease in man. This does not exclude the view of a few workers who believe that a significant degree of protection against the occurrence of massive worm burdens may, however, be related to the biological characteristics of the schistosomes in addition to some ecological and even sociological factors. The subject of immunity in schistosomiasis is an aspect of the disease which has been, and still is, a subject that draws the attention of several investigators. Accordingly, there have been many advances. For example, we now know much more about the immunopathology of the disease; we have become aware that the parasite can circumvent the immune defenses of the host, and we have an in vitro system which may elucidate the mechanism of the immune response. The worms which can circumvent the immune defenses of the host live for a number of years. Reports on some immigrants from endemic areas indicate that the three species of schistosomes can live for 20 to 30 years, and that egg output continues over the entire period. However, based on some data from endemic areas (the Philippines and Egypt), it was estimated that the mean life span was 4.5 years for *Schistosoma japonicum*, 3.8 years for *S. haematobium*, and 3.5 years for *S. mansoni*.

Going over the life cycle of the schistosome, transmission of the parasite seems quite simple, indeed. On the contrary, transmission is a complex process that requires timing of contact of humans and other mammals with the water and the larval stages contained in it. Accordingly, in recent years attention has been given to the study of the dynamics of transmission. Both the parasites and the snail intermediate hosts must be considered, together with the complexities of the host-parasite relationships through which each may influence the other in a variety of ways. Another problem is that, especially in Africa, several species of parasites and of snails are involved, and both groups show considerable infraspecific variation. As a result many field investigations raise taxonomic problems which cannot be solved readily. Studies on the dynamics of transmission from one host to the other necessitate a quantitative approach which is needed to effect measurements of the factors and the chain of events leading to transmission of schistosomiasis in a given area. Mathematical models have thus been developed to help in epidemiological research and aid in the development of a sound plan for the control of the disease.

In a few endemic areas of schistosomiasis some degree of control has been achieved, but in general the disease is thought to be on the increase, particularly in countries where water resources are being developed. The necessity to control schistosomiasis has become more eminent since its public health significance has been documented. Much information has been gathered on control measures. Among the possible control measures are use of chemotherapeutic agents to treat patients; prevention of contact between humans and water bodies likely to contain infected snails, or interrupting the life cycle of the parasites at some point or the other through sanitary measures and provision of piped water. It is believed by many experts that, at present, control or destruction of the snail hosts by various methods (environmental, chemical, or biological) is the most reliable and effective measure. Although it is recognized that there are still gaps in our knowledge of the measures applicable for control, we now have available the tools for effective control and interruption of transmission.

The interaction of environmental factors (inorganic and organic), the variety or strain of both the schistosome and the snail, and cultural and economic factors govern the geographical distribution of the disease. The population density of the human host, reservoir host, and intermediate host is related to prevalence of the disease. Where reservoir hosts exist, especially in areas of schistosomiasis japonica, a relative transmission index is necessary to assess the importance of each species in transmission.

II. HISTORICAL BACKGROUND

Schistosomiasis is an ancient disease that apparently existed since antiquity in various parts of the world. Information on historical background was included in Girgis (1934),[12] Faust and Meleney (1924),[7] Malek (1961),[17] and Farooq (1973).[6]

The disease has been known to occur in Egypt and Mesopotamia, as mentioned in Egyptian papyri, Babylonian inscriptions, medieval medical literature, and by Avicenna in the 10th century. Hematuria, an important symptom of urinary schistosomiasis, is described on certain papyri, Kahun and Ebers, of ancient Egypt, written about 1900 and 1500 B.C., respectively. Moreover, remedies to cure the disease are included in the Ebers papyrus. Ruffer (1910)[22] presented evidence of the existence of the disease in ancient Egypt by finding calcified eggs of the parasite in kidneys of two Egyptian mummies of the twentieth dynasty (1250 to 1000 B.C.). The disease was reported among Napoleon's troops during the French occupation of Egypt (1798 to 1801 A.D.). French workers stated that symptoms of the disease were frequent among the French troops.

The parasite causing the disease was not known, however, until Theodor Maximilian Bilharz encountered the worms, in 1851, at an autopsy in the Kasr-el-Aini Hospital in Cairo, Egypt. The worms were recognized as being bisexual distomes, and were named *Distoma haematobium.* Relationship was established between this trematode worm and the symptoms of dysentery and hematuria. In 1898 Manson[18] suggested, on the ground of dissimilar geographical distribution, that vesical and intestinal forms of the disease were of separate origin. In 1902, he found lateral-spined eggs in the feces of a West Indian patient in London and postulated the existence of a second species of blood fluke, but Looss opposed these views and maintained that the lateral-spined eggs were *Distoma haematobium* eggs produced parthenogenetically. Sambon (1907)[23] created the species *mansoni* for the newly differentiated type, and the work of Leiper in 1915[15] in Egypt confirmed the existence of two types of blood flukes. Leiper demonstrated the life cycle of the two parasites and determined that the snail intermediate hosts belonged not only to two different species but to two different subfamilies.

As to the generic name of this group of human blood flukes, Meckel von Hemsbach designated *Bilharzia* in 1856, and Cobbold agreed in 1859.[3] However, the generic name used at present (*Schistosoma*) was erected by Weinland in 1858.[26] The decision to use *Schistosoma* was made by the International Commission on Zoological Nomenclature in 1954.

The discovery of *S. mansoni* in the western hemisphere was made early in this century, whereby in 1904 Piraja da Silva found worm eggs with a lateral spine in patients in Bahia, Brazil and later, in 1909,[24] reported on the worms which he found at necropsy. In 1906, Soto[25] made reference to the disease in Venezuela. There is a general agreement that intestinal schistosomiasis due to *S. mansoni* was carried to the New World by African slaves who were brought from various regions in Africa.

Schistosomiasis in the Orient is a disease probably as old as Egyptian schistosomiasis. However, while there is accurate evidence of the existence of the latter during the reign of the Pharaohs, exact knowledge regarding Japanese schistosomiasis dates back only to the last century. The earliest recorded account of schistosomiasis japonica is that of Fujii (1847),[8] who described the disease in the Katayama district. Later Baelz (1883),[1] also in Japan, described the endemic area of Okayama, and in addition he listed the clinical symptoms (enlargement of the liver and spleen, bloody diarrhea, anemia, fever, ascites, and edema). However, he attributed the infection to the lesser liver fluke, *Clonorchis sinensis (Distoma hepatis endemicum).*

The causative agent of *Schistosoma japonicum* was discovered and named by Katsurada in 1904[13] in Japan. Soon after this, its pathological and clinical manifestations were reported upon in Japan and in other countries in the Far East. The first actual relation of the causative agent, the Oriental blood fluke, to the infection was secured by Katsurada (1904),[13] although Yamagiwa (1889),[28] Kurimoto (1893),[14] and Fujinami (1904)[9] had previously found eggs of an unknown parasite in various organs of individuals who had died in infected areas. By 1911 Fujinami and Nakamura[10] found that guinea pigs, monkeys, and white rats were susceptible to experimental infection via the skin when exposed to water from rice fields and that cattle, horses, dogs, and cats, as well as man, are natural hosts.

Miyairi and Suzuki,[20,21] working in an endemic area in Kyushu Prefecture, proved the existence of a snail of the family Hydrobiidae as the intermediate host, observed the penetration of the snail by the miracidia, the development inside the snail, emergence of the cercariae, and their infection of mice. Independently, Miyagawa (1911)[19] had verified the belief that *Schistosoma japonicum* has a molluscan intermediate host, and proved *Katayama* to be the mollusc by a series of exclusion experiments. Leiper and Atkinson (1915),[16] who visited China and Japan, repeated the experimental work of Miyairi and Suzuki and of Miyagawa and were able from Japanese material to verify the findings of the Japanese workers.

Investigations were also carried out in China, and whereas the infection in Japan was known as "Okayama" or "Katayama" disease, in China it was known as "Yangtze Valley Fever", "Hankow Fever", "Kiukiang Fever", and "Urticarial Fever". Infection of natives of some islands of the Philippines was reported upon by Wooley (1906)[27] and Garrison (1908).[11] Accurate and detailed work on schistosomiasis japonica was published by Cort[4,5] and especially by Faust and Meleney (1924).[7] The latter authors discovered *Oncomelania hupensis*, the intermediate host in China.

Foci of infections with *S. japonicum* were discovered at much later dates in Celebes (Sulawesi) in 1937, Laos in 1957 and 1966, Thailand in 1959 and 1960, and in Cambodia in 1968 and 1970.

III. GEOGRAPHICAL DISTRIBUTION

Countries where the disease occur and human infection rates in various parts of these countries are shown in Table 2-1.

In Europe, *Schistosoma haematobium* was found in Cyprus, and there is a focus that is dying out in southern Portugal in Algarve Province. Otherwise, urinary schistosomiasis is primarily found in Africa and in southwestern Asia. It should be noted that another schistosome of the *S. haematobium* complex, namely *S. bovis*, is found in southern Europe and some Mediterranean islands, where the snail host of *S. haematobium (Bulinus (Bulinus) truncatus)* serves as a host for bovine schistosomiasis. Schistosomiasis bovis occurs in Spain, Sardinia, and Corsica.

In Africa (Figure 2-1) urinary schistosomiasis is encountered mainly in subtropical and arid zones, and also in the dry steppe and savanna regions. It is generally absent from the woodlands of Central and West Africa and the Central Congo Basin where *S. intercalatum* is found; however, *S. haematobium* is gradually gaining ground in the edges of these regions. Along the northern coast of Africa, urinary schistosomiasis due to *S. haematobium* is present in Morocco, and several Saharan foci are encountered between northwest Africa and the Nile valley, where it extends south into the Sudan. It is absent in the Eritrean section of Ethiopia; limited foci occur in northwestern Ethiopia and northern Somalia. There is a belt of endemicity extending across Africa south of the Sahara from Upper Nile Province in the Sudan to the Atlantic. Urinary schistosomiasis occurs in East Africa and along the coast down through Mozambique and the Transvaal.

In southwestern Asia very small foci are in southern Turkey on the Syrian borders, in Syria, Lebanon, and probably also in Israel. Evidence points to the antiquity of the disease in Iraq and in the highlands of Yemen. A few scattered foci exist in Saudi Arabia, and it is found in Khuzistan Province in western Iran, probably continuous with the endemic area in Iraq; the area in Iran is the easternmost distribution of the snail *Bulinus (Bulinus) truncatus* of southwestern Asia, north Africa, and southern Europe. A small focus is in the village of Gimvi, in the state of Bombay, India, where the snail host is believed to be *Ferrissia tenuis*.

Table 2-1

WORLD-WIDE HUMAN INFECTION RATES WITH *S. MANSONI*, *S. HAEMATOBIUM*, and *S. JAPONICUM*

Country	Infection rate (%)		Year of survey	Ref.
	S. haematobium	*S. mansoni*		
Puerto Rico (overall)		1,000/10,000,+	1953—55	129
Average of Jayuya, Barranquitas, Ceiba, Comerio, Utuado and Guayama		5,117/23,262 (22)	1953	90
Overall		2,558/23,262 (11)	1956	90
Overall		325/6,780 (4.8)	1960	91
Vieques		9,047/126,244 (7)	Up to 1963	61
		1,298/10,824*,+ (12)	1962—1963	83
		21/222,+ (10)	1955	61
Dominican Republic				
Hato Mayor		52/243,+ (21.4)	1953	105
Las Palmillas		(4.6)		94
Guadeloupe		(7.8)	1948—51	59
Martinique		(6.4)	1953	59
St. Lucia overall		25,000/94,000 (26.6)		131
St. Lucia (10 localities)		(17.9—74.2)	1962	93
Venezuela				
Overall estimate (in endemic zone)		20,000/750,000 (4)		131
Aragua State		1,235/13,139 (9.4)	1964	131
Carabobo State		479/7,991 (6)	1964	131
Guarico State		102/1,833 (5.6)	1964	131
Surinam overall (in endemic area)		(12.7)	1956—57	131
Brazil				
Overall (est)		6.2	1947	95
Northeast (est)		80—95	1956	50
South and West		1-10	1956	50
Maranhao		59/12,716 (0.46),+	1950	107
Piaui		4/10,420 (0.04),+	1950	107
Ceara		379/40,314 (0.94),+	1950	107
Rio Grande de Norte		423/18,662 (2.32),+	1950	107
Paraiba		1,618/21,488 (7.52),+	1950	107

Pernambuco			
Litoral e Mata	(25.16)		
Agreste	(30.01)		
Sertao baixo	(3.13)		
Sertao Araripe	(0.93)		
Sertao Sao Francisco	(0.82)		
Sertao alto	(1.01)		
Total	12,676/50,363 (25.17)*+	1950	107
Alagoas			
Litoral	(15.75)	1950	107
Mata	(56.44)		
Sertaneja	(24.62)		
Baixo Sao Francisco	(3.81)		
Sertao do Sao Francisco	(0.43)		
Total	3,064/14,965 (20.48)*+	1950	107
Sergipe			
Litoral	(28.49)		
Central	(62.82)		
Baixo Sao Francisco	(3.19)		
Sertao Sao Francisco	(12.94)		
Oeste	(21.62)		
Total	4,422/14,676 (30.13)*+	1950	
Bahia			
Litoral Norte	(26.42)		
Sertao Sao Francisco	(2.48)		
Nordeste	(9.44)		
Feira de Santana	(18.41)		
Jacobina	(34.52)		
Matas de Orebo	(22.97)		
Reconcavo	(16.94)		
Extremo Sul	(3.79)		
Jiquie	(35.67)		
Conquista	(11.51)		
Chapada-diamantina	(13.78)		
Serra geral	(9.65)		
Cacaueira	(11.04)		
Medio Sao Francisco	(0.78)		
Planalto Occidental	(2.35)		
Total	12,344/74,590 (16.55)*+	1950	

Table 2-1 (Continued)

WORLD-WIDE HUMAN INFECTION RATES WITH S. MANSONI, S. HAEMATOBIUM, and S. JAPONICUM

Country	Infection rate (%)		Year of survey	Ref.
	S. haematobium[a]	S. mansoni[b]		
Espirito Satto		209/12,822 (1.63),*	1950	
Minas Gerais				
Mucuri		(30.02)		
Rio Doce		(12.00)		
Zona Mata		(4.76)		
Itacambira		(2.42)		
Alto Jequitinhonha		(6.66)		
Metalurgica		(5.28)		
Oeste		(1.55)		
Sul		(0.04)		
Alto medio Sao Francisco		(16.62)		
Alto Sao Francisco		(2.05)		
Uracuia		(0.00)		
Alta Paranaiba		(0.27)		
Triangule		(0.07)		
Total		6,969/158,039 (4.4%),*	1950	108
Rio de Janeiro		(0.10)	1953	108
Parana		(0.12)	1953	108
Santa Catarina		(0.00)	1953	108
Goias		(0.03)	1953	108
Mato Grosso		(0.007)	1953	108
Portugal				
Algarve	23/906 (2.5)		1948	41
(Spanish) Morocco	(13.9),*		1956	37
Gambia	(25—36)		1951	46
East	(38.8),*		1953	60
Jiborah		(71.4)	1957	123
Portuguese Guinea		13—60	1955	68
Sierra Leone				
(Boadjibu-Blama)	(70),*		1953	46
Kabala		45/215 (20.9)	1934	71

Location				
Kono		(10.9)	1970	131
Liberia				
Harbel	(56—86),+		1955	32
Suakoko	865/3,429 (25)	3/33 (9)	1956	96
Mauritania			1953	49
Estimate	40		1951	63
Estimate	31		1951	57
Atar and vicinity	50/125 (40)		1961	131
M'Bout and vicinity	231/505 (45.7)		1966	131
Senegal				
Estimate	(15)			63
Estimate	(8)			57
Thies & Kaolack	(7.5—75)			86
Basse Casamance	(15—85)			86a
Kaolack	229/997 (23)	39/997 (3.9)		86
Podor	(15—20)			86
Bakel	(22)			86
Kolda		(5—10)	1949	63
Mali				
Overall estimate	(35)	(12.0)		63
Koulikoro	(85)		1942	63
Bougouni	(2)		1942	63
Guinea				
Overall estimate	(28)	(6.4)		36
Faranah		839/1,235 (68)	1939	63
Macenta	(32)		1939	63
Kindia	(30)			63
Ivory Coast				
Overall estimate	(37)	(1.6)		56
Adzope and Agboville	3,750/150,000 (25)	4,500/150,000 (3)		131
Upper Volta				
Overall estimate	(46)	(8.0)		56
Overall estimate	(6-85)			63
Karankasso	268/672(40)			131
Silaleba	49/277 (18)			131
Bam	174/470 (37)			131
Togo				
Overall estimate	(4)	(1.0)		56
Overall estimate	(5)			63

Table 2-1 (Continued)
WORLD-WIDE HUMAN INFECTION RATES WITH *S. MANSONI, S. HAEMATOBIUM,* and *S. JAPONICUM*

Country	Infection rate (%)		Year of survey	Ref.
	S. haematobium	*S. mansoni*		
Kande	(31)			89
Tchitchao	(44.7)			89
Aveve	(30.7)			89
Dahomey				
Overall estimate	(27)	(0.2)		56
Savalou	(60)			63
Parakou	(8)			63
Porto Novo (Bekon)	(8)			63
Niger				
Overall estimate	(22)	(0.8)		56
Tillabery Cercle	6,180/13/857 (44.6)		1964	131
Magaria Cercle	211/1,008 (21.1)			131
Tessahoua	293/1,174 (25)			131
Ghana				
Overall estimate	150,000—200,000/1,000,000 (15—20)			106
Overall estimate		(0.5)		131
Nigeria				
North (Sokoto)	(10—50)	(0.2—1.0)	1947	46
(Bornu)	(30.5)	(3.6)	1953	46
West (Epe)	(63—82)	(2.0)	1953	46
(Ibadan)	(43.0)		1953	46
(Abeokuta)	(12.0)		1953	46
Ibadan	71/78 (91)			66
Epe	47/94 (50)			67
Cameroon				
Overall estimate (North)	29,250/195,000 (15)			63
Overall estimate (North)	29,250—39,000/195,000 (15—20)			58
Goulfei	2/57 (4)			63
Maltam	3/40 (8)			63
Kousseri	8/59 (13)			63
Overall estimate (North)		(1.6—10.1)	1955	36

Yaounde		(54—74)		55
Congo (Brazzaville)				
Kayes	(50)			73
Subprefecture	62/6,739 (0.9)			73
of Pointe-Noire				
Zaire				
Overall estimates	(30—69)	(3—85)	1950/1	69
Southeast (Katanga)		(0.6)	1950/1	69
West		(2—55)	1950/1	69
Northeast (Uele)		(19—86)	1950/1	69
(Kibali Ituri)		(70—100)	1950/1	69
(Faradje)		(60—94)	1950/1	69
(Lake Albert)				
Lufira lake area	362/3,019 (12.1)	190/3,019 (6.3)		116
Angola				
Salazar	(46)			81
Nova Lisboa	(11.7)			81
Cuchi	(54.1)			81
East Frontier		22/2,016 (1.1)		82
Chad				
Overall estimate	(16—87)			63
Overall estimate	(43)			56
Faya Largeau oasis	166/214.* (77.5)		1963	113
Fort Archambault		(5—15)	1965	73
Central African Republic				
Overall estimates	(22)	(40)		56
Overall estimates	(4—64)			63
Birao	(57)		1962	131
N'Dele	(86)		1960	
Paoula	(25)		1963	
M'Baiki	(80)		1962	
39 localities	1,057/18,230 (5.8)	19,912/174,674 (11.4)		120
Sudan				
North and West	(6.9)	(1.8)**	1952—53	115
Gezira	(5.5)	(7.2)	1951—52	115
Upper Nile	(2.1)**	(2.5)**	1952—53	115
Bahr-el-Ghazal	(0.25)**	(4.9)**	1949	115
Equatoria		(44.3)**	1949	115
Fung	(0.4)	(6.7)	1954—55	115

Table 2-1 (Continued)

WORLD-WIDE HUMAN INFECTION RATES WITH *S. MANSONI, S. HAEMATOBIUM,* and *S. JAPONICUM*

Country	Infection rate (%)		Year of survey	Ref.
	S. haematobium	*S. mansoni*		
Shambat (Khartoum Province)	(23.2)*,+		1959	92
Ethiopia				
Harar Province		244/1,845 (13.1)		84
		108/152 (71)*,+		
Tigre Province (Adwa)		282/459 (61.4)	1951	48
Lake Tana (Bahr Dar)		(5.5)		33
Gewani (Middle Awash)	91/189 (48.1)			119
Assayta (Lower Awash)	5/144 (3.5)	6/116 (5.2)	1965	87
Somalia				
South (2 valleys)	(41)		1951	33
Along River Juba	(15—75)*		1960	43
South	320/1,254 (25.5)*,+			102
Kenya				
Nairobi area				
Europeans		222/33,166 (0.67)	1938—44	117
Africans		965/40,503 (2.4)	1938—44	117
Asians		65/5,309 (1.2)	1938—44	117
Nairobi	(12)**	(3.2)**	1930—34	46
Mombasa	(4)**	(2.1)**	1930—34	46
Uganda				
Lango District	(50)*		1951	46
Busoga Province	(8)*,+		1953	46
Adyeda Subdistrict	(22)		1951	46
West Nile District				
Packwach		(54)*,+	1950	46
Joncum County		(100)	1954	46
West Nile District		927/7,064 (13.1)		103
Estimate prevalence rate		(31)		103
Acholi District				
Odek	(17)*,+		1960	118
Dino	(16.5)		1960	118

Location			Year	
Tanzania				
Sukumaland (Ngudu)	(33)		1951	46
Usambara foothills	(38.8)	(0.8)	1957	88
Mwanza		(30.8)+	1954	46
Handeni District	(17.4)		1959	128
Korogive District	(51.1)		1959	128
Mwanza		53/160 (33.1)+	1965	112
Mwanza		73/160 (45.6)+	1966	112
Pemba & Zanzibar Tanga area	862/2,270 (38)+			42
Zambia				
Chambezi-Luapula	384/2,617 (14.7)	178/2,575 (6.99)	1946	46
Near South of Lake Tanganyika	(25—60)			46
Lake Kariba (Siavonga)	(4—56)			78
Mankoya District		251/436 (57.6)**		76
Rhodesia				
Mashonaland	2,706/5,086 (53.2)+		1941	46
Chipoli Irrigation Estate	281/335 (84)	223/318 (70)		131
Mazoe Citrus Estates	128/251 (51)	98/251 (39)		131
Chirunda Sugar Estates	115/383 (30)	57/406 (14)		131
Iron Duke Mine	170/328 (52)	111/199 (56)		131
Malawi				
Kota Kota	263/322 (81.7)		1930—31	46
	5,189/9,861 (53)**		1935—44	114
LiKoma Island	792/3,600 (22)	163/204 (80)	1931	
(Karonga)				
Mozambique				
Overall estimate	(66.2)	(9.3)	1956	101
Niassi	(70.5)	(5.8)	1956	98
Mocambique	(80.9)	(6.4)	1956	100
Cabo Delgado	(67.8)	Nil	1956	99
Zambezia	(81.9)	(9.7)	1956	97
Manico & Sofala	(69.9)	(9.3)	1956	40
Tete	(45.7)	(17.9)	1956	40
Sul do Save	(61.5)	(11.9)	1953	39
Lourenco Marques	(50—55)	(11.85)	1956	97
Inhambane	(60—70)	(11.85)		38
South Africa				
Transvaal (East)	(74—79)	(43—57)	1955	111

Table 2-1 (continued)
WORLD-WIDE HUMAN INFECTION RATES WITH *S. MANSONI, S. HAEMATOBIUM AND S. JAPONICUM*

Country	Infection rate (%)		Year of survey	Ref.
	S. haematobium[a]	*S. mansoni*[a]		
Transvaal (North)	(43—76)	(1—77)	1955	111
Transvaal (West)	(8—88)		1955	111
Malagasy				
Majunga Province	(50)		1968	47
Maevatanana		(15)	1968	47
Tulear Province	(7—60)	(4—59)	1968	47
Diego-Suarez Province	20	0	1968	47
Tananarive Province			1968	47
Antsirabe	0	(0.1)	1968	47
Tamatave Province				
Vatomandry	0	(10—41)	1968	47
Manhanoro	0	(13—48)	1968	47
Marolambo	0	(2—25)	1968	47
Mauritius	673/7,318 (9.2),+		1951—53	53
Morocco				
Gharb (Karia ben Auda)	(37)		1939	65
(Oulad Riahi)	(35)		1939	65
Tafilalet (Gheris, Goul, Goulmina)	(9)		1943	65
Algeria				
Djanet	(27)		1935	44
Djanet	126/148 (95)		1935	131
Oran (Saint-Aime)	(44)		1935	31
Tunisia				
Gafsa	(58)		1932	70
Gafsa & Sidi Mansour	52/75 (69),+			125
	8/52 (16),+			125
Nefzaoua	572/919 (62),+			51
El-Hamma	9/88 (11),+			52
Kabili	(21),+			34

			Year	Ref.
Libya				
Fezzan	(25)		1951	127
Ghat	(32)		1951	127
Taurorga		?		45
Egypt				
Overall estimate	(40)	(20)	1937	121
Overall (perennial irrigation)	(60)		1932—36	121
Upper Egypt (flood irrigation)	(5)		1932—36	121
Nubia (5 areas)	(31)		1954—55	Egypt, MPH Report
Northeast Delta		(60)	1932—36	121
Southwest Delta		(6)	1932—36	121
14 Provinces (rural)	45,767/124,253 (38)	6,430/70,978 (9)	1955 (MPH)	131
21 Provinces, hospital returns urban and rural	(57)	(4.3)	1965	131
Southern Yemen				
Dirgag,+	up to 80	present	1963	131
Yemen				
Hodeida		12/182 (7)	1952	85
Taiz	21/218 (10)	122/218 (56)	1952	85
Ma'bar		1/26 (4)	1952	85
San'a		12/70 (18)	1952	85
Taiz	1/30 (3.3),+	14/20 (70),+	1952	33
San'a,+	3/19 (15.7)	0/9 (0)	1952	33
Saudi Arabia				
Hasa, Kharj Nejd, Aridh Nejd, Hejaz	5/204 (2.4),+	33/78 (42.3),+	1950—51	30
Israel				
Tirat-Zvi	97/451 (21.5)		1955	54
Lebanon				
Sarafand	86/591 (14.6)		1961	35
Syria				
Qubur El-Bid district	337/545 (61.8)		1951	30
Hazakeh governorate	(1.6)		1967	29
Rakka governorate	(13.5 to 25.1)		1967	29
Deir-ez-Zor governorate	(18 to 24)		1967	29

Table 2-1 (continued)

WORLD-WIDE HUMAN INFECTION RATES WITH *S. MANSONI, S. HAEMATOBIUM AND S. JAPONICUM*

Country	Infection rate (%)		Year of survey	Ref.
	S. haematobium[a]	*S. mansoni*[a]		
Turkey				
Near Syrian border	(60.9)		1956	72
Iraq				
North	(0—18),***		1949—50	30
Central	(0—84),***		1949—50	
South	(1—90),***		1949—50	
Iran				
Khuzistan overall	100,000 individuals (10)			131
				64
India				
Ratnagiri District	(20.8)		1952	62
(Gimvi village),*	(45.4)		1955	122
	53/94 (57)		1961	104
Japan*		*S. japonicum*		
Honshu (Yamanashi)		(32.0)	1950	125
(Chiba)		(3.8)		125
(Saitama)Tone R.		(2.7)		125
(Ibaraki)		(5.2)		125
(Shizuoka)		(1.8)		125
(Hiroshima)		(21.1)		125
		(47.1)		125
Kyushu (Saga-Fukuoka)				
Fujikawa-Cho		57/464 (13.2)		80
Kofu River basin (Yamanashi)		36/13,500 (0.27)	1970—71	132
Philippines				
Overall		(10—50)	1950	109
Leyte (Palo)		(20.4),*	1955	110
Mindoro (Naujan)		(40.0)	1945	79

Celebes (Sulawesi)	1,000 cases		130
Lake Lindu area	67/126 (53)	1970	74
China (Mainland)	32,777,630 cases	1950	130
Thailand (Chawang District)	(1.6—23.8)	1960	75
Laos (Khong Island) + +	47/547 (8.6)		131
	(20—35)·*	1968—1969	131
Cambodia + +			
Kratie area	(32.8)	1969	131

a + Adults only; · Children only; * Positive by skin test; ** Patients in hospital; *** Due to control efforts prevalences are now either nonexisting or are much lower than in table; + + *S. mekongi.*

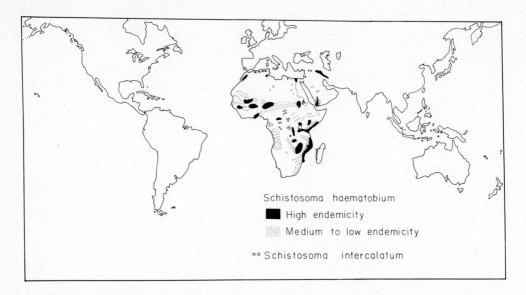

FIGURE 2-1.   Geographical distribution of *Schistosoma haematobium* and *S. intercalatum.* The very small focus of urinary schistosomiasis in India, near Bombay, is not included because no recent information has become available about it, and there is the possibility that the infection has been eradicated.

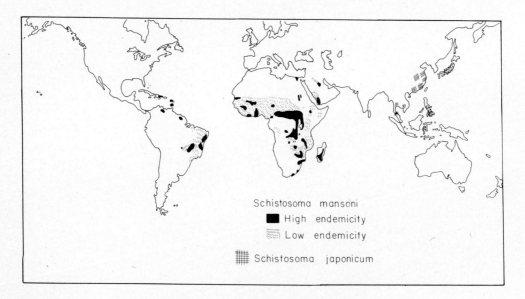

FIGURE 2-2.   Geographical distribution of schistosomiasis mansoni and schistosomiasis japonica. Note a small recently discovered focus in Malaysia, close to the borders with Thailand.

S. mansoni infections are generally characterized by being patchy, even in highly endemic areas. Intestinal schistosomiasis due to *S. mansoni* is absent in northern Africa, except for one focus in Libya and the Nile delta, but absent in the Nile valley of upper (southern) Egypt and the northern Sudan (Figure 2-2). The northernmost focus in the Sudan is in the Zeidab area, just south of the confluence of the Atbara River with the main Nile. It is present in the highlands of Eritrea and Ethiopia and occurs in the eastern part of Africa together with *S. haematobium* although to a lesser degree (Figure 2-3). There is a belt of varying degrees of endemicity extending from the south-

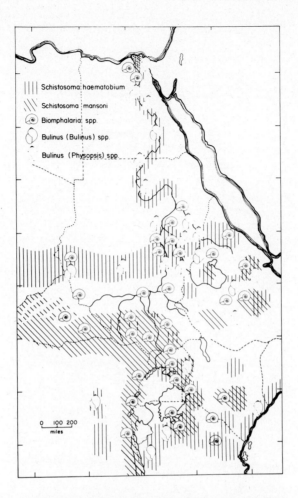

FIGURE 2-3. Showing distribution of urinary schistosomiasis (due to *Schistosoma haematobium*) and intestinal
schistosomiasis (due to *S. mansoni*) in the Nile basin and
adjacent territories in Africa. Also shown are the snail intermediate hosts for each of the two schistosome species.

ern Sudan through central Africa west to the Atlantic, running south of, and partly
overlapping, the parallel belt of *S. haematobium*. In southwestern Asia, *S. mansoni*
foci occur in the lowlands of Saudi Arabia and in the mountainous regions of southwestern Arabia from South Yemen to Asia, including all of Yemen.

There is a general agreement that *S. mansoni* was carried with the slaves to the
western hemisphere during the sixteenth, seventeenth, and eighteenth centuries. Slaves
were brought from various parts of Africa, which may account for the existence of
several strains of the parasite in the New World. The disease became established wher-
ever a suitable susceptible snail host is found, and thus at present intestinal schistosomiasis exists in Brazil, Surinam, and Venezuela in South America (Figure 2-4). In the
Caribbean it is endemic in Puerto Rico, Vieques, the Dominican Republic, Guadeloupe, Martinique, and St. Lucia (Figure 2-5). It was naturally abated in St. Kitts, and
probably also in Antigua. There is no doubt that the slaves also brought with them *S.
haematobium* from Africa, but, unlike *S. mansoni*, this parasite did not become established in the western hemisphere because of the absence of any species of the genus
Bulinus, or any other susceptible snail.

FIGURE 2-4.   Geographical distribution of schistosomiasis due to *Schistosoma man-soni* in South America.

It should be noted that in Africa *S. intercalatum*, a member of the *S. haematobium* group, is the only species of schistosome in the high rainfall and tropical forest zones of the Congo basin (Lualaba and Congo rivers) and in the region of the lower Ogowe, a river formerly connected to the Congo. Intestinal schistosomiasis due to *S. intercalatum* also exists in Gabon, Cameroon, and the Central African Republic (Figure 2-1).

Oriental schistosomiasis due to *S. japonicum* (Figure 2-2) is endemic on the mainland of China and on the large islands of a chain extending from Honshu (Japan) to the Celebes (Sulawesi), a distance of about 4000 mi. In China, the disease is very important and is present from 22°5′ to 30°7′ north latitude and 100° East longitude to

199

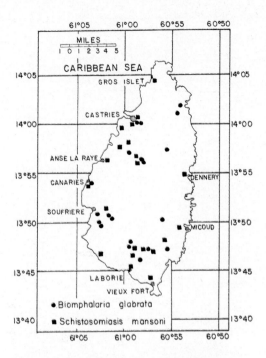

FIGURE 2-5. Map of the West Indies island of
Saint Lucia showing foci of schistosomiasis man-
soni, and its snail host, *Biomphalaria glabrata*.
(After Malek, 1963.)

the coast, primarily in the Yangtze basin (heavily infected) and along the coast from
Kiangsu in the north to Kwangtung in the south. There also are foci in the Mekong
River valley in Yunnan Province.

In Japan the disease is no longer of significance on account of a natural control
program, the exclusive use of chemical fertilizers, the improvement of life standards,
and urbanization. A survey in 1970 to 1971 revealed newly infected cases only in the
Kofu and Tone river basins, and even in the latter areas prevalences are very low.

The Philippines are next to China in importance. The disease is spread over 12 prov-
inces; most of the cases are on the islands of Leyte and Samar.

Three villages on the west shore of Lake Lindoe, in an isolated region of Central
Celebes, represent a small southern focus of the disease.

In southeast Asia, the foci so far known in which *S. japonicum* occurs are restricted
to Laos, Cambodia, Thailand, and Malaysia (Figure 2-6). The number of the population exposed to the disease in these foci is not known. Moreover, *S. japonicum* in the
latter foci has a smaller egg than the parasite in the old established foci elsewhere in
the Orient; and where the snail host has been discovered in Laos, it is not a species of
the amphibious *Oncomelania*, but an aquatic, *Lithoglyphopsis aperta* (also an hydrobiid snail like *Oncomelania*). The schistosome from Laos and Cambodia has recently
been named *Schistosoma mekongi*, a new species.

Table 2-2 shows estimates of the population exposed to the disease and the population infected (124,905,800) in the world.

IV. THE LIFE CYCLE

Figure 2-7 shows the life cycle of a schistosome. The adult schistosome worms of
man reside in the tributaries of the superior mesenteric venule (*S. japonicum*), the

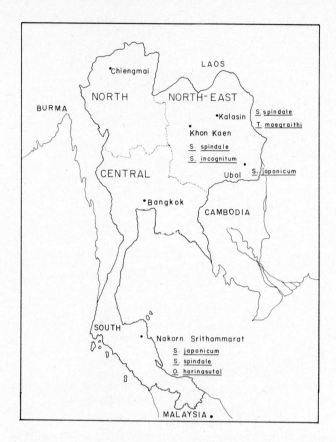

FIGURE 2-6.   Map of Thailand showing foci of the human *Schisto-soma japonicum*. Foci of the mammalian species, *Schistosoma spin-dale*, *S. incognitum*, and *Orientobilharzia harinasuti* and the bird schistosome *Trichobilharzia maegraithi* are also indicated. (Redrawn after Sornmani, 1973. The locality in Malaysia has been added.)

tributaries of the inferior mesenteric venule (*S. mansoni*), or in the systemic veins of the pelvis, especially those of the urinary bladder (*S. haematobium*). The male trans-ports the female against the current of the portal blood stream by means of its suckers into the smaller radicles of the mesenteric venules or the vesico-prostatic plexus. The female may partially or totally leave the gynecophoric canal of the male in order to migrate further into the venule. The female worm deposits the eggs in a beadlike row, often deep into the wall of the intestine or the urinary bladder, after which the female retreats into a larger venule, allowing the smaller vessels to contract about the eggs. Some of the eggs remain in the tissues while others are washed by the bloodstream into the portal circulation and are later filtered into various organs of the body, espe-cially the liver. Eggs deposited in the wall of the intestine or the bladder spend a period of maturation, the length of which depends on the species. Those of *S. mansoni* require 6 to 7 days for the development of the miracidium inside them, while those of *S. japonicum* require about 12 days. Apparently because of secretions from the miracid-ium which ooze out of the porous shell and helped by the peristaltic movements of the intestines, the eggs readily break through the submucosa and mucosa (or the wall of the urinary bladder) and are discharged into the lumen of the bowel (or the lumen of the bladder) together with extravasated blood. The majority of the eggs which are passed with the feces or urine (Figure 2-7) are usually mature when evacuated and

Table 2-2
WORLD DISTRIBUTION OF SCHISTOSOMIASIS; TOTAL REGIONAL POPULATION, POPULATION EXPOSED, AND POPULATION INFECTED[131]

Region	Total population	Population exposed	Population infected
Africa	308,021,000	226,102,740	91,200,310
Mascarene Islands (Mauritius)	741,000	370,500	66,690
Southwest Asia	87,003,000	10,745,050	2,271,020
Southeast Asia	868,531,000	337,051,500	25,223,650
The Americas	98,339,000	18,199,750	6,144,130
Totals	1,362,635,000	592,469,540	124,905,800

require only sufficient dilution of the feces or urine with water to hatch, allowing the miracidium to escape through a rent in the shell, shed its embryonic membrane, and swim in the water. Feces are either deposited directly in fresh-water bodies or are washed down into the water by rains.

Like the miracidium of other digenetic trematodes, the schistosome miracidium swims actively in search of a suitable snail to continue its development and the life cycle. It penetrates several species of snails and other organisms in the water, but only in a suitable and susceptible snail does the miracidium become transformed into a mother sporocyst which, in turn, produces a large number of daughter sporocysts. These migrate inside the snail to occupy the region of the digestive gland and gonad, where they, in turn, produce thousands of fork-tailed cercariae, the infective larvae which escape from the snail. Under optimum conditions (at about 25°C) in *S. mansoni* about 4 weeks elapse from the penetration of the miracidium to the liberation of the cercariae. This period is about 5 weeks in the case of *S. haematobium*, but about 6 to 7 weeks in the case of *S. japonicum*.

When the cercariae emerge from the snail they swim about vigorously tail first and after reaching the surface of the water, gradually sink in the water. This behavior is repeated several times, although some cercariae, especially those of *S. japonicum*, may remain temporarily near the surface. The cercariae remain vigorous for about 12 hr but their life span is about 2 days. On contact with the skin of mammals who frequent cercariae-infested waters, especially on account of their occupation (humans) or for other reasons, the cercariae penetrate down to the cutaneous capillary beds as a result of the action of their penetration glands which secrete proteolytic enzymes. Drinking cercariae-infested water can also cause infection. Penetration can occur within 5 min, but more cercariae penetrate within 15 min and the majority in ½ hr. By the end of 24 hrs the cercariae, now called schistosomula, have entered the venous circulation or the lymphatics, and are carried through the right heart to the lungs. It should be noted that a good proportion of the schistosomula cannot reach the venous or lymphatic circulation and are killed and destroyed in the skin. In the pulmonary circulation they may be delayed for several days while squeezing through the lumen of the capillaries into the veins. Once in this latter location, they can journey back through the left side of the heart into the systemic circulation. Most authors believe that the schistosomula reach the intrahepatic portal circulation by first reaching the systemic circulation, then the mesenteric artery and capillaries. This is the likely route of migration. A few authors are of the opinion that the schistosomula in the lungs break through the tissues to the pleural cavity, penetrate the diaphragm to the peritoneal cavity, and then pass through Glieson's capsule to reach the intrahepatic portal circulation. Another possible route from the lungs to the liver is via the vena cava, moving against the blood flow.

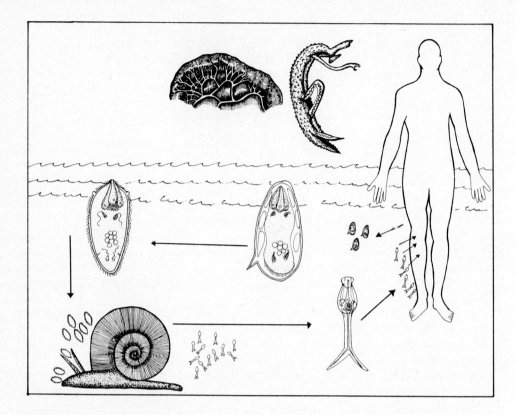

FIGURE 2-7.   Life cycle of *Schistosoma mansoni.* The adult mature male and female worms are shown
enlarged, next to a portion of the intestine with mesenteric venules where the worms reside in the mammalian host's body. The snail illustrated is *Biomphalaria* spp. which serve as the intermediate host in
Africa, the Middle East, and in some Caribbean islands and in South America (Brazil, Surinam, and
Venezuela).

Whatever route of migration the schistosomula follow from the lung to the liver
(Figure 2-8) once they reach the portal circulation in the liver, they feed on the nutri-
tious portal blood and grow rapidly to adult worms, then these move into the larger
portal venules. Embracing the female in the gynecophoric canal of the male is a requirement for the maturation of the female. Unmated females remain immature and
of a size half that of the mated female. The mature couples move down the hepatic
portal vein against the blood flow to reach the mesenteric venules (*S. mansoni* and *S.
japonicum*) or the pelvic venules via the hemorrhoidal plexus, and oviposition begins.
The period from the penetration of the cercariae to the passage of the eggs in the
excreta (prepatency period) varies according to the species. The period in humans is
50 to 55 days in *S. mansoni*, 40 to 50 in *S. japonicum*, and about 80 days in the case
of *S. haematobium.*

V. THE PARASITES

A. The Adult Worm

Blood flukes in general parasitize fishes, turtles, birds, and mammals. Those which
parasitize fishes and turtles are hermaphroditic, while those that infect birds and mammals are the true schistosomes, where the sexes are separate (dioecious). *Schistosoma
haematobium, S. mansoni, S. japonicum,* and *S. intercalatum* infect humans, while

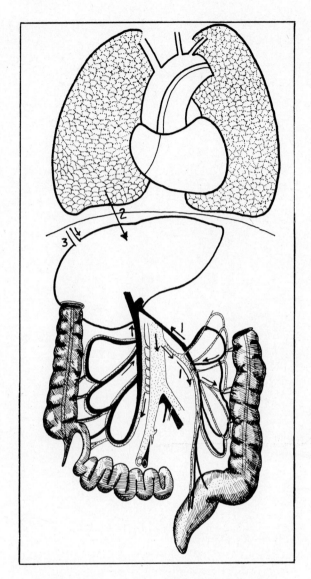

FIGURE 2-8. Diagram showing the three routes suggested for the migration of schistosomula from the lungs to the liver. (1) Migration mainly within the blood stream-aorta-mesenteric arterioles-mesenteric venules-hepatic portal-liver, (2) Migration through tissues between lungs, diaphragm, and liver, (3) Migration against the blood stream-right heart-posterior vena cava-liver.

other schistosomes are potential or incidental human parasites. The generic name *Schistosoma* is derived from "schistos", which means "split", and "soma", that is, body, since the male worm possesses a ventral and longitudinal cleft, the gynecophoric canal, in which the female is embraced during copulation and movement (Figure 2-9).

Schistosoma mansoni, S. haematobium, and *S. japonicum* agree in their basic life cycles and some other aspects, but they differ in the external and internal morphology of the adults, the shape of their eggs and larvae (miracidia) hatched from the eggs, in the specific identity and groups of snails which serve as their intermediate hosts, and in the range of mammalian hosts which they can parasitize.

FIGURE 2-9.   A large number of *Heterobilharzia americana* obtained from naturally infected raccoons and dogs in Louisiana.

Male human schistosomes have a short, preacetabular portion of the body and a wide, flattened postacetabular portion with the gynecophoric canal on the ventral surface. The female worm is filiform, round in transverse section, and the body is thinner but longer than the male. The three major human species vary in their size, tegumentary tuberculation, number of testes, shape and position of certain organs, and the number of in utero eggs. The *S. japonicum* male is the longest (12 to 20 mm) and 0.50 to 0.55 mm in greatest diameter; it possesses seven testes, the cuticle is nontuberculated, and the posterior union of intestinal ceca is caudal to midbody. The *S. mansoni* male is about 9 to 12 mm in length and 1.0 to 1.2 mm in width; it has six to nine testes, its tegument is grossly tuberculated (Figure 2-10), and the posterior union of intestinal ceca is anterior to midbody, while the *S. haematobium* male (Figure 2-11) is 12 to 14 mm in length and 0.8 to 1.0 mm in width, its tegument is finely tuberculated, it possesses four to five testes, and the posterior union of intestinal ceca is at midbody.

The *S. japonicum* female is 18 to 25 mm in length and about 0.3 mm in width, the position of its ovary is about midbody, and the in utero eggs are 50 or more; the *S. mansoni* female is 12 to 16 mm in length and 0.16 mm in width, its ovary is anterior to midbody, and there is usually only one egg at a time in the uterus; the *S. haematobium* female is 16 to 20 mm in length and 0.25 mm in width, its ovary is caudal to midbody, and there are 20 to 100 eggs at a time in the uterus. Thus the position of the

FIGURE 2-10. Scanning electron micrograph of a male *Schistosoma mansoni*, showing the tuberculated surface, typical of the male of this species, and the gyneco-phoral canal. (Prepared and presented by Mrs. Danuta M. Krotoski.)

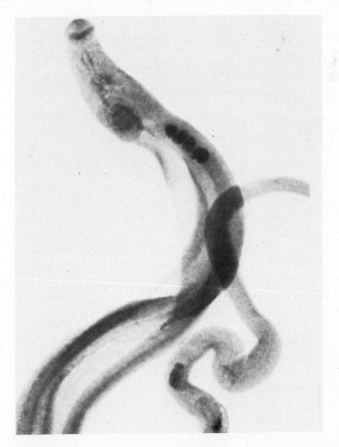

FIGURE 2-11. Adult *Schistosoma haematobium* in copula. Testes of the male, and the ovary of the female are clearly shown.

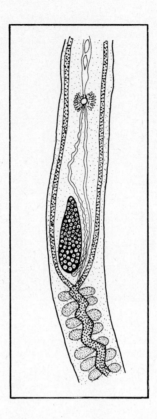

FIGURE 2-12.   Reproductive organs, and accessory glands of a female   schistosome   (*Schistosoma
haematobium*). Vitelline duct runs
more or less parallel to oviduct and
both open in the ootype. The latter
is surrounded by the Mehli's
gland. The uterus arises from the
ootype.

ovary determines the length of the uterus, the latter being the shortest in *S. mansoni*.
Figure 2-12 shows the reproductive organs and accessory structures of a female schistosome.

The ultrastructure of *Schistosoma* spp. was discussed in Chapter 1.

Scanning electron microscope studies of adult schistosomes reveal certain structures
on the body surface, such as spines in the oral and ventral suckers of both sexes, and
marked differences between various parts of the same worm, male or female. Figure
2-13 shows some of these features in the case of the male and female of *Schistosoma
intercalatum*.

The life span of the schistosomes in the human body has always been an interesting
question. Some immigrants who have resided away from endemic areas of the three
species of human schistosomes continued to have eggs in their excreta up to 20 or 30
years. This has given the impression that the worms live that long in the human body,
and it is as misleading as saying that human beings live for 120 years, which should
mean that these humans are the exceptions. The most convincing evidence for a relatively short life span of the schistosome worms, at least for female schistosomes, consists of the observed rates at which positive people become negative for schistosomia-

A

FIGURE 2-13. Scanning electron micrographs of *Schistosoma intercalatum* (Cameroon). (A) Anterior end of male and female schistosome in copula. (Magnification × 120.) (B) Oral sucker of male. (Magnification × 360.) (C) Oral sucker of female. (Magnification × 1340.) (D) Portion of (B) at higher magnification showing nature of spines on inner surface of oral sucker. (Magnification × 3200.) (E) Portion of (C) at higher magnification showing nature of spines. (Magnification × 4200.) (F) Surface spines on gynecophoral fold of male. (Magnification × 4400.) (Courtesy of Drs. Kuntz, Tulloch, Davidson and Huang.)

sis. Hairston[149] obtained information which proved that, if becoming negative reflects the death of female worms, the death rates of the worms must be higher than has been supposed in the past. Regardless of the maximum possible, the mean life of female schistosomes may be as short as 2 years, and probably is not greater than 5 years.

The mating behavior and development of schistosomes in the host's body has been another interesting aspect of schistosome infections. It has been shown in experimental animals that in unisexual infections and in bisexual infections where the female remains unmated, the female worms are stunted in growth and sometimes in sexual maturity. This has been demonstrated for *S. mansoni*, for *S. japonicum*, and for *S. haematobium*. However, Sahba and Malek[166] showed *S. haematobium* of both sexes reached maturity in unisexual infections in mice 5 months after exposure, and females and males grew to more than half the length and slightly less than the width of corresponding worms in bisexual infections. Although unisexual females did not reach the size attained by paired ones, they did produce eggs which were probably infertile, and a few appeared to contain a partly formed miracidium, but they failed to hatch in cultures. A similar incomplete parthenogenesis has been demonstrated for *Bilharziella polonica, Heterobilharzia americana*, and *Schistosoma mattheei*. Uniparental miracidia of *Schistosomatium douthitti* were obtained, an example of complete parthenogenesis among schistosome worms.

In order to learn more about mating behavior and development of the schistosomes, Armstrong[134] infected some of his mice with a mixed infection of *S. mansoni, H. americana*, and *S. douthitti*. These mixed infections with one or both sexes of the three schistosomes revealed almost random mating during the early stages of the mating

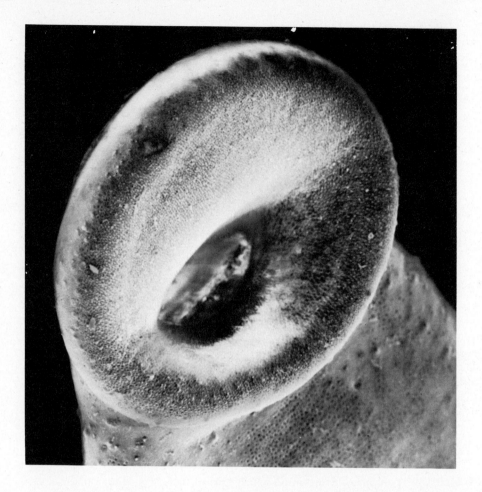

FIGURE 2-13B

process. These observations imply that, at least in the mouse host, mating occurs by
trial and error after the two sexes rendezvous in the liver. Homosexual pairing of males
occurred and further suggested that the pairings resulted from thigmotaxis rather than
chemotaxis. *S. mansoni* females, cross-mated with *S. douthitti* or *H. americana* males,
reached sexual maturity and produced a few viable eggs, but did not grow larger than
unmated females. It appears, therefore, that body growth depends more than does
sexual maturation on conspecific mating. The findings suggest that a pheromone is
produced by the male which is essential for the initiation and maintenance of the female partner's maturation, and which may inhibit the development of homosexually
embraced males.

Although the schistosomes are bisexual flukes, hermaphroditism has been encountered among both males and females of more than one species. The subject has been
reviewed by Sahba and Malek.[165]

The metabolism of the schistosomes has been reviewed by Coles.[143] The sporocysts
and worms are stages of growth and multiplication. The ability of schistosomes to
adapt to different osmotic pressures and temperatures and to penetrate and establish
themselves in two distinct host animals makes them organisms of great interest to both
the physiologist and biochemist, as well as the clinician.

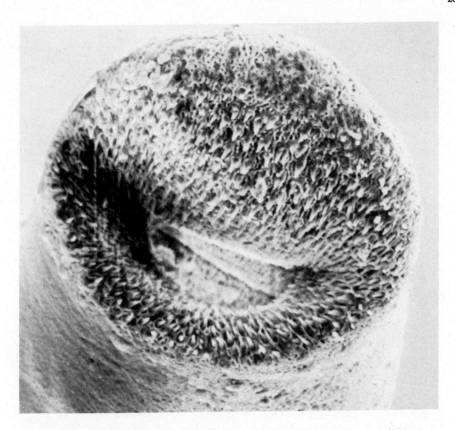

FIGURE 2-13C

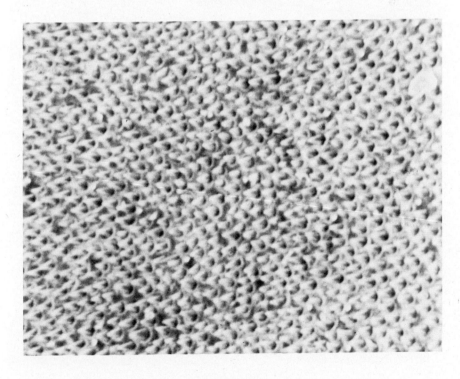

FIGURE 2-13D

FIGURE 2-13E

FIGURE 2-13F

There is an obvious difference between males and females in biochemistry and in drug response. Quantitative differences between esterase isoenzymes have been shown for mated and unmated female worms from the same mouse,[141] and there are doubtless many other metabolic differences.

Studies on the carbohydrate metabolism of adult worms demonstrated that, although oxygen is used, uptake of glucose and excretion of lactic acid is of major importance; worms using 15 to 26% of their body dry weight of glucose per hour. Glycogen is stored in large amounts in the male worm, being 13.6 to 29% of the dry weight compared with only 2.7 to 5% in the females. According to Bueding and Fisher[137] nearly all this glycogen is of low molecular weight. It is thought that the drug niridazole may kill the worms because of its effect on glycogen breakdown (Bueding and Fisher).[138] Studies of the phosphorylation of glucose, fructose, manose, and glucosamine[139] showed that there are four different hexokinases in adult schistosomes, and these were fractionated.

The lipid composition of cercariae and adult schistosomes has been reported by Smith and Brooks[170] and the metabolism of lipids investigated by Meyer et al.[159]. They found no evidence for *de novo* synthesis of fatty acids by worms. However, acetate is incorporated during chain elongation, but desaturation of oleate and stearate does not occur. The worms seem capable of synthesizing complex lipids *de novo* if supplied with long-chain fatty acids.

Certain media containing nutritive substances have been developed in which the in vitro cultivation of the schistosomula and the maintenance of the adult worms for a period of 1 to 2 months have become possible. It is evident from the habitat location of the worms, the blood vessels of the host, that they depend on the blood components for their nutrition. Schistosomula apparently begin to ingest red blood cells about two weeks after the cercariae have penetrated the host.

Although not much is known about the relationship of microorganisms to the adult worms in the human or other mammalian host, studies in experimental schistosomiasis indicate that bacteria can have a lethal effect on the adult worms. The studies also indicated that the bacteria have to be alive in order to kill the schistosomes. This observation decreases the significance of the findings, and thus bacteria should not be used as biological control agents.

B. The Eggs

Eggs of *S. mansoni* are laid singly, while those of *S. japonicum* and *S. haematobium* are laid in groups. The female pushes its way into the small venules or capillaries in the intestinal or bladder wall. As has been mentioned under "Life Cycle", these eggs are laid immature and develop in the tissues, after which they gain access to the intestinal or urinary bladder lumen, while other eggs are swept away with the blood stream which carries them to various organs and tissues of the human host.

Eggs of *S. haematobium* possess a terminal spine and measure 120 to 180 μ in length by 40 to 70 μ in width. Those of *S. mansoni* have a lateral spine and measure about 110 to 170 μ by 45 to 70 μ. Those of *S. japonicum* are the smallest, are ovoid, and measure 70 to 100 μ by 50 to 65 μ; the shell has a knob-like structure which is not always visible. Eggs of various schistosomes of humans, other mammals and birds are shown in Figures 2-14, 2-15, and 2-16. They are all drawn to the same scale to allow comparison of their size. Photomicrographs of some of these eggs are shown in Figure 2-17.

The period of maturation of the eggs in the tissues varies from one species to the other; it is 6 days in the case of *S. mansoni* and 12 days in the case of *S. japonicum*, whereas this period is not known for *S. haematobium*. The eggs which leave the body

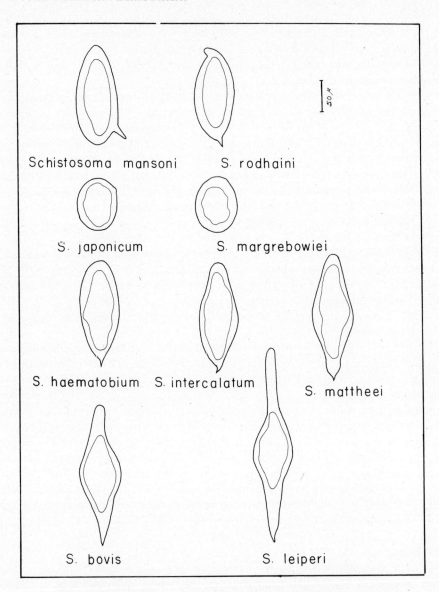

FIGURE 2-14.   Eggs of some schistosomes drawn to the same scale. The following measurements are in microns. *S. mansoni* Sambon, 1907, 110 to 170 × 45 to 68 (av. 145 × 55). *S. rodhaini* Brumpt, 1931, 120 to 170 × 48 to 70 (av. 148 × 58). *S. japonicum* Katsurada, 1914, 70 to 100 × 50 to 65 (av. 85 × 58). *S. margrebowiei* Le Roux, 1933, 70 to 97 × 50 to 68 (av. 87 × 60). *S. haematobium* (Bilharz, 1852) Weinland, 1858, 120 to 180 × 40 to 70 (av. 140 × 60). *S. intercalatum* Fisher, 1934, 140 to 240 × 50 to 85 (av. 175 × 65). *S. mattheei* Veglia and Le Roux, 1929, 145 to 250 × 40 to 70 (av. 200 × 64). *S. bovis* (Sonsino, 1876) Blanchard, 1895, 182 to 248 × 45 to 80 (av. 207 × 62). *S. leiperi* Le Roux, 1955, 210 to 305 × 38 to 65 (av. 260 × 55).

with excreta are mostly mature, each with a fully developed miracidium. Eggs in the tissues remain viable for about 3 weeks, after which they are destroyed by the host; they become black and sometimes gradually calcified. As confirmed by electron microscopic studies, the eggshell has a number of evenly spaced pores, which allow secretions and excretions of the miracidium to ooze out. Such materials are believed to help the egg in moving outside venules, and when the eggs settle in tissues these materials incite

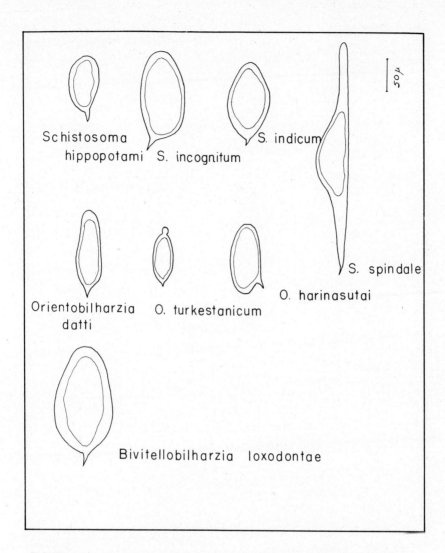

FIGURE 2-15.   Eggs of some schistosomes drawn to the same scale. All measurements in microns. *Schistosoma hippopotami* Thurston, 1963, 93 × 40. *S. incognitum* Chandler, 1926, 97 to 148 × 45 to 80 (av. 118 × 61). *S. indicum* Montgomery, 1906, 120 to 140 × 68 to 72. *S. spindale* Montgomery, 1906, 360 to 400 × 68 to 80. *Orientobilharzia datti* Dutt and Srivastava, 1952, 120 to 137 × 30 to 43. *Orientobilharzia turkestanicum* Skrjabin, 1913, 72 to 74 × 22 to 26. *Orientobilharzia harinasuti* Kruatrachu et al., 1965, 112 to 127 × 27 to 52. *Bivitellobilharzia loxodontae* Vogel and Minning, 1940, 145 to 203 × 56 to 83 (av. 168 to 71).

an inflammatory reaction around the egg. The "Circum-oval Precipitin Test", which is one of the serodiagnostic tests for schistosomiasis, is based on the secretions and excretions of the miracidium inside the egg. In addition to the pores, the surface of the schistosome egg is covered with microspines, about 3 to 5 μ in length, revealed by scanning electron micrography. It has been suggested that the microspines may help the egg to cling to the vascular endothelial lining of the blood vessels, and to penetrate into adjacent tissues.

On the bases of infections in experimental animals, estimates have been made of the daily egg production of the female human schistosomes. It has been found that the female *S. japonicum* produces from 1400 to 3500 eggs per day, that of *S. mansoni* from 250 to 350, that of *S. haematobium* 50 to 300, and that of *S. intercalatum* 150 to 400.

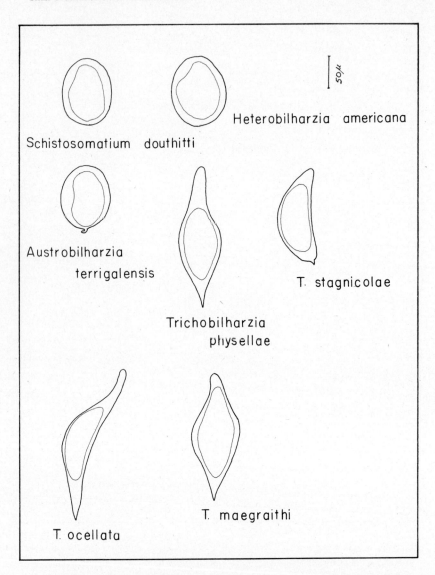

FIGURE 2-16.   Eggs of some schistosomes drawn to the same scale. All measurements in microns. *Schistosomatium douthitti* (Cort, 1914) Price, 1931, 94 to 122 ×
74 to 98 (av. 108 × 82). *Heterobilharzia americana* Price, 1929, 74 to 113 × 60 to 80
(av. 87 × 70). *Austrobilharzia terrigalensis* Johnston, 1917, 96 × 83. *Trichobilharzia
physellae* Talbot, 1936, 170 to 250 × 65 to 80. *Trichobilharzia stagnicolae* Talbot,
1936, 130 to 160 × 50 to 60. *Trichobilharzia ocellata* (La Valette, 1855) Brumpt,
1931, 160 to 220 × 28 to 60. *Trichobilharzia maegraithi* Kruatrachu et al., 1968, 190
to 200 × 75 to 85.

Unlike intestinal helminths like hookworms, the schistosomes produce eggs which
are passed out in the excreta, and some which are deposited in the tissues. Nevertheless,
it has been demonstrated that a correlation exists between the number of eggs in the
excreta and the number of worms and accordingly the intensity of infection.[158]

The metabolism of the egg and the miracidium inside it in host tissues has been
studied by some workers. Glucose is taken up, as well as some chemotherapeutic
agents. A soluble protein or proteins are released from the eggs, causing a host response. A collagenase-like enzyme leaks out of the eggs that might aid in their passage

through the host tissues (intestinal wall and urinary bladder). There is evidence for a functional citric acid cycle within the egg; this is to be expected since the infective miracidium is a free-living organism on emergence from the egg, but it is not known whether glycolysis or oxidative metabolism is of greater importance in energy formation within the egg.

Some data are available on the effect of environmental factors on schistosome eggs in the water. Eggs have been found to hatch within various temperature ranges; for *S. japonicum* these include 8 to 37°C, 25 to 30°C, and 3 to 33°C (maximum), and 13 to 28°C as optimum. By changing water temperature, the author observed a range of 17 to 35°C in which eggs of a Sudanese strain of *S. haematobium* hatched. A good proportion of the eggs of *S. mansoni* hatched at 23 to 28°C and continued to hatch for 2 or 3 days with further dilution of the liver, fecal, or intestinal cultures.[155] The process of hatching of the schistosome eggs was discussed in Chapter 1.

C. The Miracidium

Miracidia of the schistosomes are similar morphologically, but vary slightly in size. That of *S. mansoni* is the largest of the three human species, averaging about 160 by 60 μ. It is cylindrical with a cone-like anterior end, and the body surface shows a large number of cilia (Figure 2-7). The cilia are present on all the body surface of the miracidium except at the excretory pores and the lateral papillae. Transmission electron microscope studies show that there are also short simple cilia protruding from the posterior and middle portions of the folded surface of the apical papilla, but none are present at the apex (Figure 1-18).

When released out of the egg in fresh water, the miracidium swims actively, helped by its cilia; it shows some random movements but specifically aims to find and infect snail hosts.[155] The miracidium is chemosensitive and once it is in the vicinity of a snail host it moves with vigor and attempts to attach to and penetrate the snail. MacInnis[153] identified some chemicals to which the miracidium is attracted, among which are short-chain fatty acids, some amino acids, and a sialic acid. These chemicals were attractants to the miracidia when agar or starch gel pyramids were impregnated with them.

Chernin[140] demonstrated that *Biomphalaria glabrata*, the intermediate snail host of *S. mansoni*, and other unrelated snails naturally emit some substance(s) which stimulate *S. mansoni* miracidia, in the vicinity of the snail or in the absence of the snail. The miracidial stimulant, termed "miraxone" by Chernin, is thermostable, retains its activity during prolonged storage, has a mol weight of <500, and is thus probably a structurally simple chemical. According to several authors the miracidia of *S. mansoni* are negatively geotropic and positively phototropic. On the other hand, miracidia of *S. haematobium* seem to be positively geotropic as they tend to travel to the bottom of the habitat to locate and infect the snail hosts. Moreover the miracidia of *S. haematobium* seem to infect more snails in the dark (shade) than in the light.[168,174]

It has been determined by several investigators that miracidia of the schistosomes penetrate various snails but that their fate in the mollusc depends on the compatibility of certain susceptible snails. In a nonsusceptible snail they die as a result of a host-tissue reaction.[136,157,172]

In their free-living existence in the water, the miracidia are exposed to the effect of other organisms in this habitat; some are toxic, some are predators and some are enemies. Glaudel and Etges[147] demonstrated the toxic effect of certain turbellarians on the miracidia of *S. mansoni* and *Gigantobilharzia huronensis*, a bird schistosome. Khalil[150] showed that the oligochaete *Chaetogaster limnaei* captures and destroys the miracidia of *Fasciola gigantica*, an observation which was later confirmed in the case of *S. mansoni* miracidia by Michelsen.[160] It has also been demonstrated that certain species of fish ingest a large number of schistosome miracidia.

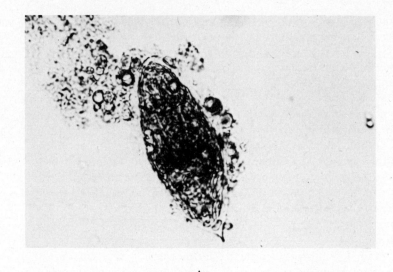

A

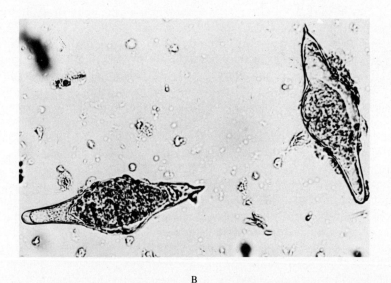

B

FIGURE 2-17.   Eggs of some schistosomes. (A) *Schistosoma haematobium*, from
human urine, (B) *S. bovis*, from Nile rat liver, (C) *S. leiperi* from antelope feces,
(D) *Heterobilharzia americana* from dog feces, (E) *Schistosoma japonicum*, from
rabbit feces, (F) *Orientobilharzia turkestanicum* from cattle feces, (G) *Schistosoma
mansoni*, from mouse liver, (H) *S. rodhaini*, from rat liver. (*S. leiperi* and *S. rodhaini* courtesy of the late Dr. P. L. LeRoux.)

D. The Mother Sporocyst

Mother sporocysts are usually found near the surface of the head-foot of the snail,
on the mantle, in the tentacles, and the pseudobranch of planorbid snails. Occasionally
they may be located on the surface of the lung, and near the nerve ganglia. A 7-day
old sporocyst is a simple elongate sac, measuring about 0.25 mm long and is filled
with germinal cells. As it becomes more developed the sac contains many germ balls
of various sizes (Figure 1-26). The germ balls become elongated as they form daughter
sporocysts.

217

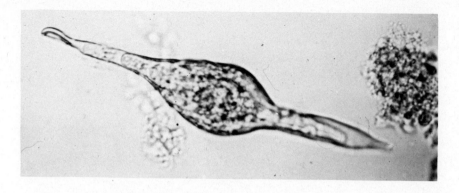

FIGURE 2-17C

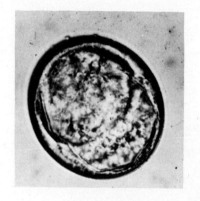

FIGURE 2-17D

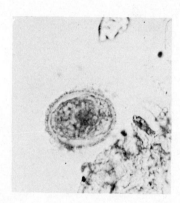

FIGURE 2-17E

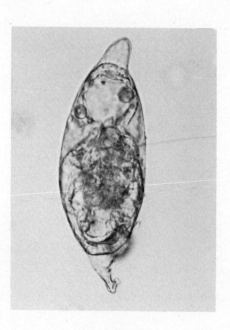

FIGURE 2-17F

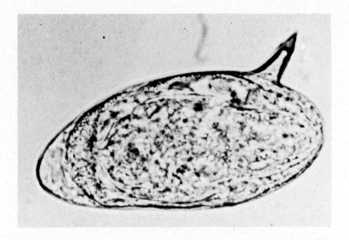

FIGURE 2-17G

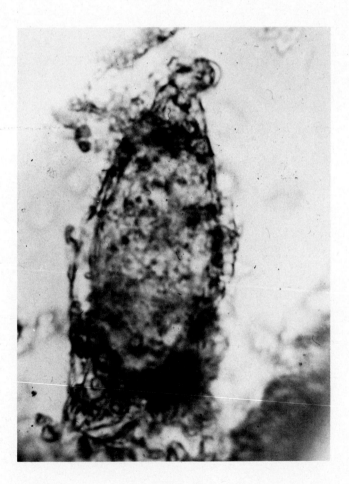

FIGURE 2-17H

E. The Daughter Sporocyst

A daughter sporocyst about to leave the mother sporocyst is in the form of an elongate thin-walled sac, containing numerous single germinal cells varying in number from 50 to 130. The surface of the daughter sporocyst is covered with spines which are quite obvious anteriorly but are smaller and very inconspicuous posteriorly. A daughter sporocyst still inside the mother measures about 0.15 to 0.24 mm long, and the width is about one sixth of its length.

Daughter sporocysts of *Schistosoma mansoni* and *S. haematobium* developing in *Biomphalaria* spp. and *Bulinus* spp., respectively, are usually found in the digestive gland-ovotestis area. In some infections, however, and particularly in heavy ones, daughter sporocysts are found in other locations such as in the wall of the lung, the rectal ridge, on the kidney, albumen gland, and other organs. Such sporocysts might have first reached the digestive gland area and then again moved anteriad, or on migration from the head-foot failed to reach their destination, that is, the digestive gland area, and thus settled in various organs.

F. The Cercaria

Cercariae of the three human schistosome species are similar in morphology, *S. haematobium* and *S. mansoni* tend to be distributed evenly in the water, while *S. japonicum* tends to settle at the surface. The cercariae are furcocercous, lack eyespots and a pharynx, and are about 1 mm long (Figure 2-18). The surface of the cercaria consists of a mucous film over a tegument with scattered minute spines and hairs believed to be sensory in function. Dominating the internal organs of the cercariae are the large unicellular glands whose processes serve as ducts. Two pairs are preacetabular and three or four pairs are postacetabular. These glands occupy the posterior two thirds of the body. Their contents are emptied through ducts which open at the anterior rim of the oral sucker. A sixth pair of unicellular glands, called escape glands, situated in the anterior portion of the body, can be seen only in unemerged cercariae. The preacetabular and postacetabular glands can be differentiated histochemically. The postacetabular glands are the smaller of the two types, and they contain finely granular, periodic acid-Schiff staining contents. The preacetabular glands contain alkaline, macrogranular, alizarin staining material, and they contain calcium. Information on ultrastructural details of the cercarial tegument, of the penetration glands, and of sensory structures, are in Chapter 1.[171]

In their free-living existence in the water the cercariae are exposed to the predatory action of some organisms, members of the biota. The predatory activity of the guppy (*Lebistes reticulatus*) on cercariae of *S. mansoni* was mentioned by Oliver-Gonzalez[161] and Rowan.[164] On the basis of laboratory and field experiments Pellegrino et al.[162] confirmed the role of the guppy as an active predator on *S. mansoni* cercariae, and Knight et al.[151] confirmed the same role by using cercariae labeled with radioselinium.

Rising temperatures stimulate the emergence of cercariae, as does a sudden change in temperature. Cercariae can, of course, withstand a temperature of 37°C, that of the human host, but schistosomiasis has also been contracted from waters which are slightly warmer.

Effect of other environmental factors on the schistosome cercariae was dealt with in Table 1-2, Chapter 1.

G. The Schistosomulum

MacInnis[154] carried out experiments to show that certain chemicals present on mammalian skin cause cercariae of *S. mansoni* to initiate penetration responses and discharge the contents of the penetration glands. The chemicals were identified by MacInnis to be certain short-chain fatty acids and amino acids. Austin et al.[135] reported

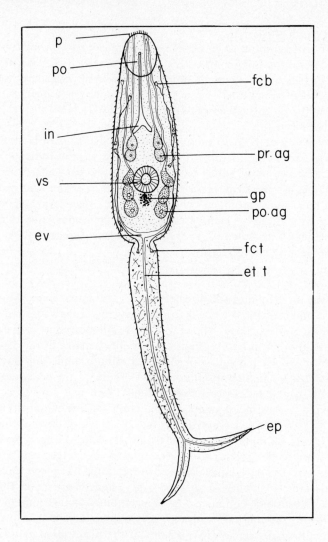

FIGURE 2-18.   Cercaria of *Schistosoma mansoni.* ep, excretory pore; et t, excretory tubule in tail; ev, excretory vesicle; fcb, flame cell in body; fct, flame cell in tail; gp, genital primordium; in, intestine; po, penetration organ; po.ag, postacetabular gland; pr. ag, preacetabular gland; and vs, ventral sucker.

that cercariae of *S. mansoni* are stimulated to penetrate the mammalian skin by a polar lipid fraction containing fatty acids, and Shiff et al.[169] concluded that penetration is initiated by unsaturated and saturated fatty acids. Histochemical studies indicate that cercarial penetration of mammalian skin is brought about by enzymes which are lost during penetration.

Usually the cercariae shed their tail while penetrating the skin, and after penetration they are called schistosomula (Figure 2-19). Now it is as if a new parasite has evolved which is different morphologically and physiologically from the cercaria. Now the schistosomulum has no precise shape, and, differing from the cercaria, it assumes a worm-like appearance. Looking like a nematode, instead, it cannot live in water anymore but is adapted to saline and serum. The Cercarienhüllen Reaktion, which is used as a serologic diagnostic tool with the cercariae, is no longer formed around the schistosomulum. In a few hours there is a change from the trilaminate structure of the

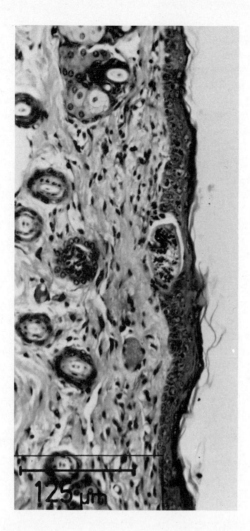

FIGURE 2-19. Section in mouse skin, showing schistosomulum of *Heterobilharzia americana*, during the first day after penetration of the cercaria.

outer membrane of the cercarial tegument to the heptalaminate structure of the adult membrane. However, the schistosomula coat lacks the large amounts of glycolipid found in the surface coat of adult worms. Depending on the mammalian host, a varying percentage of the schistosomula are killed in the skin; others move and gain access to a peripheral venous or lymphatic vessel within one to several days. In this way they can reach the heart and then the lungs and migrate further to reach the portal system.

In the liver they grow rapidly to the adult stage and egg laying commences. The prepatency in the case of *S. mansoni* in man is about 7 to 8 weeks, in *S. japonicum* it is 5 to 6 weeks, but in the case of *S. haematobium* this period is longer, up to 3 months.

The schistosomula excrete lactic acid as well as use oxygen.[142] However, oxidative phosphorylation is still of importance in energy formation. The mechanism of the switch from the fully aerobic metabolism of the cercaria to the excretion of lactic acid by the schistosomulum is not known. Change in environment may play a part in the change of metabolism.

H. Schistosome Groups

Schistosome species of the genus *Schistosoma* that infect man and animals are considered as belonging to "complexes" or "groups",[144,152] centered around the main human species *S. haematobium, S. mansoni,* and *S. japonicum.*

The *S. haematobium* group comprises schistosomes with a terminal-spined egg, and these are: *S. haematobium* (Bilharz, 1852) Weinland, 1858; *S. bovis* (Sousino, 1876) Blanchard, 1895; *S. mattheei* Veglia and LeRoux, 1929; *S. intercalatum* Fisher, 1934; and possibly also *S. spindale* Montgomery, 1906; and *S. incognitum* Chandler, 1926. The dimensions of the eggs, especially the length-breadth ratio, are the usual criteria for differentiating these species, rather than the snail intermediate host or the morphology of the adult and the cercariae. The breadth of the egg at a distance of 50 μm from the nonspined end was used for differentiating the species.[133] However, for an accurate method for differentiating the eggs, and accordingly the species, another measurement, namely, the breadth at a distance of 40 μm from the tip of the spine, became of value.[158,163]

S. haematobium is a parasite of man and a few other mammals over a large part of Africa and southwestern Asia. Mammals which were found naturally infected are certain species of monkeys, baboons, the chimpanzee, some rodents, two species of Artiodactyla, and one species of Pinnipedia. Several mammalian species have been experimentally infected with this schistosome.

S. bovis is a parasite of cattle, sheep, goats, and equines in central and northern Africa, southern Europe, and southwest Asia, as well as of the camel in Africa. It was first described from cattle in Egypt by Sonsino.

S. intercalatum was first described from several human cases in Zaire.[146] The parasite causes an intestinal form of the disease and does not inhabit the vesico-prostatic and uterine plexuses of veins, as is the case with *S. haematobium.* In addition to Zaire, the parasite also occurs in Gabon, the Cameroon, and the Central African Republic. Its biology and maintenance in the laboratory have been treated by Wright et al.[175]

S. mattheei is a bovine species infecting cattle, sheep, and goats in South Africa and Rhodesia, and in general in an area of Africa south of the zone of distribution of *S. bovis.*

S. spindale is an Asiatic species from India, Sumatra, and southeast Asia, infecting cattle, sheep, goats, and the water buffalo.

S. incognitum normally parasitizes pigs and dogs in India, and rodents and other mammals in Thailand, and in Sulawesi and West Java, Indonesia.

The *S. mansoni* group comprises schistosomes whose adults are similar morphologically, with lateral-spined eggs; these are *S. mansoni* Sambon, 1907 and *S. rodhaini* Brumpt, 1931.

S. mansoni is common in man in many parts of Africa, southwestern Asia, some islands of the West Indies, and some South American countries. It infects various laboratory animals and was found naturally parasitizing certain primates and some rodents in Africa, and rodents and opossums in the western hemisphere.

S. rodhaini was first described in 1931 by Brumpt and was redescribed by Schwetz and Stijns.[167] It is primarily a parasite of wild rodents in Zaire, but has been reported in man[148] and in the dog.[145] In addition to Zaire, this schistosome has been reported from Uganda.

A variety of *S. mansoni, S. mansoni* var. *rodentorum* Schwetz, 1953, is now regarded by many investigators as a synonym of *S. mansoni.*

The *S. japonicum* group includes *S. japonicum* Katsurada, 1904 and *S. margrebowiei* LeRoux, 1933.

S. japonicum parasitizes man and several reservoir hosts in China, Japan, the Philippines, Taiwan, Celebes, and Thailand, and there are about ten recent cases in Malaysia. The schistosome in Laos and Cambodia is regarded as *S. mekongi.*

S. margrebowiei possesses an egg which is similar to that of *S. japonicum* and occurs in equines, cattle, sheep, and various species of antelopes in Zambia and South Africa. Reports of *S. japonicum* in man in South Africa may be traced to this species.

VI. THE SNAIL HOSTS

A. Species Involved and Their Systematics

Table 2-3 shows the snail species (Figures 2-20, 2-21, and 2-21a) which are involved in the transmission of the three main human schistosomes in the various countries where the disease is endemic. The capacity of the species to act as transmitter is based on natural, experimental or epidemiologic evidence. The table has been updated from Malek.[208]

Problems continue to arise concerning the taxonomic relationship of these fresh-water, medically important snails. Such problems are due to the development of infraspecific variations in fresh-water snails, resulting from the fact that their habitats exist for only a relatively short time geologically. However, the time is not sufficient for the evolution to proceed to higher levels. Moreover, geographical isolation of small local populations of the snails results in the development of divergent forms, and variability of fresh-water snails at the infraspecific level. Another problem in the systematics of the actual and potential snail hosts of the schistosomes has been the fact that the majority of the original descriptions of African and American fresh-water snails were based on shell characters.

With regard to the African forms, a framework for the basic identification of snails has been provided by the introduction of the species-group concept into their taxonomy, by Mandahl-Barth.[213] This approach has clarified the large numbers of synonyms among the various species of the genera *Biomphalaria* and *Bulunus.* The arrangement has required some later revisions.[214,215] The process will no doubt continue, but a more or less reliable framework now exists within which taxonomic problems can be handled.

1. The Intermediate Hosts in Africa

Species groups of *Biomphalaria* in Africa include:

1. *Pfeifferi* group: *B. pfeifferi pfeifferi* (Kraus); *B. pfeifferi bridouxiana* (Bourguignat); *B. pfeifferi rhodesiensis* (Mandahl-Barth); *B. pfeifferi nairobiensis* (Dautzenberg); *B. pfeifferi gaudi* (Ranson); and *B. germaini* (Ranson).
2. *Choanomphala* group: *B. choanomphala choanomphala* (Martens); *B. choanomphala elegans* (Mandahl-Barth); *B. smithi* (Preston); and *B. stanleyi* (Smith).
3. *Alexandrina* group: *B. alexandrina alexandrina* (Ehrenberg); *B. alexandrina wansoni* (Mandahl-Barth); and *B. angulosa* (Mandahl-Barth).
4. *Sudanica* group: *B. camerunensis camerunensis* (Boettger); *B. sudanica sudanica* (Martens); and *B. sudanica tanganyicensis* (Smith).

Mandahl-Barth[214] found that it is more practical from the medical point of view not to divide *B. pfeifferi pfeifferi* into subspecies.

Among the *pfeifferi* group, *pfeifferi pfeifferi* occurs in Africa south of the Sahara; *rhodesiensis* is found in Zambia and Malawi; *germaini* is found immediately south of the Sahara and in Lake Chad; *gaudi* occurs in the western Sudan and in west Africa; and *nairobiensis* in east Africa.

Table 2-3
SNAIL INTERMEDIATE HOSTS OF *SCHISTOSOMA HAEMATOBIUM*[254]

Country	Intermediate host	Cap.[a]
Portugal	*Planorbarius metidjensis (= dufourii)*	Exp.
Cyprus	*Bulinus (Bulinus) truncatus*	
Morocco	*B. (B.) truncatus*	Nat.
Gambia	*B. (Physopsis) jousseaumei*	Nat.
	B. (Ph.) globosus	Nat.
	B. (B.) guernei	Nat.
	B. (B.) senegalensis	Nat.
Portuguese Guinea	*B. (Ph.) globosus*	Nat.
	B. (Ph.) jousseaumei	Epi.
Sierra Leone	*B. (Ph.) globosus*	Nat.
Liberia	*B. (Ph.) globosus*	Nat.
Guinea	*B. (Ph.) globosus*	Nat.
Mauritania	*B. (B.) guernei*	
Senegal	*B. (B.) senegalensis, B. (Ph.) jousseaumei, B. (B.) guernei*	Epi.
Mali	*B. (Ph.) jousseaumei, B. (B.) truncatus rohlfsi*	Epi.
Chad	*B. (Ph.) jousseaumei, B. (B.) truncatus rohlfsi*	Nat.
Ivory Coast	*B. (Ph.) globosus*	Epi.
Cameroon	*B. (Ph.) globosus, B. (B.) camerunensis*	Nat.
Ghana	*B. (Ph.) globosus*	Nat.
Nigeria	*B. (Ph.) globosus*	Nat.
Zaire	*B. (Ph.) globosus*	Nat.
	B. (B.) coulboisi	Epi.
Angola	*B. (Ph.) globosus*	Nat.
Algeria	*B. (B.) truncatus*	Nat.
Tunisia	*B. (B.) truncatus*	Nat.
Libya	*B. (B.) truncatus*	Epi.
Egypt	*B. (B.) truncatus*	Nat.
Sudan	*B. (B.) truncatus*	Exp. Nat.
	B. (Ph.) ugandae	Exp.
	B. (Ph.) globosus	Epi.
Ethiopia	*B. (B.) sp. (?)*	
	B. (Ph.) abyssinicus	Epi.
Somalia	*B. (Ph.) abyssinicus*	Epi.
Kenya	*B. (Ph.) globosus*	Nat.
	B. (Ph.) nasutus	Nat.
	B. (Ph.) africanus	Epi.
Uganda	*B. (B.) coulboisi*	Epi.
	B. (Ph.) globosus	Exp.
	B. (Ph.) nasutus	Exp.
Tanzania	*B. (Ph.) nasutus*	Nat.
	B. (Ph.) globosus	Epi.
Malawi	*B. (Ph.) globosus*	Epi.
	B. (Ph.) africanus	Epi.
Zambia	*B. (Ph.) africanus*	Epi.
	B. (Ph.) globosus	Nat.
Rhodesia	*B. (Ph.) globosus*	Nat.
	B. (Ph.) africanus	Nat.
South Africa (southern)	*B. (Ph.) africanus*	Nat.
South Africa (northern)	*B. (Ph.) globosus*	Nat.

Table 2-3 (continued)

SNAIL INTERMEDIATE HOSTS OF *SCHISTOSOMA HAEMATOBIUM*[254]

Country	Intermediate host	Cap.[a]
Mozambique	*B. (Ph.) globosus*	Nat.
	B. (Ph.) africanus	Nat.
	B. (Ph.) globosus	Nat.
	B. (Ph.) africanus	Nat.
Mauritius	*B. (B.) cernicus*	Exp.
Madagascar	*B. (B.) obtusispira*	Nat., Exp.
Turkey	*B. (B.) truncatus*	Epi.
Syria	*B. (B.) truncatus*	Nat.
Lebanon	*B. (B.) truncatus*	Nat.
Israel	*B. (B.) truncatus*	Exp.
Saudi Arabia	*B. (B.) truncatus*	Nat.
Yemen	*B. (B.) truncatus*	Nat.
Iraq	*B. (B.) truncatus*	Nat.
Iran	*B. (B.) truncatus*	Nat., Exp.
India	*Ferrissia tenuis*	Exp.
Snail Intermediate Hosts of *Schistosoma mansoni*		
Puerto Rico	*Biomphalaria glabrata*	Nat.
Dominican Republic	*B. glabrata*	Epi.
St. Kitts	*B. glabrata*	Epi.
Guadeloupe	*B. glabrata*	Nat.
Martinique	*B. glabrata*	Nat.
St. Lucia	*B. glabrata*	Nat.
Dutch Guiana	*B. glabrata*	Nat.
Venezuela	*B. glabrata*	Nat.
Brazil	*B. glabrata*	Nat.
	Biomphalaria straminea	Nat.
Gambia	*Biomphalaria pfeifferi gaudi*	Nat.
Portuguese Guinea	*B. pfeifferi*	Nat.
Sierra Leone	*B. pfeifferi*	Nat.
Liberia	*B. pfeifferi*	Nat.
Guinea	*B. pfeifferi*	Nat.
Mauritania	*B. pfeifferi*	Epi.
Senegal	*B. pfeifferi*	Nat.
Mali	*B. pfeifferi*	Exp.
Ivory Coast	*B. pfeifferi*	Epi.
Cameroon	*B. sudanica, B. camerunensis*	Epi., Nat.,
Ghana	*B. pfeifferi, B. sudanica*	Nat.
Nigeria	*B. pfeifferi, B. sudanica*	Nat.
Zaire (north)	*B. rüppellii*	Nat.
	B. choanomphala	Nat.
	B. sudanica	Nat.
	B. sudanica tanganyicensis	Nat.
	B. stanleyi	Exp.
Zaire (south)	*B. bridouxiana*	Nat.
Zaire (east)	*B. nairobiensis*	Epi.
Angola	*B. pfeifferi*	Nat.
Egypt	*B. alexandrina*	Nat.
Sudan	*B. sudanica*	Nat.
	B. rüppellii	Exp.
	B. rüppellii	Nat.
	B. pf. gaudi	Epi.
Ethiopia	*B. rüppellii*	Nat.

Table 2-3 (continued)

SNAIL INTERMEDIATE HOSTS OF *SCHISTOSOMA HAEMATOBIUM*[254]

Country	Intermediate host	Cap.[a]
Kenya	*B. sudanica*	Nat.
	B. sudanica tanganyicensis	Nat.
	B. choanomphala	
	B. rüppellii	Epi.
	B. nairobiensis	Nat.
Uganda	*B. rüppellii*	Exp.
	B. choanomphala	Nat.
	B. sudanica	Nat.
	B. sudanica tanganyicensis	Nat.
	B. pfeifferi	Nat.
	B. smithi	Epi.
	B. stanleyi	Epi.
Tanzania	*B. rüppellii*	Epi.
	B. sudanica	Epi.
	B. sudanica tanganyicensis	Epi.
	B. pfeifferi	Nat.
	B. bridouxiana	Epi.
Malawi	*B. sudanica*	Epi.
	B. rhodesiensis	Epi.
	B. angulosa	Epi.
Zambia	*B. pfeifferi*	Nat.
	B. rhodesiensis	Epi.
	B. angulosa	Epi.
Rhodesia	*B. pfeifferi*	Nat.
Union of South Africa	*B. pfeifferi*	Nat.
(Natal & Transvaal)	*B. angulosa*	Epi.
Mozambique	*B. pfeifferi*	Nat.
Madagascar	*B. pfeifferi*	Nat.
Israel	*B. alexandrina*	Exp.
Saudi Arabia	*B. pfeifferi*	Nat.
Yemen	*B. pfeifferi*	Nat.
Snail Intermediate Hosts of *S. japonicum*		
Japan	*Oncomelania nosophora*	Nat.
China	*O. hupensis*	Nat.
Philippines	*O. quadrasi*	Nat.
Formosa	*O. formosana*	Nat.
Laos	*Lithoglyphopsis aperta*	Exp.

[a]   The snails' capacity as a transmitter is indicated in this column. Nat. indicates natural infection; Exp. is experimental infection; and Epi. indicates a host on epidemiological grounds.

Members of the *choanomphala* group occur in the east-central African lakes. Among the *alexandrina* group, *alexandrina alexandrina* is found in the Nile delta while *angulosa* occurs in eastern and southern Africa. in the *sudanica* group, *sudanica* is found from southern Sudan southward through Uganda, Kenya, Tanzania, and westward to Zaire and the Lake Chad region, and *sudanica tanganyicensis* is prevalent in central Africa, Uganda, and Tanzania.

All species of *Biomphalaria* in Africa could be considered as potential intermediate hosts of *S. mansoni*. However, all species and strains are not equally susceptible to all strains of *S. mansoni*, and in some cases they may prove nonsusceptible. Several stud-

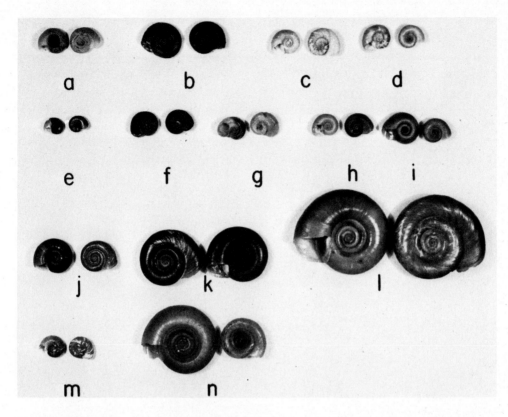

FIGURE 2-20.   Biomphalarid snails, hosts of *Schistosoma mansoni*. (a) *Biomphalaria alexandrina*, Egypt, (b) *B. sudanica*, Sudan, (c) *B. pfeifferi*, Senegal, Tanzania, (d) *B. pfeifferi*, South Africa, (e) *B. rüppellii*, Kenya, (f) *B. angulosa*, Tanzania, (g) *B. choanomphala*, Lake Victoria, Uganda, (h) *B. smithi*, Uganda, (i) *B. stanleyi*, Lake Kivu, (j) *B. glabrata*, St. Lucia, (k) *B. glabrata*, Brazil, (l) *B. glabrata*, Brazil, (m) *B. straminea*, Brazil, (n) *B. tenagophila*, Brazil.

ies have been conducted in various countries of Africa to test the susceptibility of local species of *Biomphalaria*, and identification of the cercariae shed from naturally infected snails as *S. mansoni* cercariae have been carried out. In addition to *S. mansoni*, *B. pfeifferi* and *B. sudanica* have been found to act as hosts for *Schistosoma rodhaini*.
   Species groups of *Bulinus* include:

1.   *Africanus* group: *B. africanus* (Krauss); *B. abyssinicus* (Martens); *B. nasutus* (Martens); *B. ugandae* (Mandahl-Barth); *B. globosus* (Morelet); and *B. jousseaumei* (Dautzenberg).
2.   *Tropicus* group: *B. tropicus* (Krauss); *B. angolensis* (Morelet); *B. zanzbaricus* (Clessin); *B. sericinus* (Jickeli); and *B. obtusispira* (Smith).
3.   *Truncatus* group: *B. truncatus* (Audouin); *B. trigonus, B. truncatus rohlfsi* (Clessin); *B. coulboisi* (Bourguignat); and *B. guernei* (Dautzenberg).
4.   *Forskalii* group: *B. forskalii* (Ehrenberg); *B. senegalensis* (Muller); *B. cernicus* (Morelet); *B. reticulatus* (Mandahl-Barth); *B. beccarii* (Paladilhe); and *B. scalaris* (Dunker).

   As to geographical distribution, *B. africanus* is found is eastern and southern Africa; *B. globosus* in Africa south of the Sahara; *B. ugandae* in east Africa; *B. tropicus* in equatorial and southern Africa; *B. obtusispira* in Madagascar; *B. truncatus* in Africa

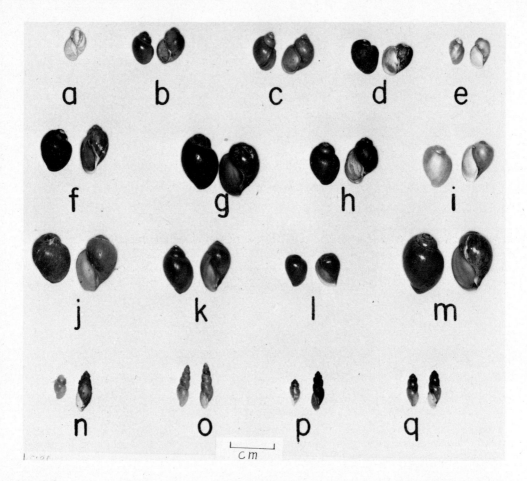

FIGURE 2-21.   Bulinid snails hosts of *Schistosoma haematobium* and related species. (a) *Bulinus (Bulinus) truncatus*, Egypt, (b) *B. (B.) truncatus*, Sudan, (c) *B. (B.) truncatus*, Iran, (d) *B. (B.) truncatus trigonus*, Tanzania, (e) *B. (B.) guernei*, Senegal, (f) *B. (Physopsis) globosus*, Tanzania, (g) *B. (Ph.) globosus*, Angola, (h) *B. (Ph.) globosus*, Senegal, (i) *B. (Ph.) globosus*, Ghana. (j) *B. (Ph.) ugandae*, Sudan, (k) *B. (Ph.) nasutus*, Tanzania, (l) *B. (Ph.) jousseaumei*, Mali, (m) *B. (Ph.) africanus ovoideus*, Kenya, (n) *B. (B.) senegalensis*, Senegal, (o) *B. (B.) forskalii*, Tanzania, (p) *B. (B.) forskalii*, Senegal, (q) *B. (B.) forskalii*, Sudan.

north of the equator, in the Middle East, Mediterranean islands, and southwest Asia; *B. forskalii*, cosmopolitan in Africa; *B. senegalensis* in west Africa; *B. reticulatus* in east and central Africa and southern Yemen; *B. cernicus* in Mauritius; and *B. beccarii* in southern Yemen.

Some species of the genus *Bulinus* are intermediate hosts of *Schistosoma haematobium* and other related species with terminal-spined eggs, viz., *S. bovis*, *S. mattheei*, *S. intercalatum*, and *S. leiperi*. The problem, however, is complicated by the fact that there are certain species which are not susceptible to these schistosomes and that the geographical distribution of such species overlaps that of the susceptible species. Moreover, several strains of the same species occur, some are susceptible while others are nonsusceptible, and the number is more than is found in the case of species of the genus *Biomphalaria* in Africa.

It seems likely that all members of the *truncatus* group are potential hosts for *S. haematobium* and *S. bovis* north of the Sahara, whereas members of the *tropicus* group are not hosts, although some reports are in the literature about their capacity

FIGURE 2-21a.   Habitat of the snail *Oncomelania quadrasi*, intermediate host of *Schistosoma japonicum* in the Philippines.

to transmit mammalian schistosomes. It is difficult to differentiate members of these two groups on a morphological basis, and cytological studies have proved of value in this respect. In all species of *Bulinus* the haploid number of chromosomes is 18, typical of the family Planorbidae. However, polyploidy (36 and 72) is common among members of the *truncatus* group.

South of the Sahara in Africa, members of the *africanus* group are the main transmitters of the mammalian schistosomes with terminally spined eggs. Members of this group can be differentiated from the other groups of *Bulinus*, but it is difficult to differentiate the individual species of *africanus.*

In general the *forskalii* group comprises small forms, with high-spired shells. Four species are known as transmitters of *S. haematobium* in Africa and southwest Asia, viz., *B. senegalensis* (west Africa), *B. reticulatus* (east and central Africa, southern Yemen), *B. cernicus* (Mauritius), and *B. beccarii* (southern Yemen). Early reports on this group have incriminated *B. forskalii* as a natural intermediate host for *S. haematobium*, but this was due to misidentification and confusion of *B. forskalii* and other members of its group. Experimental infections of certain populations of *B. forskalii* with certain strains of *S. haematobium* showed very low infection rates of short durations, with emergence of very few cercariae, which indicate that this species is not important from an epidemiological point of view in the transmission of urinary schistosomiasis. It has been proven, however, that *B. forskalii* transmits *S. intercalatum* in Gabon.

2. The Intermediate Hosts in Africa

On the basis of morphological studies and cross-breeding experiments it became evident that the snail hosts of *S. mansoni* in Africa and the Americas are congeneric. The International Commission on Zoological Nomenclature (Opinion 735, 1965) ruled that the generic name *Biomphalaria* is to be given precedence over the generic names *Planorbina*, *Taphius*, and *Armigerus* by any zoologist who considers that any or all of these names apply to the same taxonomic genus. The generic names *Australorbis* and *Tropicorbis*, which have been used for the snail hosts in the Americas, are both of more recent origin and also fall into synonymy. In view of this ruling all intermediate snail hosts in the Americas fall under *Biomphalaria*, the same genus in which are included the snail host species in Africa. Revision of the nomenclature and keys to the species in the Americas are included in *A Guide for the Identification of the Snail Intermediate Hosts of Schistosomiasis in the Americas*, Pan American Health Organization, Scientific Publication No. 168, 1968, Washington, D.C. Table 2-3 shows the species involved in each endemic area. Three species have been found naturally infected, viz., *Biomphalaria glabrata, B. straminea*, and *B. tenagophila*. The three species are hosts in Brazil, and the first two are in Venezuela, otherwise *B. glabrata* is the recognized host in all the other areas. Susceptibility studies have indicated that populations of *B. glabrata* vary considerably as to their capacity to act as hosts. For example, a population from Dique, Bahia, Brazil is very poorly susceptible to certain strains of *S. mansoni*. Laboratory studies indicate that certain populations of *B. riisei, B. albicans, B. havanensis*, and *B. peregrina* are potential hosts, but they have never been found naturally infected.

It is of interest to note that although *B. tenagophila* is considered as a poor host, this species acts as a highly susceptible host for the strain of *S. mansoni* occurring in the Paraíba River valley in the state of São Paulo, Brazil. Also, populations of *B. tenagophila* from areas outside the Paraiba valley have usually proved to be susceptible to the strain of *S. mansoni* from the Paraiba valley.

There are several other species of *Biomphalaria* in the Americas which are nonsusceptible to infection. Nevertheless, they are important because they occur in habitats similar to those of the susceptible hosts and often are confused with them. The systematics and grouping of these species have been treated by Malek.[212]

3. The Intermediate Hosts in the Orient

The snail hosts of *Schistosoma japonicum* are members of the family Hydrobiidae. *Oncomelania* spp. are amphibious snails in China (mainland), Taiwan, Japan, Celebes, and the Philippines, whereas *Lithoglyphopsis aperta* is an aquatic snail in Laos and Cambodia. Various generic names have been used for the amphibious snails: in China (mainland), *Oncomelania, Hemibia, Katayama*, and *Schistosomophora*; in China (Taiwan), *Blanfordia, Katayama, Tricula*, and *Oncomelania*; in Japan, *Katayama, Blanfordia*, and *Oncomelania*; and in the Philippines, *Oncomelania, Blanfordia*, and *Schistosomophora*. Abbott[176] considered that all the known intermediate hosts of *S. japonicum* in the Orient belong to one genus, *Oncomelania*.

Tricula chiui, from Alilao, Taiwan, was later reported as a new hydrobiid snail, and it was proven in the laboratory that it is susceptible to four of the *S. japonicum* strains. Taxonomic studies proved, however, that it is either a geographical race of *Oncomelania formosana* or a subspecies of *Oncomelania hupensis*.

It has been recognized that in each geographical area of *S. japonicum* there is a different oncomelanid species, viz., *O. hupensis* in China (mainland), *O. formosana* in China (Taiwan), *O. nosophora* in Japan, and *O. quadrasi* in the Philippines. The four geographical oncomelanian species seem to be distinct biologically in many respects. However, they were interbred successfully in different species-crossing combinations. Moreover, the chromosome number of these four species was found to be the same, and in their F_1 hybrids no apparent anomalies occurred in chromosome pairing that were not as prevalent in the normal parent. Therefore some malacologists consider that there is only one species of *Oncomelania, O. hupensis*, with four subspecies which had been considered as species.[186] Another subspecies has been described from the schistosomiasis focus at Lake Lindoe, Sulawesi, Indonesia, and called *Oncomelania hupensis lindoense*. Some schistosomiasis workers, however, are of the opinion that the specific names should be retained as species and not as subspecies.

The aquatic hydrobiid snail *Lithoglyphopsis aperta* is the snail host of *S. mekongi* in Laos and Cambodia. The snails have been found on Khong Island, Laos, and were successfully reared and infected in the laboratory. Although the snail host in Thailand remains unknown, it is believed that it is *L. aperta*, or a hybrobiid snail related to it.

B. Ecology and Habitat Characteristics

The planorbid snail hosts in the Americas and Africa are found in fresh-water bodies, large and small, flowing and standing; in waters with pH in the range of 5.8 to 9.0; in tropical forest regions and in arid situations; at low or at high altitudes; and at water temperatures from 20 to 30°C. Usually they inhabit shallow waters; however, one species, *Biomphalaria choanomphala*, in Lake Victoria in Africa, lives in deep water. The habitats are usually with moderate organic content, little turbidity, a muddy substratum rich in organic matter, submerged or emergent aquatic vegetation, and abundant microflora. They have been collected from rivers, lakes, marginal pools along streams, borrow-pits, marshes, flooded areas, irrigation canals, and aqueducts. They are not present in habitats with high tidal fluctuations or in reaches of streams having a fall steeper than 20 m/1000 m of length, though protected pools or swampy areas alongside these steep streams often harbor them. The life span of the planorbid hosts does not normally exceed 12 to 15 months. Aside from the known physical deterrents, such as stream gradient, the water quality in certain habitats may be largely responsible for their spotty distribution in otherwise suitable environments. The snails occur in habitats with high content of bicarbonate, carbonate, sulphate, chloride, magnesium, and calcium. Lethal concentrations of these ions as determined by laboratory studies are not usually encountered in nature.

Oncomelania spp. hosts of *S. japonicum* in Japan, China, the Philippines, Taiwan, and Indonesia are amphibious snails which favor the moist soil at the margins of slow-flowing streams, ditches, narrow irrigation canals, and overflow areas, where vegetation is dense. They are not frequently found in rice fields, contrary to the impression often given. Ditches that have running water throughout the year rarely support colonies, but those with water for only a few months are the most common habitat, especially if they contain abundant humus on the soil.[245] Irrigation canals represent secondary adaptation for *Oncomelania*. In Japan it is reported that broad marshlands between dikes and channels[230] and in the Philippines, flood-plains, forests, and swamps[203] represent the original habitat. Fluctuation in water level affects oviposition of *Oncomelania*. *O. quadrasi* prefers to lay its eggs at or above water level on a solid substratum, whereas *O. nosophora* in Japan lays its eggs in the water, with soil as a substratum. In Japan there is winter hibernation from November to March. Sexual dimorphism is observed in *Oncomelania*, and more females than males have been found to be infected with *S. japonicum* in the Philippines. The life span of *Oncomelania* is from 6 months to 3 years.

In Laos and Cambodia the snail host of *S. mekongi* is *Lithoglyphopsis aperta*, a hydrobiid like *Oncomelania* but totally aquatic. Populations of *L. aperta* are often restricted to several small patches within a larger area. This is probably due to discontinuity of suitable substrates and of relatively sheltered areas characterized by a slow current and zone of deposition. For example, one population at Ban Khee Lek of Khemarat was reported located within an area of 4000 m², and another population was not found until ½ mi upstream. In other regions, such as Khong Island (Mekong River), the population may be continuous for 2 to 3 mi along the shore of the island with nodes of increased density reflecting a greater abundance of suitable substrates in protected shallow water areas. The snail lives on solid substrata, particularly wood, shells, and leaves in the Mekong River. The species is a colonizer in a river with severe annual floods, and it occurs from Khemarat, Thailand, to the Cambodian border, 200 river mi downstream. The range probably extends another 100 river mi downstream to Kratie, Cambodia.

Some of the chemical, physical, and biological factors which affect the planorbid and oncomelanid snails are discussed below.

1. Ionic Composition of Water

In east Africa and Puerto Rico the snails are not found in waters with low electrical conductivity, indicating very low concentrations of dissolved salts. Malek[207] showed that in the Sudan both *Biomphalaria* and *Bulinus* were found where the ratio of sodium to calcium in the water was 1:2, but only *Biomphalaria* was found where it was 0:2. In South Africa the *Africanus* group and *Biomphalaria pfeifferi* were found where the ratio was from 0:5 to 2:0, but when it rose above 2:4 the *africanus* group was present and *B. pfeifferi* was most uncommon.[232]

It has been shown by laboratory tests[198] that the egg laying of *Biomphalaria pfeifferi* was significantly reduced and then inhibited by increasing the magnesium content of the water. It has also been observed in Rhodesia that small populations of pulmonate snails in certain streams occurred on account of a high magnesium content. In other Rhodesian streams a high natural turbidity caused by a mixture of kaolin and illite or sericite, or both, also affected the reproductive activity of *B. pfeifferi* and *B. globosus*.[197]

2. Temperature

It seems that temperature affects the distribution of the intermediate host snails in Africa.[182] *B. pfeifferi* is absent from the low-lying coastal areas in the equatorial region but is present at similar altitudes to the north and south, which suggests that high temperatures may be unsuitable for this species.[236] In southern Tanzania the daily maximum temperature in standing water at Mahiwa (altitude 200 m) can rise to over 30°C at most times of the year and at Nachingwea (altitude 350 m) this can occur between October and February, whereas at Songea (altitude 1050 m) such temperatures are unlikely to occur.[181,182] *Bulinus globosus* is abundant at all three localities but *B. pfeifferi* is found only at Songea and is not known to occur below Tunduru (altitude 600 m) in this region. Brygoo[184] showed that *Bulinus obtusispira*, the intermediate snail host of *S. haematobium* in Madagascar, will not reproduce at temperatures fluctuating up to 25°C but reproduces intensively when exposed to maximum temperatures of 35 to 37°C, which cannot be tolerated by *Biomphalaria pfeifferi*. This explains the fact that on Madagascar *B. obtusispira* occurs in the western and southern parts of the island. On the other hand, conditions in the forested central plateau and eastern areas provide shaded conditions in which *B. pfeifferi* (and accordingly *S. mansoni*) can thrive almost down to sea level. Also in South Africa, the distribution of *Bulinus africanus*, *B. globosus*, *B. tropicus*, and *Biomphalaria pfeifferi* is determined by altitude and by temperature.[183,239]

Temperature also affects reproduction of the snails and snail population dynamics. Under controlled laboratory conditions the growth and fecundity of *B. pfeifferi* and *B. globosus* were recorded and life tables constructed for a range of temperatures.[199] The latter study, carried out in Rhodesia, showed that in each case the peak value was 25°C, but *B. globosus* had a much higher peak than *B. pfeifferi*. Other studies support the conclusion that the optimum temperature for *B. pfeifferi* and for *B. glabrata* is around 25°C.[218,236] The high intrinsic rate of natural increase in *B. globosus* in Rhodesia appears to be an adaptation to adverse conditions such as the drying up of the habitat in temporary environments. This allows a rapid expansion of the population to exploit favorable conditions when these occur.[199] The same might well be true for *B. globosus*, *B. nasutus* in Tanzania, and *B. glabrata* and *B. straminea* in northeast Brazil.[177,182,241]

3. Rainfall

In the natural habitats (i.e., not including irrigation canals and drains) of the intermediate snail hosts, it is generally possible to relate seasonal cycles in the density and reproduction of the snails to the climate of the region, particularly to the rainfall. Most habitats undergo considerable seasonal changes in several related factors such as size, water level, rate of flow, flora, water chemistry, oxygenation, and pollution.[182] Optimum conditions for the snails will be produced by certain combinations of factors, while other combinations may create very adverse conditions. In certain habitats there is an increase in the snail population during the rainy season, but in general an increase is evident early in the dry season. In still others, the optimum size of the population is reached towards the end of the dry season.

One of the important characteristics of the planorbid intermediate hosts is their ability to withstand drought for long periods of time. It has been found that if desiccation takes place gradually in a habitat where the humidity in the immediate vicinity (microhabitat) of the snails is high, the level of subsoil water is high, and the snails placed among vegetation and debris, a good proportion of them survive. Snails exhibit species, strain, and individual differences in their ability to survive during dry periods. This occurs in natural habitats as well as man-made habitats (irrigation canals and drains). When the water returns, survivor estivating snails repopulate the area within a short period, usually 1 to 2 months.

It should be noted that the seasonal cycles of snails in irrigation canals are influenced by the frequency of fluctuation in the level of the water and by the presence or absence of silt and accordingly the presence or absence of aquatic weeds. Figures 2-22, and 2-23 show seasonal fluctuation in snail population density of the planorbids.

4. Dissolved Oxygen

The snail intermediate hosts of *S. mansoni* and *S. haematobium*, members of the family Planorbidae, being pulmonates, are provided with a lung which enables them to make use of atmospheric oxygen for their respiration. They also depend on dissolved oxygen in the water when they remain submerged. They make use of dissolved oxygen throughout their entire surface, and also through a pseudobranch (a false gill) which is highly vascular and is considered as an outgrowth from the respiratory wall.

In the Sudan in certain habitats which harbored both *Biomphalaria rüppellii* and *Bulinus truncatus*, dissolved oxygen averaged 4.7 to 7 ppm. However, in habitats where algae are abundant, and where aquatic weeds are present in moderate numbers, but not dense enough to stop light penetration, supersaturation of the habitat, due to photosynthetic activities, is noticed. In addition to aerobic respiration the planorbid snails can undergo anaerobic respiration. Investigations have been carried out on the resistance to anaerobic conditions shown by various species of snails, including planorbids, and it was found that they had a higher resistance than lymnaeid and physoid snails.[240] In other experimental work it was shown that, under anaerobic conditions, *Biomphalaria glabrata* infected with *S. mansoni* ceased to shed cercariae or only shed them in very small numbers, and that infected snails were less able to withstand anaerobic conditions than were uninfected ones.

Oncomelania spp. snail hosts of *Schistosoma japonicum* in the Orient are amphibious and possess a gill to obtain dissolved oxygen. The young snails are mostly aquatic and depend on oxygen dissolved in the water, while the adult snails which live at the margin of the water utilize dissolved oxygen periodically.

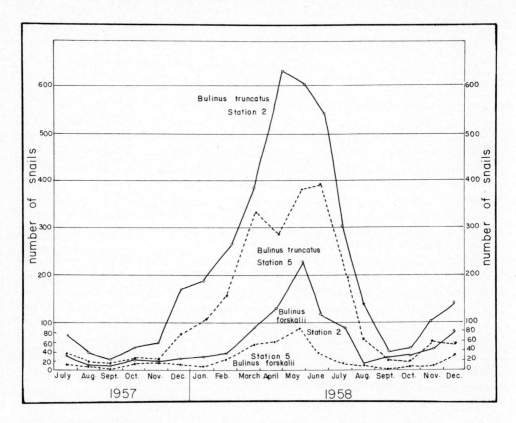

FIGURE 2-22.   Seasonal fluctuation in snail populations as indicated by the monthly collections of *Bulinus (Bulinus) truncatus* and *Bulinus (Bulinus) forskalii* at two stations (2 and 5), in irrigation canals, Ministry of Agriculture Farm, near Khartoum, Sudan, during 1957—1958. (After Malek, 1962).

Lithoglyphopsis aperta is the snail host of *S. mekongi* in Laos and Cambodia. It is an aquatic snail throughout its life history. Dissolved oxygen is essential to its survival and is utilized by a well-developed gill in its mantle cavity.

5. Pollution

Pollution of a water body is brought about by several means: by the dumping of industrial wastes, by sewage in the form of animal or human excrement, or by the accumulation of large quantities of organic matter of vegetable or animal origin.

Refuse from factories pollutes the water and makes it unsuitable for schistosomiasis intermediate host snails. This industrial waste usually contains large quantities of oils, acids, and a high mineral content, far in excess of the limit of tolerance of these animals. Pollution also has an indirect effect on the snails through harming the vegetation, thus reducing the oxygen supply, and through altering the nature of the bottom.

The effect of animal and human excrement on schistosomiasis intermediate hosts has been reported by several investigators. In general, the snails are abundant near human habitations, which pollute the water with excrement. Some workers consider the excrement as a requirement for the existence of the snails in several African countries. It appears, however, that although the snail hosts thrive better in the presence of animal and human excrement, they do occur in unpolluted habitats, as long as their food, mainly in the form of microflora, is plentiful.

Pollution caused by decaying vegetable and animal matter seems to be beneficial to the snails when these are deposited on the substratum of the water course. The type

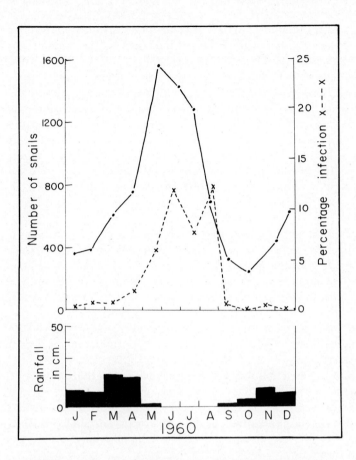

FIGURE 2-23. Seasonal changes in snail numbers (*Bulinus nasutus* productus), and infection rate with *Schistosoma haematobium*, in small pools in an area near Mwanza, Tanzania. The continuous line shows the numbers of snails and the broken line the infection rates. (Adapted after Webbe, G., *Bull. W.H.O.*, 27, 59, 1962.)

of substratum almost always associated with schistosomiasis snails is a firm mud bottom, rich in decaying organic matter. This provides support for aquatic plants, a surface on which flourish algae and other microorganisms on which the snails feed, and a crawling surface for the snail.

6. Food

Food in the habitat seems to influence the reproduction of the snail hosts and their population dynamics. The snails browse and feed more or less continuously as they move, and there are reports in the literature about their food.[182,185,201,202,207] Algae seem to be the commonest diet component, with green algae and diatoms being particularly favored. Blue-green algae are also components of the diet and *Oscillatoria formosa* has proved very satisfactory in the laboratory;[188] but some other species may be directly harmful. According to laboratory observations, it seems that filamentous algae only cause mechanical obstruction rather than being poisonous to the snail hosts. Leaves of aquatic weeds are usually not eaten by the snails unless they are in a state of decomposition. The snails, however, scrape the periphyton accumulating on the aquatic weeds. Periphyton is the assemblage of organisms growing on free surfaces of objects submerged in water and covering them with a slimy greenish-brown coat. The organisms composing the periphyton belong to several plant and animal phyla, and thus certain microscopic sedentary animals are sometimes also taken in by the snail in the process of ingesting the microflora. This is indicated by the examination of stomach contents of the snails. It should be noted that a diet of this type is found in the majority of habitats, whether they contain well-established colonies or only a few individuals. Consequently it is not the type of diet but its quantitative composition which is important in influencing the habitat and the reproduction and population dynamics of the snails. We have no quantitative information about the food requirements of a population of snails or about the factors which may influence the production of suitable food in natural habitats. Berrie[182] is of the opinion that until such studies are carried out on the trophic relationships of the snails we remain unable to assess exactly the importance of the food supply in the population dynamics of the snails.

7. Predators

Laboratory experiments indicate that a number of predators feed on the snail hosts. Various carnivorous leeches, certain crustaceans (crabs, crayfishes, ostracods), aquatic Coleoptera, fishes (*Gambusia* spp., *Lebistes* spp., and *Tilapia* spp.), amphibians (certain species of salamanders), birds (ducks), and mammals (water rats) feed on freshwater snails, some to a very considerable degree. Other snails (e.g., *Marisa cornuarietis*) compete with the snail hosts for the food available in the habitat and accidentally consume eggs and young of the medically important snails, whereas the snails *Helisoma* spp. are believed to affect the medically important snail species through their secretions or excretions.

Although it is expected that these predators might affect the number of the medically important snails, it is unlikely that they exterminate them under natural conditions. This has to be substantiated through detailed field investigations. Only the effect of *Marisa cornuarietis* on populations of *Biomphalaria glabrata* has been demonstrated in Puerto Rico,[231] and that of *Helisoma duryi* on *Biomphalaria pfeifferi* has been demonstrated in Tanzania.[226]

8. Pathogens

A number of pathogenic bacteria (*Bacillus pinottii, Mycobacterium* sp.) have been isolated from snails, and were successfully transmitted to other snails of the same species or other species.[190,217] No doubt other bacteria, pathogenic and nonpathogenic, infect snails. A number of fungi have been reported to kill adult snails and their eggs, under laboratory conditions. Effects of these pathogens under natural conditions have not been thoroughly evaluated, especially as it concerns the snail population dynamics.

Among the protozoa Richards[227] has reported the occurrence of two species, *Hartmanella biparia* and *H. quadriparia* in *Bulinus globosus* and *Biomphalaria pallida*, respectively. Both amoebae cause pathologic reactions within their hosts. Parasitized snails may become moribund, and the presence of *H. biparia* is believed to affect the growth and reproduction of *Bulinus globosus*. Whether these amoebae have an effect on snail population dynamics remains to be tested.

Similarly, several species of digenetic trematodes infect the snail hosts of the schistosomes. It is known that these trematodes have mechanical and physiological effects on the snails, but whether they considerably affect the snail population is not known.

C. Snail-Schistosome Interrelationships

Various aspects of the snail-schistosome relationships are of significance in the epidemiology and transmission of schistosomiasis. Some of these aspects are discussed here.

1. Specificity

There is a high degree of specificity of schistosomes, as well as of other trematodes, to their intermediate snail hosts. A schistosome miracidium might penetrate several species of snails, but its fate in the tissues of the snails is determined by a biochemical adaptation to certain species. They might develop and produce cercariae in some species, whereas they are walled off in others as a result of a strong host reaction which is the expression of an innate cellular internal defense mechanism.[204,211,220,237] In general such a reaction is not produced in a susceptible snail. However, slight and restricted encapsulation may also occur around larvae in their natural hosts at an advanced stage of the infection,[223] but these extremely light reactions usually inflict little or no damage upon the parasites (Figure 2-24). On the other hand, the extensive capsules that surround the larvae (miracidium and/or mother sporocyst) in totally or partially incompatible snails usually result in destruction of the parasites. The chemical basis for this destruction remains undetermined; nevertheless, it may be generalized that encapsulation by leukocytes and/or fibroblasts resulting in death is by far the most effective form of innate resistance in molluscs against incompatible trematode larvae. Newton[220] considered that susceptibility or resistance of a snail to infection is an hereditary character, and recently Richards[228,229] confirmed this contention by extensive studies on the genetics of *Biomphalaria glabrata*.

Members of the *Schistosoma haematobium* complex in Africa and in the Middle East develop only in certain species of the genus *Bulinus*. On the other hand, *S. mansoni* and the related species in Africa (*S. rodhaini*) and in the western hemisphere develop in species of *Biomphalaria*. *S. japonicum* matures in hydrobiid species of the genera *Oncomelania* and *Lithoglyphopsis* in the Orient and southeast Asia.

Adaptation is very pronounced among species of *Bulinus* in Africa to the local strains of *S. haematobium*, and of animal schistosomes of its complex, viz., *S. bovis*, *S. intercalatum*, and *S. mattheei*. *B.(B.) truncatus* from Egypt was found susceptible to the Egyptian strain of *S. haematobium* but was not susceptible to *S. haematobium* strains from Rhodesia and Zambia, to *S. mattheei* from Zaire, from Rhodesia and from the Transvaal, or to *S. intercalatum* from Zaire. Moreover, *B. (Ph.) africanus* from Durban was incompatible with the Egyptian strain of *S. haematobium* but susceptible to Rhodesian strains of *S. haematobium* and *S. mattheei*.

2. Effects of Environmental Factors

Certain environmental factors have pronounced effects on the schistosome-snail interrelationships. Temperature affects the penetration of the miracidia into the snail, the larval development inside the snail, and the emergence of the cercariae from the snail. The higher the temperature, the shorter the incubation period of the parasite inside the snail. Gordon et al.[194] give 32 to 33°C as optimum temperature for the development of both *S. haematobium* and *S. mansoni* inside their snail hosts in Sierra Leone, while Standen[234] considers 26 to 28°C as the optimum temperature for *S. mansoni* in *Biomphalaria glabrata*. Stirewalt[235] reports an 18 day incubation period at 31 to 33°C for *S. mansoni* in the same species of snail. The author's observations showed that at 23 to 24°C the incubation period of *S. mansoni* was 30 days, at 28°C it was 26 days, and at 30 to 31°C it was 22 days. For *S. haematobium* (Sudanese strain) in *Bulinus (B.) truncatus* there was a difference of 16 days in the incubation period in snails maintained at 20 to 22°C and those maintained at 28°C. Moreover, rising temperature stimulates the emergence of cercariae, as does a sudden change in temperature. Barbosa and Coelho[178] record 19 to 37°C as the most favorable range for shedding of *S. mansoni* cercariae from *B. glabrata*; shedding was inhibited at temperatures below 13°C, and above 41°C. For *S. japonicum*, 20 to 28°C was reported as favorable for the emergence of cercariae; on both sides of this range the emergence was reduced.[196]

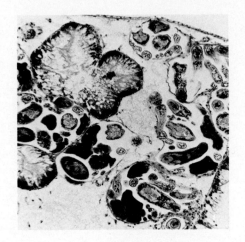

A

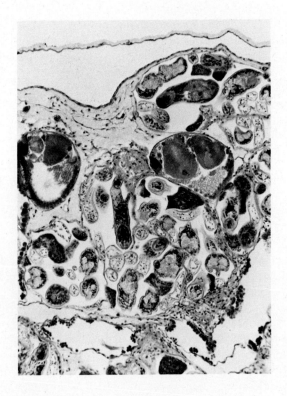

B

FIGURE 2-24.   (A) Section in digestive gland of
the snail *Biomphalaria alexandrina*, showing
sporocysts of *Schistosoma mansoni*, containing
cercariae, between the tubules, (B) Section in
ovotestis of same snail, also showing sporocysts
and a large number of cercariae.

The permanence of the water is essential for the snail and more so for the schistosome. Eggs remain viable for some time on moist soil, for example, around unhygenic relief places. Winter conditions, such as prevail in some parts of China and Japan, keep the eggs of *S. japonicum* viable for a long time. Drought is detrimental to the miracidium and the cercariae. Drought also has its effects on the schistosome inside the snail. In Brazil, mature infections of *S. mansoni* in *B. glabrata* are less resistant to the desiccation conditions to which the snail is exposed than are immature infections. After the drought period, immature infections are able to complete their development into cercariae, whereas the mature infections die out.[180]

Wright[243] confirmed a suggestion by several authors that overcrowding of snails and the consequent stunting of their growth and reduction in their fecundity is caused by accumulation of excretory products in the water. Such substances are very effective at very low concentrations and may be of a hormone-like nature. Apparently the excretory products of the snail, and probably other substances in the water also reduce the susceptibility of the snail to infection with schistosomes. *Biomphalaria alexandrina* produced by overcrowded conditions in the aquaria showed considerably less susceptibility to infection with *S. mansoni*; moreover, the incubation period of the parasite in the snail was increased, as compared to infections in snails grown under uncongested conditions.[205] Whether crowding in natural colonies of snails would result in a reduction of schistosome infection rates, and the size of the infection in each snail, as judged by the number of sporocysts and cercariae produced, has to be investigated. The outcome of such studies would be of significance in a better understanding of the schistosome-snail relationships in general.

3. Effects of the Schistosome on the Snail
This subject was discussed in some detail in Chapter 1.

4. Snail Infection Rates
Table 2-4 shows the correlation between schistosome infection rates in snails collected in the field and prevalence of infection in the human population. Variation is observed, and this depends on the species of snail, the species of schistosome, and the endemic area within the geographical region of distribution of the schistosome. In general, relatively few infected snails are responsible for high prevalence rates in humans. This is especially the case with *S. haematobium* in *Bulinus (Bulinus) truncatus* in irrigation schemes in the Nile valley and in the Middle East, and with *S. mansoni* in *Biomphalaria straminea* in Brazil. The figure quoted in the table for the area of the Zoba tribe in Iraq is very exceptional for *B. (B.) truncatus*, and might be due to animal schistosomes (*S. bovis*) in the area. On the other hand, the bulinid hosts of *S. haematobium* in west, central, and South Africa show higher infection rates. Bovine schistosomes are sometimes responsible for these high rates. Those quoted from Gambia, however, were proven by Smithers by passage into mice to be those of *S. haematobium*.[233]

Infection rates of *S. mansoni* in *Biomphalaria* spp. in Africa and in *B. glabrata* in the western hemisphere usually are much higher than in *B. straminea* or *B. tenagophila*. In the western hemisphere certain populations of *B. glabrata* are highly susceptible and the host-parasite relationship is well balanced, in such a way that very high infection rates are encountered, as shown in Table 2-4 for St. Lucia, for Agua Branca and Catolandia, Brazil.

Table 2-4

CORRELATION BETWEEN NATURAL INFECTION RATES WITH SCHISTOSOMES IN SNAILS AND HUMANS

Locality	Snail host	No. snails examined	No. +ve	% +ve	Human infection rates[a]			Ref.
					mansoni	*haematobium*	*japonicum*	
Japan								
Koya-Mura, Ibaraki Prefecture	*Oncomelania nosophora*	534	71	13.3			11/97 (11.3)	221
Numazu area, Shizuoka Prefecture	*O. nosophora*	315	2	0.6			14/155 (9.0)	245
Philippines								
Palo, Leyte	*O. quadrasi*	34,198 ♂	1 383	4.0				225
		48,861 ♀	2 521	5.2				
		83,059 ♂ + ♀	3 904	4.7				
Brazil								
Pernambuco	*Biomphalaria straminea*	3,515	4	0.1	231/747 (30.9)			179
Pernambuco	*B. straminea*	5,919	1		143/431 (33.2)			179
Pernambuco	*Biomphalaria glabrata*	685	18	2.6	96/687 (14.0)			179
Pernambuco	*B. glabrata*	6,619	582	8.8	86/286 (30.1)			179
Lake Santa, Minas Gerais	*B. glabrata*	2,306	230	10	40/290 (13.9)			224
Bambuí, Minas Gerais	*B. labrata*	6,349	260	4.1	151/1600 (9.44)			189
Rio de Janeiro	*B. glabrata*	2,905	10	0.3	7/129 (5.4)			187
Dominican Republic								
Hato Mayor	*B. glabrata*	1,044	216	20.7	52/243 (21.4)			222
Las Palmillas	*B. glabrata*	153	3	2.0	3/65 (4.6)			216
Puerto Rico	*B. glabrata*	5,185	4	0.1	(27.4)			242
Puerto Rico—Guayama	*B. glabrata*	624	73	11.7	High prevalence			192
Puerto Rico—Sabana	*B. glabrata*	50	0	0	High prevalence			192

Location	Species						Ref.
Llana Puerto Rico—Sabana	*B. glabrata*	10	1	10.0	High prevalence		192
Llana Puerto Rico—Caguas	*B. glabrata*	88	2	2.3	High prevalence		192
Yemen Ta´izz	*Biomphalaria alexandrina*	75	27	35.0	Very high prevalence		200
Iraq Zoba tribe, north of Baghdad	*Bulinus (Bulinus) truncatus*	769	71	9.2		14/14 (100)	219
Gambia—Daru	*B. (B.) senegalensis*	625	21	3.1		25/40 (62.5)	233 191
Gambia Simoto bolon	*B. (Physopsis) jousseaumei*	143	25	17.5		High prevalence	233
Kumbiji	*B. (B.) guernei*	132	2	1.5		High prevalence	233
Egypt Qualyub	*B. (B.) truncatus*	14,239	57	0.4		(50%)	238
Qualyub	*Biomphalaria alexandrina*	3,263	19	0.6	0		238
Sudan Gezira scheme	*B. (B.) truncatus*	5,121	3	0.1	7110/81027 (8.8)	7179/81027 (8.9)	195
	Biomphalaria ruppellii	6,601	80	1.2			195
Shambat	*Bulinus (B.) truncatus*	400	7	1.7			207
Shambat	*Bulinus (B.) truncatus*	3,655	17	0.5		(23.2)	210
White Nile, Kosti	*Biomphalaria sudanica*	587	13	2.2	High prevalence		Malek (unpublished)
Zaire[b] Lake Albert	*Biomphalaria* spp.		37	94			193
Kibali Ituri	*Biomphalaria* spp.		58	86			193

Table 2-4 (continued)
CORRELATION BETWEEN NATURAL INFECTION RATES WITH SCHISTOSOMES IN SNAILS AND HUMANS

Locality	Snail host	No. snails examined	No. +ve	% +ve	Human infection rates[a]			Ref.
					mansoni	*haematobium*	*japonicum*	
Lomami Kasai	*Biomphalaria* spp.		91	25				193
Elizabethville	*B. (Physopsis) africanus*	10,728	75	0.7		(58.9)		193

[a] Number infected/number examined and ( ) percentage.
[b] Infection rates in snails and humans range up to levels given.

VII. EPIDEMIOLOGY

A. Prevalence and Incidence

In an epidemiological survey, the usual method employed to determine the prevalence of the disease is to make a urine examination (*S. haematobium*) or a fecal examination (*S. mansoni* and *S. japonicum*) of the respective population groups. For convenience, the population sampling usually consists of school-age children, who represent a relatively uniform group and provide opportunity to assess changes which may occur in incidence of the disease in a relatively short span of years. Relatively few eggs are laid per female *S. mansoni* per day (only about one sixth the number produced by *S. japonicum*). For this reason direct fecal films in the case of *S. mansoni* fail to reveal the characteristic eggs unless the infection has just matured or the number of worms is considerable. Concentration techniques for fecal examination are necessary in such cases. These techniques include gravity sedimentation, sodium sulphate-HCl-Triton-ether method, and the formalin-ether technique of Ritchie.[259] The Kato thick-smear method has been used with some success for *S. mansoni*.[257]

Various immunological tests have been employed for the diagnosis of schistosomiasis. A critical review of these tests has been provided by Kagan and Pellegrino.[252]

The distribution of prevalence by age groups shows a pattern for each species which is more or less uniform in various endemic areas where conditions of transmission are similar. Figure 2-25 shows the prevalence rates for *S. mansoni* by age groups in the town of Pontezinha near Recife, Pernambuco, Brazil. The uniform pattern shows low prevalence rates in the early years of life, a maximum at adolescence, and a decrease during the next age groups. The figure also shows higher rates when the skin test was used in the diagnosis as compared to feces examination, and it is to be remembered that this test is more sensitive for adults as compared to children. Similar findings were reported from the endemic area in the Philippines (Figure 2-26).

Prevalence rates for both sexes show different results in different endemic areas. This is to be expected as social customs differ between endemic areas. In some, women are equally exposed to infected waters as the men, and thus the infection rates are almost similar. In other endemic areas, however, there are restrictions as to the degree of exposure of women and their use of streams for swimming, washing utensils and clothes, and bathing.

Incidence is the rate at which people become positive. Incidence provides a much more sensitive indication of epidemiological events than does a simple statement of prevalence, the latter being the proportion positive at the time of survey. Incidence is measured by the repeated examination of persons initially negative for the parasite, the proportion becoming positive being recorded per unit of time; conventionally, the time unit is 1 year. If the observations are carried out at intervals other than 12 months, the annual incidence may be calculated as

$$I = 1 - X^{12/Y}$$

where X is the proportion remaining negative after Y months. An example was given in Farooq et al.[249] where the number negative at first survey for *S. haematobium* was 175, the number negative at second survey was 130, and the observation period was 18 months:

$$\text{Incidence} = 1 - \frac{130}{175}^{(12/18)}$$

$$= 1 - 0.820$$

$$= 0.18 \text{ or } 18.0\%$$

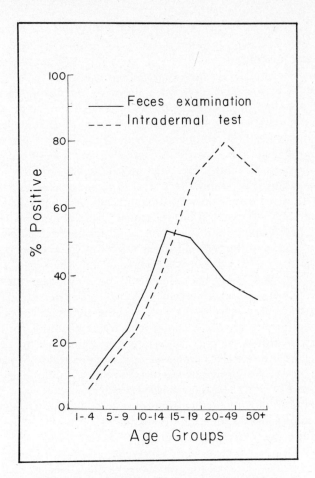

FIGURE 2-25.   Infection rates of schistosomiasis mansoni in various age groups, in Pontezinha, Pernambuco,
Brazil. (Redrawn after Barbosa, 1970.)

B. Intensity of Infection

Several authors are of the opinion that prevalence does not appear to be completely
satisfactory as an index of intensity of the infection. Much attention therefore is being
given in schistosomiasis to the use of egg counts as a more accurate reflection of the
worm load. In schistosomiasis the difficulty is realized that the schistosome eggs are
excreted with the stools or urine as well as being deposited in the tissues. In this respect,
infections with the schistosomes differ from some intestinal helminthic infections for
which correlation of egg counts and intensity of infection has been established.

The pioneering work by Scott[262] on the quantitative aspects of schistosome infections
was only resumed some years later by Stimmel and Scott[264] and Scott.[263] Scott[262] had
shown that the egg output of *Schistosoma mansoni* was no more irregular than that
of hookworm in the same person or than that of hookworm and other intestinal worms
reported by other authors. There was also a regularity in the egg output of *S. haematobium* among the patients which they used for their studies.[263,264] Although there was
a variability in the daily egg counts, the coefficients of variation were no higher than
similar values for intestinal nematodes for which quantitative estimates have been successfully used over some decades. The authors concluded that quantitative estimates
of the average intensity of infection in groups of people and not of an individual seem
warranted.

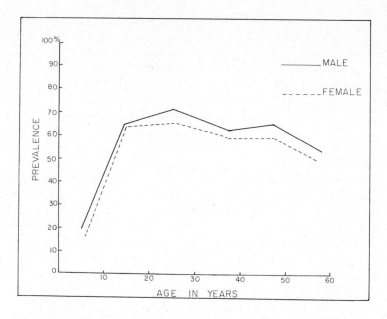

FIGURE 2-26. Age and sex distribution of individuals infected with *Schistosoma japonicum* in Palo, Leyte, Philippines. (Redrawn after Pesigan, T. P., Farooq, M., Hairston, N. G., Jaurequi, E. G., Garcia, B. C., Santos, B. C., and Besa, A. A., *Bull. W.H.O.,* 18, 345, 1958.)

Correlations of egg content of stools and worm burden in man was conducted in a post-mortem study by Cheever.[247] The latter author, after examining stools from the rectum or the sigmoid colon of a number of cadavers infected with *S. mansoni* in Brazil showed that the number of eggs per gram of feces reflected the worm burden in a nearly linear relation. However, the examination was done in relatively few "asymptomatic" cases with more than 80 worm pairs.

Kloetzel[253] in Brazil, and Jordan[251] in Tanzania and St. Lucia have found that prevalnce of clinical manifestations of disease is positively correlated both with prevalence of disease in population and with intensity of infection as measured by fecal egg excretion. It follows, then, that studies carried out in an area of high prevalence and intensity will uncover correspondingly greater morbidity, and vice versa. A controlled study of morbidity of schistosomiasis mansoni in St. Lucian children 7 to 16 years of age, based on quantitative egg excretion, was conducted over 4 years and included a comparable uninfected control group.[248] Infected subjects were divided into three levels of intensity of infection: heavy, 400 or more eggs/mℓ of feces; moderate, 100 to 300 eggs/mℓ; and light, 10 to 75 eggs/mℓ. The results showed that hepatomegaly and splenomegaly were significantly more frequent in the heavy-moderate infection group, and extension of the liver below the costal margin was found to increase with intensity of infection. Serum globulin level and skin-test reactivity were directly related to intensity of infection. The study provided a method for assessing morbidity of schistosomiasis based on quantitative egg excretion and demonstrated the relation of objective morbidity to intensity of infection.

Among animal schistosomiasis a study was carried out using *Heterobilharzia americana* in dogs, which were infected with varying number of cercariae; the eggs in feces were counted over a period of up to 14 months.[256] When the dogs were sacrificed and the worms counted there was a correlation between the number of female worms recovered and the mean number of eggs passed in the stools. The results of this experiment are of interest because the dog is one of the natural hosts of *H. americana*, and thus this schistosome in the dog is a good model for quantitative studies on schistosomiasis.

C. Socio-Economic Aspects

A number of habits common to many parts of the world are very significant, even vital, in the transmission of schistosomiasis. Such habits are tied up with the culture of many communities, and to change them will require great effort for generations to come. The habit of defecating and urinating in the nearest body of water is common in endemic areas of the disease. This pollution of the water does not always result from lack of sanitary facilities; latrines have been constructed in some of these areas, but the population ignores them. It may be that such latrines are sometimes not properly constructed or are inconveniently located, but this is certainly not always the case. Pollution occurs most often in streams that pass through or near villages; agricultural laborers also pollute the canals going through the fields and those near the roads to and from the village. In certain parts of Africa, South America, and the Caribbean islands the natives urinate in streams but defecate a short distance way. They choose shady and protected sites and this enhances the survival of the eggs until the rain comes and washes the excreta down into the water. Domestic and wild animals also track human feces down to the water when they go to drink. The washing of commodes in the canals of China washes fecal matters into the water. Fecal material also reaches the water through laundering of soiled clothes in the streams and through children playing and swimming in the water whereby perianal fecal material drops in the water.

The differences in infection rates between the sexes in various endemic areas are determined by the social customs of the people involved. During childhood both sexes are usually equally exposed to infection; this is the case on Leyte, the Philippines, Venezuela, and the Dominican Republic. Beyond childhood, sex differences become apparent, but the amount of sex differences depends on the position of women in the society, as well as to the amount and kind of work assigned to women. In some communities where women are secluded, for example, in certain countries in Africa, the Middle East, and southwestern Asia, they come into contact with the water rather infrequently and thus they are less infected than men. In one country, the Sudan, the following rates illustrate this contrast: in the Gedaref area of the east, which is very conservative, infection rates with *S. haematobium* were 4.02% in men and 0.56% in women; in Shambat village (also conservative) among school children the author found infection in 102 of 439 boys (23.2%) and only 3 of 395 girls;[255] and in the south (Juba Hospital), where women are emancipated, the infection rate with *S. mansoni* was 50% among men and 33.3% among women. In many parts of southern Sudan women not only bring drinking water from swamps, pools, and rivers and use these water bodies for laundering clothes and washing household utensils, but also habitually bathe in these waters just as the men do. Several other African countries, Caribbean islands, and South America exhibit the same social structure as the southern Sudan.

Schistosomiasis is mainly an occupational disease. Individuals whose occupation brings them in prolonged contact with infested waters contract the infection from these waters (Figure 2-27). Such occupations are farming under irrigation, with Egypt, Sudan, and other parts of the Middle East, certain Caribbean islands, and Venezuela as important examples; the growing of rice under the paddy system, as in the Orient; fresh-water fishing; the leaching-out of bitter manioc (cassava root) in various parts of tropical Africa; and the extensive washing carried out in gold and diamond mining. Inland fishermen in Palo, Leyte showed a higher infection rate than coastal ones.[258] Inland fishermen in Zaire and Ghana, and ferrymen working on central African lakes and rivers, are often infected. Members of the community who have the lowest infection rates are office workers and professional people.

A

B

FIGURE 2-27. Some activities which bring humans in contact with cercariae-infested waters. (A) The use of irrigation instruments in Egypt which is accompanied by immersion of the legs and arms in water for long periods (*S. haematobium* and *S. mansoni*). (B) Seining in Egypt, (C) Washing and fishing in Zaire (*S. mansoni*), (D) Swimming in the Volta Lake, Ghana (*S. haematobium*), (E) Swimming in irrigation canals near Khartoum, Sudan (*S. haematobium*), (F) Watering animals, and fetching water from small bodies of water in western Sudan (*S. haematobium*), (G) Washing clothes and kitchen utensils in the Sudan, (H) Boating in the White Nile reservoir, Sudan (*S. mansoni* and *S. haematobium*), (I) Planting rice in northern Iran (only cercarial dermatitis, due to *Orientobilharzia turkestanicum*).

FIGURE 2-27C

FIGURE 2-27D

FIGURE 2-27E

FIGURE 2-27F

FIGURE 2-27G

FIGURE 2-27H

FIGURE 2-271

However, schistosomiasis is only partly occupational; it is also to a greater degree social. Individuals of low economic status whose occupations do not necessarily bring them into contact with water would still be forced to use infected waters for bathing, washing, laundering clothes, and sometimes drinking.

The long-established endemic areas of schistosomiasis in Africa, the Middle East, and the Americas have apparently been undergoing natural expansion and many areas suitable for transmission have been occupied by the parasites. However, it appears that human interference, through present economic developments, has created new sites suitable for propagation of the infections.

While water resource developments are essential for economic progress, they unfortunately prove to be health hazards in countries where schistosomiasis is endemic. The construction of dams (Figures 2-28 and 2-29) to impound water for human and domestic animal consumption, for irrigation systems, and for power production increases the number of breeding places of the snail intermediate hosts, resulting in increased prevalence and intensity of the disease. This is because the same factors which make an area more satisfactory for man usually also make it more suitable for the snail hosts. Some areas once very inhospitable for man and the snail now have dense populations of both. This, combined with human contacts and with water polluted by excretions, makes them suitable for the transmission of schistosomiasis.

The most important economic factors incriminated in the aggravation of the disease are water storage and man-made lakes, irrigation, fish farming, industrialization, and population movements.

D. Schistosome Transmission Foci

The presence of infected snails in a certain area is not evidence that there will be cercariae in the surrounding water. However, in addition to the survey for snails, the study of cercarial populations in the field can provide a useful means of locating schis-

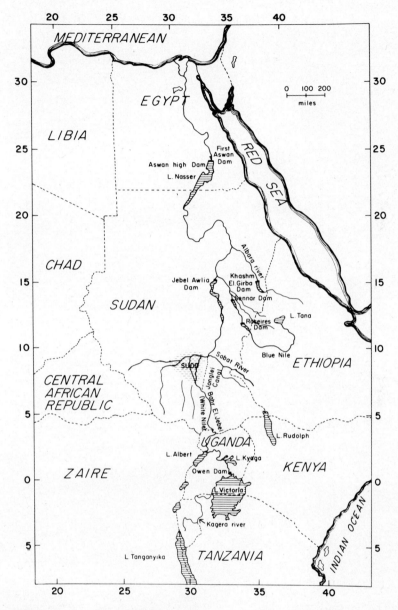

FIGURE 2-28.   The Nile basin in Africa, serves as an example of major economic developments in large river basins, accompanied with disease repercussions. As shown on the map the dams which have been constructed on the Nile are: the Owen dam at Lake Victoria in Uganda; the Roseires, Sennar, Khashm El-Girba, and Jebel Awlia dams in the Sudan; and the Aswan High dam and the First Aswan dam in Egypt.

tosome transmission foci. Several methods have been developed for cercariometry, and among these methods are the following:

1.   Rowan's paper filtration method: the water containing the cercariae is forced through filter paper which will retain them
2.   Mechanical vacuum filtration[260]
3.   Modification of Rowan's methods by employing a battery-operated pumping system[261]
4.   Centrifugation for concentration of the cercariae

A

B

FIGURE 2-29.   Some of the major dams in Africa. (A) The Owen dam, Lake Victoria, (B) The Sennar dam on the Blue Nile, Sudan, (C) and (D) The first Awan dam, Egypt, (E) The Akosombo or Volta dam, Ghana, (F) The main canal (a manmade river) originating from the Sennar reservoir, and irrigating about one and a half million acres of the Gezira agricultural scheme in the Sudan. (Photograph of the Akosombo dam, courtesy of Mr. R. Klumpp.)

253

FIGURE 2-29C

FIGURE 2-29D

FIGURE 2-29E

FIGURE 2-29F

Sandt compared in the laboratory some cercariae recovery techniques. The modified Rowan's technique[261] ranked first for mean recovery accuracy (79.4%), followed by continuous centrifugation (53.8%), and Rowan's technique (25.9%).

Exposure of Mice in Natural Habitats — The mice are usually placed in natural waters in wide-mesh wire cages (Figure 2-30), and left for about 1 hr and examined about a month later for their infection. Such immersion has the advantage of proving the infectivity of the cercariae and identification of their species. For better results for qualitative and quantitative purposes in determining transmission foci it seems advisable to carry out cercariometric determination, rodent exposure, and location of the infected snails.

E. Concomitant Diseases

The occurrence of significant degrees of pathology in schistosomiasis is related to the intensity of infection, but also to the intensity of host response. This is modified in the course of natural infection and by other factors, including malnutrition and concomitant diseases. It is often difficult to determine the relative contribution of each of these diseases to the lowered health of the individual and the community. Table 2-5, from a stool examination conducted by the writer[256a] on the Caribbean island of St. Lucia, shows a number of intestinal helminthiases in the same patients with schistosomiasis mansoni. The findings are similar in the majority of endemic areas. Added to the intestinal helminths are Chagas disease in South America, African trypanosomiasis in Africa, and some other protozoal infections including amoebiasis.

VIII. IMMUNITY

It has already been stated above that in the majority of endemic areas age prevalence rates of infection increase to a peak, usually in the second decade of life, and then decline. This is the case with infections caused by the three main schistosomes of man. The extent of the decline in prevalence in the older age groups varies in different endemic areas and in all three species, and is possibly greatest in *S. haematobium*; this is thought by some investigators to reflect different immunological responses. The in-

A

B

FIGURE 2-30.   Mouse exposure in transmission foci of schistosomiasis mansoni in Brazil, to detect the presence and intensity of the cercariae in natural snail habitats.

tensity of infection also varies with age and shows a similar pattern to prevalence. The fall in prevalence and intensity of infection in adults is attributed to acquired resistance to reinfection, and the gradual spontaneous death of the worms which had been acquired during childhood.

The acquired immunity in adults is supported by the observation that aged individuals in the majority of endemic areas are resistant or much less susceptible to infection with the disease than healthy persons moving in from a nonendemic area. Likewise, when the disease has been recently introduced or intensified in an area, the adult infection rate will, because of lack of acquired immunity, be higher, compared to that of children, than in the area where the disease has long been endemic.

Experimental infection of domestic and laboratory animals has demonstrated that some of them are obviously resistant, others are slightly susceptible, and still others show high susceptibility. These findings could be explained only by the "natural" or "innate" immunity of certain animals. That man is not infected by certain animal schistosomes is another evidence of genetic or natural immunity.

FIGURE 2-30C

FIGURE 2-30D

Both cellular and humoral host factors are evidently involved in the development of such immunities. The development of immunity to schistosomes involves antigenic stimuli. The various stages of the parasite (penetrating cercariae, schistosomula, adult worms, and eggs) within the human body or in other mammals are all potentially antigenic.

Certain studies suggested that there are soluble innmune complexes formed in schistosomiasis, which may be deposited in the kidney glomeruli, causing the nephrotic syndrome. Madwar and Voller[271] reported the finding, by means of gel-diffusion tests, of circulating soluble antigen and antibody in the sera of people infected with *S. mansoni* and *S. haematobium*. Moreover, Houba et al.[272] were able to demonstrate the presence of soluble antigens and antibodies in sera from baboons infected with *S. mansoni*, by countercurrent immunoelectrophoresis. The first antigens detected were

FIGURE 2-30E

gut-associated antigens (2 to 4 weeks after infection), followed by membrane antigens (3 to 5 weeks), and soluble egg antigens (7 weeks). Corresponding antibodies were detected 1 to 2 weeks later. There is a possibility that the simultaneous detection of both components in serum strongly suggests the presence of circulating immune complexes.

The presence of at least two *S. mansoni*-derived antigens has been demonstrated in the trichloroacetic acid (TCA)-soluble fraction of concentrated hamster serum 45 days after exposure to 1500 cercariae.[269] One of the two antigens (an anodically migrating antigen) was also detected in the TCA-soluble fraction of concentrated urine from infected hamsters.

Antigens are produced in secretions, especially the histiolytic secretions of the penetrating cercaria and the escaping egg; digestive enzymes released during the expulsion of the digested blood from the gut of the adult; excretory products, especially those from adult worms; proteins released during turnover of the tegument; and the breakdown products of all stages that may die within the mammalian host. Species-specific antibodies are thus produced in the body of the host. Evidently, sufficient time has to elapse for immunity to develop, during which time heavy and repeated infections may break down the developing immunity. Immunity in schistosomiasis has recently been reviewed.[276,285,288,290]

An immunological state exists in some animals and perhaps in man, where, although the host is immune to reinfection, it is unable to kill off the established population of worms from a primary infection. It should be noted that immunity to schistosomiasis in man is best regarded as a delicately balanced relationship between the immunological forces of the host and the circumvention of these forces by the parasite.

A. Mechanism(s) of Immunity

A wide variety of experimental animals acquire immunity to schistosomiasis in response to previous infection with the parasite. This has been known for many years but only recently has some light been shed on the mechanism(s) of this immunity. It has been expected that schistosomes are able to provoke a multitude of different humoral and cellular responses. The difficulty exists in relating these immune responses induced by the parasite to the process whereby schistosome infections are eliminated

Table 2-5

RESULTS OF STOOL EXAMINATION FOR THE DIAGNOSIS OF HELMINTHIC INFECTIONS IN ST. LUCIA NOVEMBER—DECEMBER 1962[256a]

School or town	No. examined	Age	S. mansoni		Ascaris		Trichuris		Necator		Strongyloides		Enterobius	
			No. +ve	%	No. +ve	%	No. +ve	%	No. +ve	%	No. +ve	%	No. +ve	%
Marchand school	81	5—14	30	37.0	66	81.4	75	92.6	25	30.9	1		2	
Ciceron school	86	4—14	34	39.5	55	63.9	77	89.4	29	34.9	2		2	
Bexon school	89	5—15	66	74.1	66	74.1	60	67.4	42	48.3	4		2	
Joyeaux school	106	5—14	26	24.5	92	87.7	89	83.9	65	61.3	1		1	
Soucis school	18	3—60	11	61.1	14	77.8	13	72.2	9	50.0	4		2	
Banse	23	3—20	12	52.1	18	78.2	15	64.2	11	47.8				
Castries	28	4—25	5	17.9	22	78.6	20	71.4	13	46.4	3		2	
Gros Islet	16	2—60	0	0	10	62.5	12	75.0	9	56.2	4			
Ti Colon	5	8—41	3	60.0	4	80	3	60.0	1	20.0	1			
Ravine Poisson	9	3—49	2	22.2	7	77.8	6	66.7	6	84.4	1			
Bexon	13	4—30	7	53.8	11	53.8	13	1.0	4	30.6	2			

in resistant animals. The evidence now available suggests that, in some experimental animals, acquired resistance during its most efficient phase employs an effect or mechanism involving the cooperation of three components: antibody, complement, and unsensitized cells. It is possible that these elements mediate the rejection of challenge infections through a direct cytotoxic effect against the surface of the schistosomulum, mediated by an antibody-dependent cellular mechanism, and/or damage or trapping by an inflammatory process generated by the interaction of antibody with parasite antigens released into local tissue sites.

Investigations by the London School of Hygiene and Tropical Medicine (Nelson and co-workers) confirmed and extended observations that immunity can develop between species of schistosomes. This type of immunity was coined "heterologous immunity" by the London group. In mice and rhesus monkeys immunity to *Schistosoma mansoni* will develop after exposure to *S. mattheei* or *S. bovis*,[265.266] and in calves, immunity to *S. mattheei* will develop after exposure to *S. mansoni*.[273] These observations have confirmed a suggestion that exposure to animal schistosomes in some parts of Africa may diminish the susceptibility of the population to *S. mansoni*. The writer also showed that an immunity develops against *S. mansoni* in white mice as a result of previous patent infection with *Heterobilharzia americana*.[279]

As to homologous immunity, the major stimulus to its development against *S. mansoni* in the rhesus monkey is the presence of the living adult worm.[286] This was demonstrated by the direct transfer of adult worms from other hosts into the hepatic portal system of normal monkeys, thus inducing an adult infection without prior exposure to cercariae or migrating schistosomula. Monkeys treated in this way with about 80 pairs of worms were almost completely resistant to challenge with 2000 cercariae 8 to 14 weeks later. Worms killed by snap freezing immediately before transfer did not induce immunity, nor did half a million viable eggs injected into the mesenteric veins. From experiments with irradiated cercariae (which will penetrate the host but die in the liver 3 to 4 weeks later), it appears that the schistosomulum provides a much poorer stimulus to immunity, although, if this stimulus is prolonged, as in multiple exposures to large numbers of irradiated cercariae, then some degree of immunity may be induced. These findings apply to the *S. mansoni*-rhesus monkey model but do not apply to all host-parasite systems; for example, irradiated cercariae are as immunogenic as normal cercariae, in the rat.[286]

Acquired immunity in laboratory animals can be demonstrated at an early stage of the infection by the lung recovery method.[283] The technique was first reported by Perez et al.[281] After a challenge exposure to *S. mansoni*, fewer schistosomula are recovered from the lungs of laboratory mice that have had a previous infection than from the lungs of animals that have had no previous contact with the parasite. This is because a significant percentage of the invading schistomula is eliminated in the lungs or at an earlier point, i.e., in the skin. Accordingly, recovery from the lungs is a valid and convenient assay for analysis of the immunologic basis of acquired resistance to schistosome infections. In experimental schistosomiasis strains of the host animal might show different levels of acquired immunity to schistosomes.

In acquired immunity there is apparently a cross resistance to *S. mansoni* and *S. haematobium*. In the hamster, primary infection with *S. mansoni* or *S. haematobium* conferred a high level of immunity to reinfection with either species of schistosome, judged by the perfusion assay, involving recovery of adult worms 6 to 10 weeks following challenge.[284]

An interesting finding with regard to the development of immunity is that protection against cercarial challenge might be produced by anti-snail antibodies.[274] The latter author demonstrated the existence of intermediate host antigens associated with the cercariae of *S. haematobium*. Immunofluorescence studies, using cercariae treated with fluorescein-labeled rabbit antisera to snail *Bulinus (Physopsis) africanus*, showed the snail antigen on the cercarial tegument. The Cercarienhüllen Reaktion was positive with cercariae in rabbit antisera to snail but not in normal rabbit serum. The invading cercariae may thus sensitize the mammalian host to snail antigen as they penetrate the host's skin, and thus resistance to further invasion by cercariae could be induced in this manner. Later, a survey of the prevalence of anti-snail antibodies in various population groups was carried out by Jackson and De Moor.[275] The authors used an extract of the snail *Bulinus (Physopsis) africanus*, the intermediate host of *S. haematobium*, as the antigen in hemagglutination tests. It was found that sera from known schistosomiasis-infected individuals and randomly selected individuals from schistosomiasis endemic areas had significantly higher prevalences as well as higher titers of antibodies to this snail antigen than noninfected individuals and individuals from nonendemic areas.

B. Mammalian Host Antigens

The ability of adult schistosomes to survive in the bloodstream of their mammalian hosts for months or even years implies that the worms are able to escape the consequences of the host's immune response. One of the important findings in the immunology cf schistosomiasis in recent years has been the explanation of the mechanism by which the worms survive. Apparently the presence of host material on the surface of the worm protects it against the host's immune response. Antigens are evidently acquired from the host and are incorporated into the surface of the schistosome, and these antigens serve in some way to block or mask the sites that are vulnerable to the host's immune response. Moreover, it appears that the adult worm liberates antigens that induce a state of immunity against the migrating schistosomula of a challenge infection; but this immunity is ineffective against the adult worms. In this way, the parasite would not only avoid destruction by the host but, through the agency of the host's immunity, would create a barrier against continued reinfection that might otherwise lead to overcrowding and death of the host.

The studies which led to the findings about the acquisition of host antigens by the adult worms were conducted by Smithers and Terry[287] in their worm transfer experiments. Although adult worms transferred from mouse, monkey, or hamster donors into monkey recipients stimulate immunity to reinfection, the behavior of the worms from each of these three hosts is different on transfer.

C. Acquired Immunity in Man Through Infection

Investigators are of the opinion that studies on animals will not be the final answer to the problems of schistosomiasis immunity in man; they can only suggest how such immunity might operate by analogy between effects in animals and in man. The only solution is to expose human volunteers, which involves ethical problems. There are, however, such attempts in the literature. Fischer,[270] in Zaire, infected six presumably immune human volunteers (fishermen over 35 years of age) in a hyperendemic focus of *Schistosoma intercalatum*. Only half of the volunteers passed a few eggs after eight months. Mice exposed to infection from the same lot of cercariae contained adult worms. Clarke,[268] in Rhodesia exposed volunteer adults to cercariae of what was probably *S. mattheei*. The volunteers had lived all their lives in an area endemic for this schistosome. There was some dermatitis and eosinophilia following exposure, however, neither individual passed eggs or showed any other symptoms of infection. Aside from these two schistosomes, Gothe[271] exposed himself to cercariae of *S. haematobium* which had been treated with D.D.T. and finally with untreated cercariae of the same species. No eggs or any other symptoms of the disease were observed.

There is some evidence that a condition of "premunition" exists in schistosome infections. This is an immunty that persists as long as the infecting agent is present and subsides when the host is completely cured. Human old infections (middle-aged individuals), mainly a few adult worms but also eggs, survive and produce resistance. The existence of immunity in the presence of an active adult infection is considered by Smithers and Terry[288] to be an example of "concomitant immunity", a phenomenon occurring in tumor transplants in which animals bearing one tumor are resistant to a second graft, although the first tumor continues to grow progressively. The two terms "premunition" and "concomitant immunity" are being used by schistosomiasis workers. Earlier, Chandler[267] state that premunition in parasitic diseases in general is quickly developed and is highly effective, but usually is of short duration unless continually boosted by reinfection or survival of a few parasites in controlled infection.

D. Acquired Immunity Through Vaccination

Since the development of immunity in schistosomiasis depends on host mechanisms similar to those employed against bacterial or viral infections, one would expect then that an injection of metabolic antigens in the body of a potential host would produce a degree of protection against subsequent infections. Trials have been made, with a certain degree of success, by injection of a saline suspension of whole worms or cercarial antigens. Ozawa[280] "vaccinated" dogs by injection with either whole worms or cercarial antigens and reported that a second infection with *S. japonicum* produced small and very poorly developed worms. Lin et al.[277] succeeded in producing a barely detectable degree of protection by previous injections of saline suspensions of whole *S. japonicum* worms. Similar results were obtained by Sadun and Lin.[282] However, several other authors were unsuccessful in immunizing various species of laboratory animals against *S. japonicum* or *S. mansoni*.

Attenuated parasites can be obtained by irradiation of the cercariae. The latter can penetrate the skin of the host, migrate but fail to reach maturity. Several groups of investigators have reported varying degrees of protection in experimental hosts induced by irradiated cercariae. Taylor et al.[289] recently reported the successful immunization of sheep against *S. mattheei*, using either irradiated schistosomula or irradiated cercariae.

In summary, the degree of resistance induced in schistosomiasis varies greatly with the species of schistosomes, with the host species, and with the technique used in the process of vaccination. Best results have been obtained by previous exposures to homologous or heterologous parasites either intact or attenuated by X-irradiation. Conversely, vaccination attempts with nonliving parasite materials or passive transfer of serum have given somewhat disappointing results.

IX. PATHOLOGY AND CLINICAL MANIFESTATIONS

With regard to pathological damage and clinical aspects there are four stages:

A. Cercarial dermatitis
B. The Katayama Syndrome
C. Stage of egg deposition and extrusion
D. Stage of tissue proliferation and chronic infection

A. Cercarial Dermatitis

Cercarial dermatitis is a sensitization reaction and it was demonstrated on exposure of Egyptians with known schistosome infections to cercariae of *S. mansoni* and *S.*

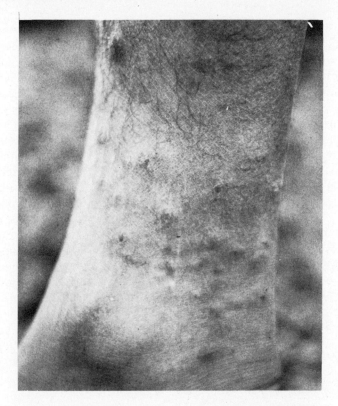

FIGURE 2-31.   Dermatitis on the leg, caused by cercariae of *Orientobilharzia turkestanicum* in rice plantations in northern Iran. (After Sahba, G. H. and Malek, E. A., *Am J. Trop. Med. Hyg.*, 28, 912, 1979.)

haematobium.[295] The dermatitis consists primarily of pruritic papules (Figure 2-31). The chances of development of skin lesions increases with an increase in the number of cercariae which die or are killed in the skin. Most of the enzymes of the cercariae are secreted immediately on penetration through the outer layer.[323] Cercarial secretions and excretions produce immunologic and toxic reactions. Mechanical damage due to penetration and movement of the cercariae is believed to be highly localized.[298] While penetrating the skin of different hosts a varying number of cercariae die, the number depending on the degree of susceptibility of the host.[303,308,324] With regard to the human schistosome species it is believed that more cercariae of *S. mansoni* and *S. haematobium* die in the skin than cercariae of *S. japonicum*. That most of the loss of the parasite in the host is due mainly to the loss of the invading cercariae in the skin was shown by Lewis and Colley.[312] The latter authors showed that with incubation of lung fragments for up to 48 hr the recovery of schistosomula of *S. mansoni* from the lung correlated closely with 7-week worm recoveries.

B. The Katayama Syndrome

This term was first used in infections with *S. japonicum* but later on was also used in infections with *S. haematobium* and *S. mansoni*, and it usually commences 3 to 6 weeks after penetration of the cercariae. In endemic areas people are exposed continuously to the cercariae and thus the onset of penetration of the cercariae and the Katayama syndrome are not known with accuracy. However, the syndrome was described in American tourists where the date of exposure and penetration of the cercariae on St. Lucia, West Indies was known[318] and in Africa by Rabinowitz[320] and Clarke et al.[302] One of Most's[318] patients was exposed to the cercariae on January 29, 1960. On the evening of March 9, 1960, there was some malaise noted, with an oral temperature of 90°F (37.2°C). Beginning with the morning of March 10th and continuing for about 1 week, there were symptoms very similar to those of "flu" with marked malaise, aches and pains all over, as well as chilliness and fever ranging from 101°F (38.3°C) in the morning to 104°F (40.0°C) in the evening. Headache was also a symptom. A persistent symptom was profuse sweating, associated with sleep. During the first few weeks the stools were somewhat frequent and semiformed, but no parasite eggs were observed. On March 16th the eosinophiles-blood smear was 28%, and on March 19th the eosinophil count rose to 34%. On March 24th stool examination and rectal biopsy showed eggs of *S. mansoni*. Thus the prepatency of this species of schistosome in the human body is 55 days. In a few cases death may occur before onset of egg laying. In Brazil post-mortem examination of four cases dying in the acute stage of *schistosomiasis mansoni*[297] revealed massive egg dissemination in the liver and intestines, with granulomatous inflammation. The spleen was congested and large numbers of histiocytes and eosinophils were observed. Warren,[326] in his excellent and thorough review of the subject, believed that while the mechanism of the Katayama syndrome (fever) remains unknown, it is possible that it may be a form of immune complex disease or serum sickness.

Pathology due to immature and adult worms has also been studied. The role that migrating and adult schistosomes play in the pathogenesis of schistosomiasis has been a matter of dispute. The worms affect the host by utilization of host materials including metabolites and red blood cells, output of toxic and antigenic excretions, and secretions and formation of embolic lesions when the worms die.

When the worms die naturally, or when killed by chemotherapy, they pass up in the bloodstream as emboli to the next capillary bed and eventually are trapped in the liver and incite an inflammatory response. However, Cheever et al.[301] found no gross manifestations of liver disease after three cycles of heavy infection and treatment in mice.

The spleen undergoes a transient enlargement early in the course of infection, during the Katayama syndrome. Marked reticuloendothelial activation is believed to be the cause of the involvement of the spleen. Migrating immature worms as well as adult worms cause this activation accompanied by an increase in gamma-globulin-producing plasma cells.[292]

The lungs are involved in the early stages of development. Schistosomula, presumably of all species which infect man and animals, are carried to the lungs via the pulmonary circulation. Migration through the lungs, especially on first exposure, produces no significant damage to the lungs but subsequent exposures to cercariae bring about a marked inflammation around the schistomula, at times with hemorrhages and local accumulation of eosinophils, epithelioid cells, and giant cells around pulmonary blood vessels.

C. Stage of Egg Deposition and Extrusion

S. haematobium and *S. japonicum* are known to deposit a larger number of eggs than *S. mansoni*. The eggs of the first two species accumulate in minute, short chains in radiating branches of the smaller mesenteric venules within the submucosa in the case of *S. japonicum*, and in the minute branches of the vesical venules in the wall of the urinary bladder in the case of *S. haematobium*. In addition to eggs passing to the lumen of the intestinal canal or the cavity of the urinary bladder, eggs become free in the venules, are carried with the blood to the liver, where they filter out perivascularly in the liver and various other organs. It is believed that worms occasionally migrate out of the portal into the caval venous system, with deposition of nests of eggs in pulmonary vessels, or into the vertebral venous system, and deposit eggs in many ectopic locations.

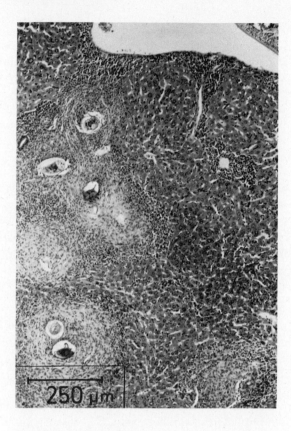

FIGURE 2-32.   Section in liver of white mouse infected
with *Schistosoma mansoni*, showing granuloma formation
around eggs of the parasite.

A large proportion of the eggs which are produced in the host reach the liver in the
case of *S. mansoni* and *S. japonicum*. According to Cheever[300] the proportions of the
eggs which reach the liver are as follows: 33% in man, 25% in rhesus monkeys, 15%
in chimpanzees, and 63% in mice with long-term infections.

In contrast to the migrating schistosomula and the mature adult worm, the egg is
the principal cause of pathology in schistosome infection. The egg incites an inflammatory reaction around where it is deposited in the tissues. The reaction is in the form
of concentric layers of cells and is termed a pseudotubercle. Continued development
of pseudotubercles produces fibrosis (Figures 2-32 and 2-33) and normal function of
the organ where the eggs are deposited is reduced. It should be noted that in Rhodesia
eggs of schistosomes have been found by digestion of every organ in the human
body.[291]

The secretions of the schistosome eggs, regardless of species, are both toxic and
antigenic. Intact or broken eggs release a substance which has a toxic and suppressive
effect on the growth of monkey heart tissue in vitro.[309]

The so-called Hoeppli phenomenon is related to the secretion of substances by intact
schistosome eggs. Hoeppli[310] described eosinophilic ray-like material around *S. japonicum* eggs in tissue sections. Hoeppli speculated that the eosinophilic substance might
consist of antigen-antibody complexes, which was later confirmed in this laboratory
by the use of the fluorescent antibody test[315] and later by Lichtenberg et al.[314]

FIGURE 2-33. Section in mouse liver infected with *Schistosoma mansoni*, showing granulomas around eggs in the parenchyma, and cross section of male and female worms in intrahepatic blood vessel.

D. Stage of Tissue Proliferation and Chronic Infection

The toxic-sensitizing reaction continues throughout the life of the schistosome worms, accompanied by continuous development of pseudotubercles and fibrosis, especially in the liver, and papillomatous growth in the intestinal tract resulting in poor digestion, thickening of the large veins of the liver and periportal fibrosis, portal hypertension, blockage of portal blood flow, and eventually ascites and esophageal and subcutaneous varices. Symmers'[325] original description of the chronic liver disease of schistosomiasis, "clay pipe-stem cirrhosis" (now called fibrosis) implicated the egg in its pathogenesis. Experimental studies using mice confirmed the important role of the egg in the development of liver disease. It was observed that mice with *S. mansoni* infections developed a disease syndrome characterized by hepatomegaly, splenomegaly, portal hypertension, and esophageal varices that was similar in most respects to the human disease.[304,327] There were also studies conducted using chimpanzees as primate models,[313,321] and in some of these animals there were gross lesions in the liver resembling Symmers' clay pipe-stem fibrosis.

In humans the liver shows a characteristics pathological picture of advanced hepatosplenic schistosomiasis mansoni or japonica. The left lobe is disproportionately enlarged, the cut surface (Figure 2-34) reveals pathognomomic dense periportal fibrosis (Symmers' clay pipe-stem fibrosis). The liver parenchyma maintains its lobular structure, but focal areas of nodular regeneration can be seen. Occasionally Symmers' fibrosis is seen in the absence of splenomegaly. Andrade and Cheever[294] reported Sym-

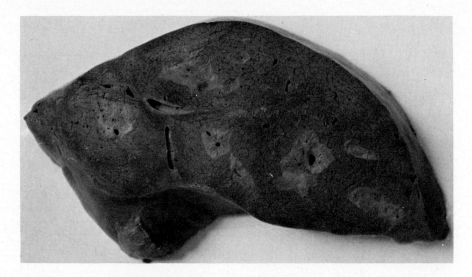

FIGURE 2-34.   Cut surface of a human liver infected with *Schistosoma mansoni*, showing some
of the consequent pathology. (Specimen courtesy of Dr. Shukri Abdel Malek, Alexandria, Egypt.)

mers' fibrosis of the liver in all of approximately 100 cases of hepatosplenic
schistosomiasis autopsied in their hospital in Brazil. Microscopically, the fibrosed portal spaces are infiltrated by lymphocytes and plasma cells. Schistosome eggs are most
often abundant but may be scarce or absent. Vascular lesions are frequent; these include occlusion of small vessels by granulomas, by dead worms, thrombophlebitis of
larger portal branches, narrowing of portal venules, and intimal thickening of arteri-
oles.

With regard to the human liver in urinary schistosomiasis it is believed that the eggs
of *S. haematobium* in the liver are too few to cause significant pathology. However,
there are reports from time to time involving the liver in this form of the disease.[299]
On the experimental side Sadun et al.[322] showed heavily infected chimpanzees with
scattered egg deposition and granuloma formations of variable degree with some com-
posite lesions. Diffuse infiltration of the portal fields with lymphoid cells and eosinophils was also evident.

In human schistosomiasis the spleen is greatly enlarged (Figure 2-35), averaging 1000
g in weight. The usual changes of congestive splenomegaly are seen and there is a
marked reticuloendothelial hyperplasia and infiltration by gamma globulin-producing
plasma cells. The venous sinuses are dilated and may function as venules permitting
rapid blood flow.[293]

In Brazil intestinal lesions are generally not seen,[294] although occasional cases show
isolated polyps or areas of marked serosal or periintestinal fibrosis. In cases with Symmers' fibrosis of the liver, large numbers of eggs are found in the proximal colon and
small intestine, while the rectal mucosa contains few eggs. In cases without Symmers'
fibrosis, eggs are most numerous in the rectal mucosa and distal colon. Rectal fibrosis
is uncommon. Mucosal congestion, edema, and petechiae may be seen at proctoscopy.
Schistosomal polyps are not commonly seen in Brazil by clinicians or pathologists, in
contrast to their apparently high prevalence in Egypt.[305]

In *S. haematobium* infections, about 70% show rectal involvement, and the remainder of the large intestine and appendix may also be affected. Gross lesions are rare,
but granular areas in the mucosa resembling sandy patches are common.[306]

Schistosome eggs have been found in the lungs in many humans and animals with
patent infections. This is very common in *S. haematobium* infections, but also occurs

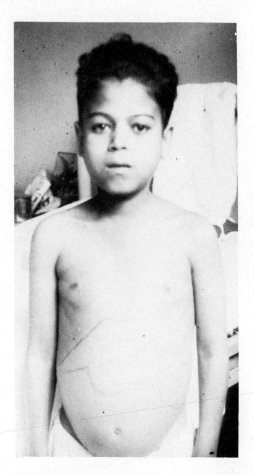

FIGURE 2-35. A 14-year-old boy in Belo Horizonte, Brazil, with an enlarged liver and spleen due to schistosomiasis mansoni.

in *S. mansoni* and *S. japonicum* infections. Granuloma formation in the lungs appears to be similar to that in the liver and other organs. The eggs produce acute necrotizing arteriolitis, after which they escape outside the vessels and produce small parenchymatous lesions. Eggs from such lesions may reach the alveoli and appear in the sputum.[307] Widespread occlusion of the pulmonary vessels occurs in cases of heavy schistosomal infection. Pulmonary hypertension results in dilation of the pulmonary artery and its main branches and hypertrophy of the right ventricles. Cor pulmonale is a syndrome which is usually fatal. It has been reported in schistosomiasis patients in several countries, among which are Egypt, Puerto Rico, and Brazil.

The gross lesions in the urinary and genital systems produced by *S. haematobium* have long been recognized in Egypt, summarized by Makar[317] and Elwi.[307] During the last decade, however, investigations on this disease in Tanzania, Nigeria, Rhodesia, Zambia, etc. demonstrated the importance of schistosomiasis haematobia infections. Moreover, recent investigations on primates revealed the evolution and involution of the gross lesions in the bladder and ureters. Gross lesions in the form of sandy patches occur mostly in the posterosuperior wall of the bladder and near the ureteral openings (Figure 2-36). Schistosomiasis polyps of the urinary bladder have the same pathogenesis as their colonic analogs but they are fewer in number and are more restricted in distribution. Extensive fibrosis of the whole wall of the bladder may lead to marked

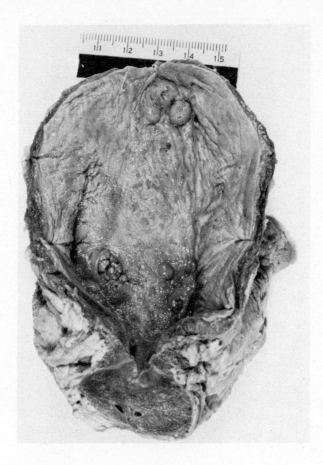

FIGURE 2-36.   Human urinary bladder opened to show pathology of the epithelial lining of the lumen due to schistosomiasis haematobia. (Specimen courtesy of Dr. Shukri Abdel
Malek, Alexandria, Egypt.)

contraction of the organ, whereas fibrosis of the neck obstructs the flow of urine and
leads to the development of hydroureters and hydronephrosis (Figure 2-37). The latter
may be associated with bacteruria, most often involving *Salmonella* species.[311] Schistosomiasis carcinoma of the bladder is seen on the lateral wall, posterior wall, anterior
wall and vault. Of 217 cases of carcinoma of the bladder in the Department of Pathology, the University of Zambia, over a 5-year period, 65% had concomitant schistosomiasis.[296]

Perquis et al.[319] reported 20 patients with stones in the renal pelvis or ureter associated with infection with *S. haematobium* in Africa and Madagascar (Figure 2-38).

Maged[316] considers the L-shaped ureter a rare complication of schistosomiasis. The
anatomical course of the abdominal portion of the ureter, apparently due to schistosomiasis fibrosis of the wall, is completely altered, taking the shape of the letter L,
and instead of lying on the medial side of the psoas major it lies on the quadratus
lumborum muscle.

In humans with schistosomiasis haematobia the genital organs are frequently affected. Lesions may be found in the prostate and seminal vesicles in males and the
cervix and Fallopian tubes in females.

Reports are in the literature about infection of the central nervous system with the
three main human schistosomes, *S. japonicum* infecting the brain and *S. haematobium*
and *S. mansoni* the spinal cord.

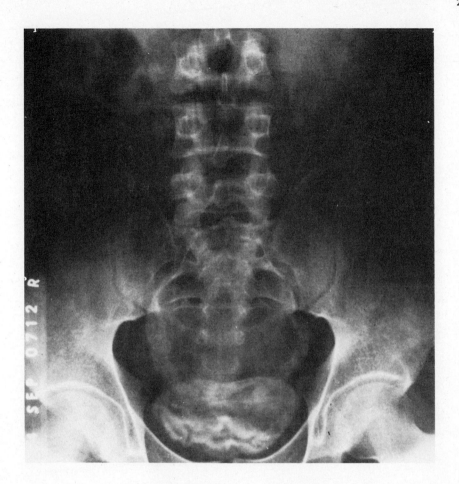

FIGURE 2-37. Schistosomiasis haematobia: (A) Abdominal flat plate without contrast me dium. The marked calcification and distortion of the urinary bladder and distal ureters are clearly visible. (B) Intravenous pyelogram in another patient, showing gross dilatation and distortion of the right distal ureter. (From radiographs demonstrated by Capt. W. Miner, U.S., NAMRU 3. Photographs of Dr. Wojciech A. Krotoski.)

Cerebral schistosomiasis might cause focal Jacksonian epilepsy and occasionally there are signs and symptoms of a generalized encephalitis. Only nests of eggs have been found in brain lesions, indicating that a female or a pair has been there. Eggs of *S. mansoni* or *S. haematobium* have been found in the brain of some individuals on digestion.

Thus at the symptomatic stage the patients present physical signs and laboratory evidence of the disease at its chronic stage and where symptoms have been present for six months to several years. Symptomatic cases of the disease may be divided into two groups: one in which the clinical manifestations are mainly gastrointestinal and meta-bolic, and the other, where there are evidences of involvement mainly to liver and lungs. Gastrointestinal manifestations of the chronic stage include frequent generalized abdominal pain, associated with diarrhea, intermittent with constipation. Stools may be liquid, soft, or solid and are mixed with streaks of red blood or with mucus. Tenes-mus may be present and severe. The colonic and rectal mucosa is congested and may show punctate hemorrhages that can be seen on proctoscopic examination. There is

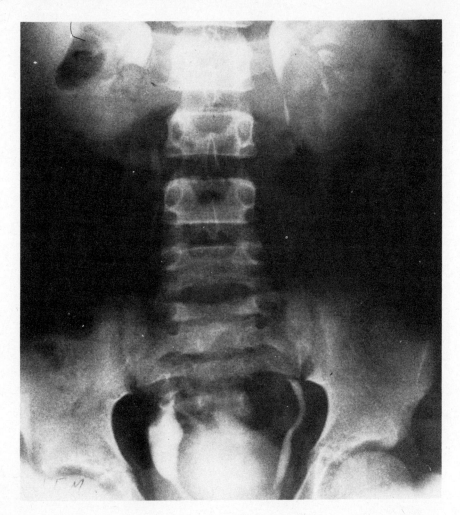

FIGURE 2-37B

thickening of the intestinal wall due to edema. A piece of the rectal valves of Houston may be obtained through the proctoscope, and when this is examined microscopically eggs can be seen. This rectal biopsy is a diagnostic tool for both intestinal and urinary schistosomiasis.

When hepatosplenomegaly appears, the late manifestations of the disease frequently give rise to a new clinical picture. The latter are emaciation, hepatosplenomegaly, ascites, (Plate I) and evidence of fibrosis of the liver and of portal hypertension. Uncontrollable massive hematemesis from ruptured gastro-esophageal varices is a frequent cause of death.

Occasional reports mention other complications involving miscellaneous lesions such as vulvitis, cervicitis, and lesions of the conjunctivae and the skin. All such lesions frequently contain many eggs.

E. Asymptomatic Disease

In endemic areas such as Puerto Rico asymptomatic cases account for 50 to 60%. It is believed that such cases are due to the strain of the parasite or the nutritional status and general health of the patients. In such endemic areas, for many patients

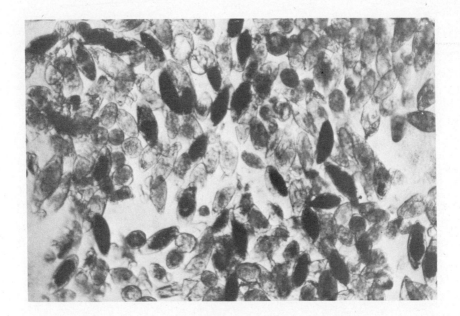

A

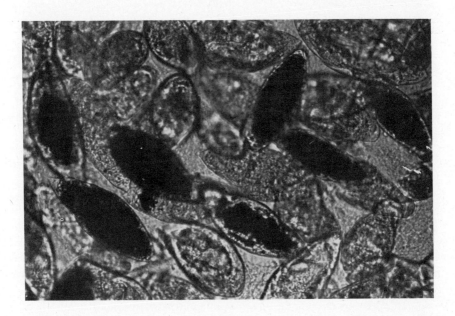

B

FIGURE 2-38.   Schistosomiasis haematobia: (A) Unstained bladder biopsy specimen from a 34-year-old Egyptian seaman presenting with gross hematuria. Complete replacement of the vesical tissue by schistosome ova in this portion of the biopsy is well demonstrated. (B) High magnification of (A). Both viable and calcified (dark) *S. haematobium* ova are present. (C) Alkaline digest of the bladder wall biopsy from same patient, performed by Dr. James Davis, III, showing *S. mansoni* ovum with lateral spine among the *S. haematobium* ova. A rectal mucosal scraping obtained from this patient also demonstrated *S. mansoni* ova, corroborating the presence of a mixed infection. (Material courtesy of Dr. Wojciech A. Krotoski.)

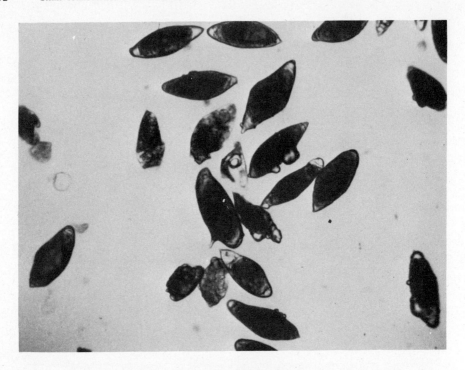

FIGURE 2-38C

who are hospitalized for disease other than schistosomiasis, a diagnosis of that infection is established as a result of routine fecal examination, particularly when stool concentration methods are used. These infections with schistosomes are considered incidental findings not related to the primary disease for which the patient had been admitted to the hospital. The finding of schistosome eggs in the stool of individuals who do not complain of or show clinical manifestations of the disease is more commonly observed in well-nourished persons belonging to high socioeconomic strata of the society.

X. DIAGNOSIS

Several methods are known for the diagnosis of schistosomiasis and these are clinical, parasitological, and immunological. These techniques are required not only for the diagnosis of the disease in individual patients but also for epidemiological surveys and for evaluation of therapy and other control measures.

Some clinical symptoms might aid in the diagnosis of the early incubational stage of the disease. Individuals in endemic areas with a history of exposure to infected waters, in some cases with cutaneous lesions, or with urticaria, eosinophilia, and pulmonary disorders, fever, and abdominal pains are suspect for schistosomiasis. Differential diagnosis from typhoid fever and angioneurotic edema is necessary.

In intestinal schistosomiasis dysentery, enlarged or fibrotic liver and splenomegaly are suggestive, but differential diagnosis from amoebic or bacillary dysenteries, other splenomegalies, and hepatic cirrhosis is necessary. Rectal examination, manual or by sigmoidoscope, may show papillomata, and rectal biopsy snips or scraping may reveal the presence of the schistosome eggs. In the vesical form of the disease hematuria (Plate I), cystitis, urinary calculi, and other vesical symptoms are suggestive. However, differential diagnosis from renal calculi, nephritis, and hemoglobinuria is required.

Cystoscopic examination may reveal grayish elevations, hemorrhagic papules, sandy patches, and papillomata. X-ray examination is helpful for the diagnosis of urinary schistosomiasis where the disease is shown by calcified demarcations or cloud-like shadows in any part of the urinary tract when there is a sufficient deposition of calcified eggs. Rectal biopsy snips or mucosal scrappings may also reveal the eggs of *S. haematobium.*

Parasitological methods of diagnosis have as objective the demonstration of the characteristic eggs of the schistosomes in urine or feces of infected individuals. There are qualitative and quantitative parasitological methods. The qualitative methods are useful for routine diagnosis when only simple laboratory facilities are present. For *S. mansoni* and *S. japonicum* the direct fecal smear method is commonly used and is accurate in cases of high intensity infections. However, concentration techniques are necessary for moderate or light infections or for infections of long duration. One of these methods is gravity sedimentation. Other qualitative techniques include the formol-ether and the merthiolate-iodine-formol concentration methods in which there is a removal of fecal detritus, mucus, and fat with acid and ether.

For *S. haematobium* the specimen of urine is collected, preferably in early afternoon, in a urinalysis or sedimentation glass and the eggs allowed to sediment at the bottom, together with erythrocytes and pus cells. Also, dilution of the urine and hatching of miracidia from eggs is a reliable means of diagnosis, as these miracidia will gather on exposure to light. The same method can also be used for hatching eggs of *S. mansoni* and *S. japonicum* in stools, but requires washing and sedimentation two or three times before exposure to light.

Quantitative parasitological methods are necessary for a chemotherapeutic relationship between prevalence and intensity of infection and morbidity based on egg output. Filtration of the urine and staining the eggs with ninhydrin is a reliable quantitative method for *S. haematobium.* The egg output in the case of *S. mansoni* and *S. japonicum* is determined by one of three methods. They are the formalin-ether sedimentation technique,[338] the Bell filtration and ninhydrin staining method,[328] and the cellophane thick smear technique of Kato and Miura[332] for the diagnosis of helminth infections. The formol-ether concentration technique has recently been modified for increased sensitivity in detecting *S. mansoni* eggs.[334] The thick-smear technique has been adapted, improved, and evaluated in the case of stools containing eggs of *S. mansoni.*[333,335,336] The latter method is based on the rapid clearance of the fecal background with a glycerin-malachite green solution (the square cellophane cover slips are soaked in this solution prior to use), and the method has the additional advantage of having an adequate sample size.

Immunodiagnostic methods, unlike parasitological demonstration of infection, are based on indirect evidence (antibodies formed against the schistosome), and thus should only be used when they are highly sensitive and very specific. Moreover, immunodiagnostic techniques of schistosomiasis should not have priority over coprological or urine examinations.

There are several available immunodiagnostic techniques for schistosomiasis which have been and are still being used. These include immediate and delayed intradermal tests using different stages of the schistosome as antigen, the miracidial immobilization test, the circumoval precipitin test, the cercarial agglutination test, the Cercarienhüllen Reaktion, several tests involving complement fixation, hemagglutination, flocculation, immunodiffusion, and immunoelectrophoresis, and several varieties of fluorescent antibody tests. Many of these procedures are reviewed[330,331,341] with a discussion of the performance of these tests in some recent field projects.

The intradermal test is highly sensitive, but it is not reliable when performed on children. Warren et al.[339] used the intradermal test in a schistosomiasis area in St. Lucia in the Caribbean. The intradermal antigens showed high sensitivity but were not as reactive in children. False positive reactions in a control population from the nearby island of St. Vincent (where the disease does not occur) were very high. Delayed skin reactions, although less frequent than immediate skin reactions, were more specific, inasmuch as 66% of the St. Lucian egg passers showed delayed skin reaction. With the same St. Lucian population the complement fixation and slide flocculation tests were insensitive, but the indirect immunofluorescence was sensitive and acceptable.

Improvements in the indirect immunofluorescence technique involved the use of frozen sections of adult worms.[329] Wilson et al.[340] evaluated adult and cercarial antigen of *S. mansoni* for sensitivity and specificity and found the cryostat adult antigen superior, in that no cross reactions with sera from patients with schistosome dermatitis and trichinosis were obtained.

The double diffusion (Ouchterlony) test was evaluated with sera from patients with acute disease whose sera were positive by complement fixation test.[337] The gel diffusion test was sensitive with sera from patients with acute disease but was most insensitive with sera from patients with chronic disease.

XI. TREATMENT

There is an increasing need for effective chemotherapeutic compounds for the treatment of schistosomiasis. This is because, although the prevalence of the disease has declined in a few countries (Japan and Puerto Rico), the prevalence of the disease has been rising in other endemic areas all over the world. New development projects have increased the snail habitats, man-water contact, and pollution of these transmission foci by the ever increasing population. Treatment of schistosomiasis started with the use of potassium antimony tartrate (tartar emetic) by Christopherson.[345] Since then several antimonial and other preparations which have come into clinical use have become available (five or six trivalent antimonials and four or five nonmetallic organic compounds).

The efficacy and cure rates of available drugs have recently been reviewed by Standen[363] and by WHO.

A. The Antimonials

1. Tartar Emetic

A "standard" course of treatment in Egypt comprises 12 intravenous injections administered on alternate days. Each dose contains 2 mg/kg body weight (the maximum individual dose should not exceed 130 mg of tartar emetic). The total amount of antimony administered to adults weighing 60 kg and over varies from 420 to 450 mg. Tartar emetic is therapeutically effective against the three main species of the parasite. The cure rates recorded are as follows: *S. haematobium* — 80 to 90%; *S. mansoni* — 75 to 90%, and *S. japonicum* — 40 to 75%.

2. Stibophen (Fouadin, Repodral)

Stibophen is effective against *S. haematobium* and *S. mansoni* infections, but has little therapeutic value for *S. japonicum*. In Egypt and the Middle East the average individual dose of stibophen used is 5 mg/kg body weight, given in an intramuscular injection. The cure rates reported are 70 to 80% in *S. haematobium*, and 60 to 75% in *S. mansoni* infections.

The drug is not used for mass treatment campaigns because of the prolonged course of treatment and the occurrence of serious reactions reported in some campaigns.

3. Stibocaptate (Astiban)

This drug is administered as a 10% solution. Originally given intravenously, current preference and practice is for intramuscular administration. The individual dose is 8 mg/kg body weight, and the prescribed course consists of five injections administered on consecutive or alternate days. The cure rate for *S. haematobium* is over 75%. For *S. mansoni* it varies from 40 to 75%. However, when given intravenously or intramuscularly in daily injections, it is not well tolerated by the majority of patients. Therefore the administration of treatment in biweekly or one-weekly intramuscular injections has been resorted to in some countries.

B. Niridazole (Ambilhar)

This drug is administered orally at a daily dose of 25 mg/kg body weight for 7 consecutive days. It is therapeutically active against *S. japonicum, S. mansoni,* and *S. haematobium* infections, with the respective cure rates of 40 to 70%, 40 to 75%, and 75 to 95%. Similar or greater percentages of reduction in egg output in uncured cases are also reported. However, with *S. mansoni* and *S. japonicum* infections there is the probability of liver function impairment and this militates against the general use of the drug. Clinical experiences suggest that a history of psychotic disturbance is also a contraindication.

C. Metrifonate

This is one of the organophosphorous compounds which include several extremely dangerous substances. Their administration to man is therefore subject to the closest scrutiny. Fortunately, metrifonate is a relatively safe drug, but in spite of this a cloud of doubt will always hang over it until it has undergone extensive monitored trials in man. This drug, which is administered orally, is effective only against *S. haematobium* infections. The recommended dose is 7.5 mg/kg once every 2 to 4 weeks, to a maximum of three doses.[348] The cure rate is 70 to 80% with minor side effects only. Metrifonate also reduces egg output in those who are not parasitologically cured. However, in man there is a depression of cholinesterase activity in both plasma and red cells. Such depression returns rapidly to normal in plasma but takes several weeks in red cells. Plestina et al.[359] demonstrated these effects in 63 children, aged from 7 to 18 years, attending school in Tanga, Tanzania, and all infected with *S. haematobium.* The drug appeared to be effective; there were only four failures to clear the schistosomiasis at 1 month, and four at 3 months. A 60% cure rate was obtained in another field trial in Rhodesia by Jewsbury et al.[353] involving rural children.

D. Lucanthone (Miracil D)

This drug has been in limited use for about 25 years. At one time it was the drug of choice in Rhodesia,[344] especially for the treatment of *S. haematobium* infections. It has the advantage of being used by mouth and is an egg suppressive, but its side effects include nausea, vomiting, anorexia, and in some cases severe depression. More recent interest in lucanthone was centered around its possible role as an egg suppressant when given at spaced intervals at the generally tolerated dose of 500 mg, weekly or fortnightly. McMahon,[355] using lucanthone to treat *S. mansoni* infections, administered the drug at 500 mg weekly for 10 weeks; 5 (out of 18) of his patients were cured, and the pretreatment fecal counts were reduced by 67.4% in the remaining patients. Other egg-suppressive results obtained with *S. mansoni* was the study by Lees,[354] who obtained 89% reduction.

E. Hycanthone

Hycanthone is a schistosomicide therapeutically effective against both *S. mansoni* and *S. haematobium* infections of man. It is the 4-hydroxymethyl analog of lucanthone (Miracil D). Rosi et al.[360] reported the conversion of lucanthone to hycanthone in a fermentation system where optimum concentrations of lucanthone were incubated in growing cultures of *Aspergillus sclerotiorum*. That hycanthone is a metabolic product of lucanthone was demonstrated by finding hycanthone in the urine of animals after an oral dose of lucanthone. The effect of hycanthone on *S. mansoni* in laboratory mice has been described by Berberian et al.[343] and Pellegrino et al.[358] among others. Trials in South Africa and Brazil indicated that hycanthone is effective when given orally as enteric coated tablets for 3 to 5 consecutive days, or when administered intramuscularly at a single dose of 2.0 to 3.5 mg/kg (optimum dose 3.0 mg/kg). The intramuscular administration is of a soluble salt, for example, sulfamate or methanesulfonate. Hycanthone is manufactured in England for commercial distribution under the name "Etrenol".

Apparently hycanthone is not effective against *S. japonicum* infections.[364] In that study Japanese and Philippine strains of the parasite were used, and in addition to the drug being ineffective it did not enhance the activity of tartar emetic or stibophen.

In Egypt Shoeb et al.[361] reported a certain degree of success in treating *S. mansoni* and *S. haematobium* infections. In Brazil, Ferraz et al.[349] reported on two patients (*S. mansoni*) who after the treatment with hycanthone at a dose of 3 mg/kg single intramuscular injection developed a symptomatic psychosis.

On the basis of trials with hycanthone, Cook et al.[346] concluded that for a mass campaign, a dose of 1.5 mg/kg or 2 mg/kg presents considerable advantages: there would be increased acceptability with no loss of efficacy. In a thorough and controlled trial of hycanthone and placebo in schistosomiasis mansoni in St. Lucia more information has recently been gained. In St. Lucia 32 patients (mean age 12.03 years) were given a single intramuscular dose (2.5 mg/kg of body weight) of either hycanthone or a vitamin placebo. The principal side effect, vomiting, was limited to 4 of the 16 patients given hycanthone; 3 patients in the hycanthone group and 2 of the 16 given placebo complained of abdominal pains.

Some data has been gathered with regard to the teratogenicity, mutagenicity, and carcinogenetic potential of hycanthone. As to teratogenicity, pregnant mice were given an injection of 10 to 50 mg/kg on the 7th day of gestation; they were killed on the 17th day and the fetuses examined.[357] Doses of 10 mg/kg had no effect. After 35 mg/kg the fetal mortality per litter was 15% (normal 4 to 8%), and 31% of the fetuses per litter were abnormal. However, 50 mg/kg was very harmful; the fetal mortality was 44.6%, and 49% of the fetuses per litter were with some malformations, among which were fusion and branching of the ribs, hydrocephaly, and microphthalmia.

F. Oxamniquine (UK-4271)

This compound is produced by the Pfizer Company and it is 6-hydroxymethyl-3-isopropylaminomethyl-7-nitro-1,2,3,4-tetrahydroquinoline. Foster et al.[352] gave preliminary results with this compound. Several workers have reported high parasitological cure rates of *S. mansoni* with a single intramuscular injection at 7.5 mg/kg; oral oxamniquine in a single dose of 12 to 15 mg/kg has also been considered very promising. Oxamniquine used against *S. haematobium* and *S. japonicum* in mice, hamsters, and primates was either inactive or gave indifferent results.[350,351] A single intramuscular injection of 7.5 mg/kg in children with *S. haematobium* infections achieved few radical cures, although egg output fell. That oxamniquine is not effective against *S. haematobium* was also demonstrated by McMahon[356] who attempted to treat 30 Tanzanian adults.

High cure rates were obtained with two groups of 227 patients with *S. mansoni* in Brazil.[362] One group (112) was given a single dose of 12.5 mg/kg, and the other group (115) was given 15 mg/kg. Cure rates at 6 months were 81.1% in the first group and 82.7% in the second group. Side effects were similar in both groups.

By using an oral dose of 800 mg/m² body surface per day for 2 days (this corresponds to between 40 and 60 mg/kg) oxamniquine was effective against *S. mansoni* in children (21 out of 24); but only 6 of 25 children were cleared of *S. haematobium*.[342]

Thus it appears that oxamniquine is effective against only one species (*S. mansoni*), but it meets some of the requirements of the ideal schistosomicide, especially ease of administration, low toxicity and chemoprophylactic activity.

G. EMBAY 8440 (Praziquantel)

A very recent drug, EMBAY 8440, under the trade name Praziquantel (Droncit in veterinary medicine) has been found effective against the three species *Schistosoma mansoni, S. haematobium*, and *S. japonicum*.[347] It is orally administered in man and is used at a dose of 50 mg/kg body weight. Praziquantel is absorbed immediately through the duodenum and is excreted largely as metabolites in urine. It has also been found effective against other trematode infections, such as those of *Clonorchis sinensis, Opisthorchis felineus*, and *Paragonimus westermani*. For schistosomiasis, it has been and is still being used experimentally in Zambia, in Brazil, in the Philippines, and in Japan.

Thus the effectiveness of the drugs available at present depends on the species of schistosome involved. For *S. japonicum* there are the trivalent antimonials and niridozole; for *S. mansoni* and *S. haematobium* there are lucanthone, stibocaptate, niridozale, metrifonate, and hycanthone, and for *S. mansoni* there is oxamniquine (UK-4271).

One important point with reference to the chemotherapy of schistosomiasis is the presence of several strains of each of the human species, *S. mansoni, S. japonicum*, and *S. haematobium*. Various investigators have demonstrated geographical strains of the parasites and these differ as to their biology, pathogenicity to animals, and degrees of snail host specificity. Chemotherapeutic studies of *S. mansoni* infections in the field and laboratory have not always yielded similar results, and differences between strains have been suggested as a possible explanation.

XII. ZOONOTIC SCHISTOSOMIASIS

Schistosome species parasitizing lower animals and man insure their continued existence by adapting themselves to a broader host spectrum. Accordingly we find that certain species normally parasitizing other mammalian species have become adapted to man, posing a threat to his well being. According to the Joint WHO/FAO Expert Group on Zoonoses,[436] the term zoonoses is defined as "those diseases and infections which are naturally transmitted between vertebrate animals and man." With regard to the schistosomes the definition of the term is still, however, in a very fluid state. Some terms and classifications have been suggested for the zoonotic schistosomiases,[412] and for zoonotic helminthiases in general.[425] The author is avoiding these terms because such terms, aside from the fact that they all require definition and explanation, have not been generally accepted by many of those concerned with zoonotic helminthiases. The classification of the zoonoses and the suggested terminology cause more problems and bring up more questions than the value obtained from them. The terms do not fulfill the beneficial goals of the authors who have introduced them. The terms, however, may be helpful in indicating the direction and the degree of infection.

Table 2-6

GEOGRAPHICAL DISTRIBUTION AND HOSTS OF THE ZOONOTIC SCHISTOSOMES

Schistosome	Snail intermediate hosts	Natural definitive hosts	Geographical distribution
Species causing dermatitis in man			
Trichobilharzia ocellata	*Lymnaea stagnalis, Stagnicola palustris*	Ducks, teals	Europe, Canada, U.S.
T. physellae	*Physa anatina*	Ducks	U.S., Louisiana
T. physellae	*Physa parkeri*	Ducks	Canada, Michigan, Wisconsin
T. stagnicolae	*Stagnicola emarginata*	Canaries (exp.)	Canada, Michigan, Wisconsin
T. nasale	*Lymnaea natalensis undussumae*	Ducks	Zaire
Bilharziella polonica		Ducks	Poland, Russia
B. polonica	*Planorbarius corneus*	Mute swan	U.S., Washington, D.C.
B. polonica		Domestic and wild ducks	Germany
Dendritobilharzia loosi		Pelican	Russia
G. igantobilharzia acotylea		Gull	England, Sweden
G. huttoni	*Haminoea antillarum guadaloupensis*	White pelican	U.S., Florida
G. sturniae	*Segmentina hemisphaerula*	Starling, sparrow, wagtail	Japan
G. gyrauli	*Gyraulus parvus*	Blackbirds	U.S., Wisconsin
G. huronensis	*Physa gyrina*	Goldfinch, cardinal	U.S., Michigan
Austrobilharzia variglandis	*Littorina pintado*	Ruddy tern stone	U.S., Hawaii
A. variglandis	*Nassarius obsoletus*	Lesser scaup duck	U.S., Rhode Island
A. variglandis	*Nassarius obsoletus*	Redbreasted merganser	U.S., Connecticut
A. variglandis	*Littorina planaxis*		U.S., California
A. terrigalensis	*Pyrazus australis*	Seagulls	Australia
A. penneri	*Cerithidea scalariformis*	Chicks, pigeons (exp.)	U.S., Florida
Ornithobilharzia canaliculata	*Batillaria minima*	Royal tern	U.S., Florida
O. odhneri		Asiatic curlew	China, Peking area
Species causing nonpatent infection in primates			
Schistosomatium douthitti	*Stagnicola palustris, Lymnaea stagnalis*	Muskrat, deer mouse, meadow mouse	Northern U.S., Canada, Alaska

Heterobilharzia americana[a]			
H. americana		Bobcat	U.S., Florida
H. americana	Lymnaea cubensis, Lymnaea (Pseudosuccinea) columella	Raccoon	U.S., Texas, North Carolina
		Raccoon, dog, nutria, rabbit	U.S., Louisiana
Schistosoma spindale	Indoplanorbis exustus	Cattle, buffalo	Malaya and Sumatra
S. spindale	Indoplanorbis exustus	Goat, cattle, buffalo	India
Species causing patent infections in man			
Schistosoma bovis[b]	Bulinus (Bulinus) truncatus	Sheep, goat, cattle, equines	Southern Europe, Southwestern Asia
S. bovis	B. (B.) truncatus, B. (B.) nasutus, B. (B.) forskalii, B. (B.) senegalensis	Sheep, goat, cattle	Africa (about the northern half)
S. bovis	B. (B.) truncatus, B. (Physopsis) ugandae	Sheep, goat, cattle, pig, equines, camel, man	Sudan
S. bovis	B. (B.) truncatus	Cattle, sheep, camels	Iraq
S. mattheei	B. (Ph.) africanus, B. (Ph.) globosus	Cattle, sheep, goat, rodents, impala, blue wildebeest, zebra, man	South Africa, Transvaal
S. mattheei	B. (Ph.) globosus	Cattle, sheep, baboons, man	Rhodesia
S. margrebowiei	Bulinus tropicus, Bulinus depressus	Equines, cattle, sheep, several species of antelopes, man	South Africa
S. margrebowiei		Antelopes, zebra	Zaire
S. leiperi (= S. spindale var. Africana)		Antelopes, man	South Africa
S. leiperi	Bulinus (Physopsis) globosus, B. (Ph.) africanus	Cattle, sheep, other herbivores, man	Zambia
S. rodhaini	Biomphalaria bridouxiana, B. sudanica	Rodents, dog, man	Zaire
S. rodhaini	Biomphalaria spp.	Rodents	Uganda, Kenya

[a] Non-patent infection was produced in rhesus monkeys, but patent infections in cebus and squirrel monkeys

[b] The Table shows geographical distribution of S. Bovis, but only few human cases were confirmed.

Zoonoses have been well demonstrated in the schistosome group. Several species of blood flukes that are parasites of lower mammals and birds infect man. One of the problems often encountered is whether a species which infects man and animals started as an animal species which became later on adapted to man, or whether it had been a human species that became later on adapted to animals. In general, however, one assumes that the schistosomes were animal parasites before man existed on earth. With the appearance of man in the Pleistocene, certain animal schistosomes became adapted to man to insure their continued existence and spread.

The zoonotic schistosomes of public health importance (Table 2-6) can be treated in three categories according to whether in humans they invade the skin only and produce dermatitis, whether they invade the deeper tissues and produce a nonpatent visceral infection in the lung and liver, or whether they invade the deep tissues and cause a patent infection with the deposition of eggs in the tissues and their excretion in the feces and urine as they do in the normal host.

A. Schistosome Dermatitis

The schistosome cercariae which penetrate the human skin and cause, at least in sensitized individuals, a characteristic dermatitis with rash and appearance of papules are mostly parasites of birds and a few mammalian species. This type of infection, first described by Cort,[380] is variously known as swimmer's itch, clamdigger's itch, Gulf Coast itch, el-caribe, and koganbyo.

Species known to cause dermatitis in man include *Trichobilharzia ocellata, T. physellae, T. stagnicolae* (in the U.S. and Europe); *T. anatina, T. berghei, T. schoutedeni* (the last three in Rwanda, Africa), and *T. maegraithi* in Thailand; *Gigantobilharzia huttoni* (Florida), *G. gyrauli* (Wisconsin); *G. huronensis* (Michigan); *G. sturniae* (Japan); *Austrobilharzia variglandis* (California, Connecticut, Rhode Island, Hawaii), *A. penneri* (Florida); *Ornithobilharzia canaliculata* (Florida, Brazil); and *O. pricei* (Virginia).

In a review of the subject of "cercarial dermatitis," Cort[381] summarized the records of such cases all over the world. There have been several more recent papers of cases observed at fresh-water lakes in the U.S., Canada, and Argentina.[385,392,393,395,430,431] In Europe similar cases were reported in the Netherlands,[371,423] in Switzerland,[410] and in Austria.[390] In Africa, cases have been reported from South Africa,[382] and Zaire (the former Belgian Congo).[386,387,388] In Australia, Bearup and Langsford[367] reported on cercarial dermatitis contracted in rice fields in Northern Territory. In addition to the above cases, which were all reported in fresh-water bodies, some have also been reported in brackish and sea water, and it has been discovered that the intermediate hosts are brackish-water and marine snails. The cases reported from continental U.S. and Hawaii were by Stunkard and Hinchliffe,[429] Chu,[379] and Leigh.[401,402] In Australia, Bearup[366] studied many cases of dermatitis caused by the cercariae of *Austrobilharzia terrigalensis*, which develops in marine snails. Marine schistosome dermatitis has also been reported from northern Queensland, Australia.[421]

As indicated by skin biopsies, cercariae of this group of parasites usually are immobilized in the skin. However, in some cases a few may penetrate to the deep tissues and cause lesions in the lungs; this is known to occur in the albino mouse and rhesus monkey.[413] Man's reaction to cercariae of this group of bird schistosomes is a sensitization phenomenon. First exposures elicit mild and transitory responses, but on subsequent exposures there is a reaction to the cercarial antigenic substances with the production of larger papules and often other symptoms of an allergic response. Recently, Hunter[394] reviewed the pathology of the primary and secondary response and the clinical course of cercarial dermatitis.

On primary exposure the episode may pass unnoticed; the cercariae penetrate rapidly, and in about 30 min they are down to the malpighian layer of the epidermis. In individuals who have been sensitized by previous exposures, skin biopsies show that the cercariae push the epidermal cells apart, probably aided by enzymes secreted from the penetration glands of the cercariae. However, the parasites are limited to the epidermis, and are not found below the stratum germinativum. Clinically, a prickling itch occurs within about ½ hr; macules then appear, a pruritis becomes intense in 6 to 7 hr, and in 24 hr large papules or nodules 8 to 10 mm in diameter develop. These are usually accompanied by erythema, edema, and intense pruritis, with the subsequent formation of vesicles. By 52 hr after penetration of the cercariae, the cercariae are completely destroyed and the tunnel in which the cercariae rested is filled with exudate, leukocytes, lymphocytes, and an amorphous eosinophilic mass. The entire dermis shows much edema, indicative of a well-marked inflammatory response. The marked erythema and even large urticarial wheals which are produced in some individuals are the result of a strong allergic response. Repair starts about 70 to 100 hr and is virtually complete by the fifth day. Some melanin-pigmented areas remain after the papules have disappeared. However, rubbing or scratching complicates the course of the dermatitis, whereby the vesicles rupture, and become pustular when they are secondarily infected. In such cases the repair process is delayed a few more days.

B. Nonpatent Visceral Schistosomiasis

The cercariae of *Schistosomatium douthitti* (Cort, 1915), Price, 1929, occurring in northern U.S., Canada, and Alaska, produce dermatitis in sensitized persons and in the rhesus monkey. In the monkey the schistosome reaches the lungs and develops to adult worms in the liver.[396.414] Sexually mature and young normal *S. douthitti*, smaller in size than worms of a corresponding age in white mice, were recovered from the liver and intestinal venules of rhesus monkeys necropsied 10 to 25 days after exposure to the cercariae. No worms were found in other monkeys when they were necropsied 30 to 92 days after exposure. Earlier, Fairley[389] infected *Macaca sinicus* with *Schistosoma spindale* Montgomery, 1906; the infection terminated, however, in 11 to 15 days after exposure. The cercariae of *S. spindale*, an Asiatic schistosome, had been incriminated in dermatitis cases known as "sawah itch" in Malaya.[373] *S. spindale* is also present in Thailand and is responsible for dermatitis cases. A survey of animals in Kalasin Province, northeast Thailand, showed a high prevalence of this schistosome in water buffaloes and cattle.[392] No mature infections were found in indigenes, although in some inhabitants the complement-fixation and skin tests, using adult *S. japonicum* antigen, were positive. These reactions were considered to result from the effects of abortive infections arising from exposure to *S. spindale* cercariae. In one human subject, and in laboratory animals, dermatitis was produced after exposure to *S. spindale* cercariae. In animals, however, adult worms were subsequently found.

Heterobilharzia americana Price, 1929 (Figure 2-39) and *S. douthitti* (Figure 2-40) are the only mammalian schistosomes known in North America. *H. americana* is found in the coastal and near coastal areas from Texas to the Carolinas (Figure 2-41). In this enzootic area *H. americana* parasitizes raccoon, nutria, marsh rabbit, opossum, and dog. In the laboratory this schistosome can be maintained in mice, hamsters, and dogs. In rhesus monkeys it produces dermatitis and a nonpatent infection. Complement-fixing antibodies can be demonstrated in infected monkeys accompanied by a very pronounced eosinophilia. In man it produces dermatitis and a nonpatent infection with peripheral eosinophilia.[409] Preadult worms and advanced schistosomulae were recovered from the livers of the rhesus monkeys together with lesions in the liver and lungs.[409] The infection is terminated by the death of the schistosomulae and the pre-

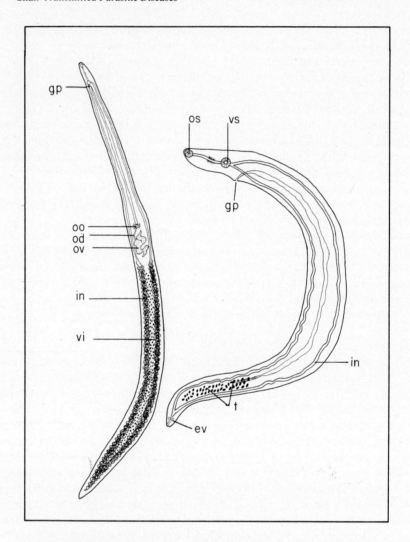

FIGURE 2-39.   *Heterobilharzia americana* from the raccoon and dog in Louisiana; adult flukes, female (left) and male (right). ev, excretory vesicle; gp, genital pore; in, intestinal cecum; od, oviduct; oo, ootype; ov, ovary; t, testes; vi, vitellaria; and vs, ventral sucker.

adults. It is quite possible that the schistosome meets the same fate in man as it does in the rhesus monkey. The fate of the schistosome is different in certain South American monkeys. In the capuchin and squirrel monkeys the infection becomes patent with excretion of viable eggs in the stools.[408] The coatimundi also proved to be a good host, and its egg (miracidia)-producing capacity parallels that of the raccoon.[408] *H. americana* is probably responsible for the dermatitis cases often reported among hunters, trappers and oil rig workers in Louisiana.

C. Patent Zoonotic Schistosomiasis

To this group belong subprimate mammalian schistosomes which may or may not produce dermatitis in humans but do develop to maturity in man and certain other primates. Among these schistosomes are: *Schistosoma bovis, S. mattheei, S. margrebowiei*, and *S. rodhaini*.

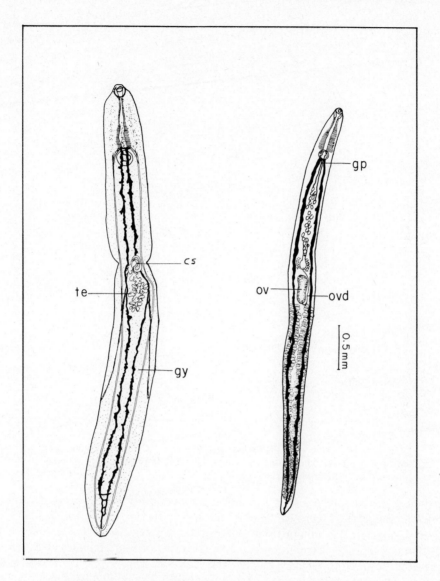

FIGURE 2-40. Adult *Schistosomatium douthitti*, male (left) and
female (right). gp, genital pore; gy, gynecophoric canal; ov, ovary;
ovd, oviduct; and te, testes.

1. Schistosoma bovis (Sonsino, 1876) Blanchard, 1895

This species, originally described from cattle in Egypt, has been reported later from
a variety of domestic animals in Africa, the Middle East, southern Europe, and western
Asia. It inhabits the mesenteric venules and has not been reported as a urinary infection
in its hosts (cattle, sheep, goats, camels, horses, and pigs). There are some records of
this bovine schistosome in man. Raper[420] reviewed the literature, and reported on a
natural infection which he himself contracted in Uganda. Other human cases were
reported in South Africa,[397] in Egypt,[424] in the Sudan,[406,407] and Rhodesia.[370] With
regard to the reports from South Africa and Rhodesia, whether these eggs are those
of *S. bovis* or of another bovine schistosome, namely *S. mattheei*, still has to be elucidated. Dinnik and Dinnik,[383] after compilation of material and data, believe that *S.
bovis* does not occur in Rhodesia or South Africa.

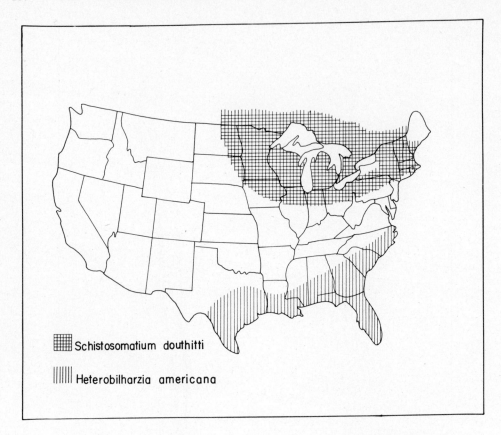

FIGURE 2-41.   Map of the U.S. and adjacent part of Canada, showing known distribution of the North American mammalian schistosomes, *Schistosomatium douthitti* and *Heterobilharzia americana.*

It should be remembered that it is not an impossibility that eggs of *S. bovis* are excreted in the urine of man. In the cases the author reported from the Sudan[406,407] of eggs of *S. bovis* in the human urine, there were also eggs of *S. haematobium*. It is likely that a male of *S. haematobium* might carry one or more females of *S. bovis* on its migratory route to the urinary bladder and the vesico-urinary plexus. It is also likely that *S. bovis* worms have an equal chance or affinity to the bladder and gut veins of man as *S. haematobium* worms.

With regard to infection of subhuman primates with *S. bovis*, Brumpt[372] recovered mature adults, but without eggs, from experimentally infected rhesus monkeys. The writer[407] recovered adult *S. bovis* (Sudanese strain) with in utero eggs from an African green monkey, *Cercopithecus aethiops*. There was deposition of some eggs in the liver and intestinal wall. Certain strains of *S. bovis* produce only dermatitis in man in the Mediterranean islands and do not develop further to mature adults.[369]

2. Schistosoma mattheei Veglia and LeRoux, 1929

This species was described from sheep in South Africa. It has been found naturally in the horse, zebra, sheep, cow, and a wide range of antelopes.[403,416] It occurs in the mesenteric venules of the bovine hosts; however, Pitchford[415] found high prevalence rates (23%) of infection with *S. mattheei* among humans in the eastern Transvaal. Viable eggs were found in both urine and feces. In later surveys, Pitchford[416] reported even high prevalence in man, 35 to 40%, among inhabitants of some villages in the Transvaal, where the schistosome frequently was found together with *S. haematobium*.

Interbreeding between these two schistosomes was thought to occur in man, producing a hybrid which also infects rodents, cattle, and sheep. In post-mortem examinations of 200 Africans in Rhodesia, Alves[365] found eggs of *S. mattheei* in the appendix, bladder, brain, liver, lung, rectum, spleen, and uterus.

3. *Schistosoma margrebowiei Le Roux, 1933*

Of cattle, sheep, antelopes in Rhodesia and South Africa, it produces an egg which is similar to that of *S. japonicum*. It is included in this group because it is probably Cawston's[377] *S. japonicum* which he reported from man in South Africa.

Le Roux[404] recovered *S. margrebowiei* from naturally infected cattle, leche (*Kobus leche*), puku (*K. vardoni*), sitatunga (*Tragelaphus spekei*), reedbuck (*Redunca arundinum*), roan antelope (*Hippotragus equinus*), wildebeest (*Connochaetes taurinus*), *dinum*), roan antelope (*Hippotragus equinus*), wildebeest (*Connochaetes taurinus*), and zebra (*Equus* sp.) from southern Zambia. *S. margrebowiei* was recovered from leche and zebra from Katanga Province, Zaire by van den Berghe,[432] and possibly Walkiers[434] who reported viable eggs of *S. faradjei* in five patients with bloody diarrhea from northeast Zaire. Lapierre and Hein[399] reported eggs possibly of *S. margrebowiei* mixed with *S. mansoni* and *S. haematobium* in a rectal biopsy from Mali. Pitchford[418] reported eggs of *S. margrebowiei* in the stools of two individuals in eastern Caprivi, South Africa. The known distribution of *S. margrebowiei* thus extends from Chobe Game Reserve in Botswana, through eastern Caprivi to southern Zambia and Katanga in Zaire.

4. *Schistosoma leiperi Le Roux, 1955*

This species was found by Le Roux[405] in several species of antelopes together with *S. margrebowiei*. It was recovered in a total of 21 herbivore species, including cattle and sheep in Zambia. Various reports of *S. spindale* in southern central Africa prior to 1955 are all believed to refer to *S. leiperi*. One of these reports was by Porter,[419] who stated that *S. spindale* var. *africana* eggs occurred in the urine of man in South Africa.

Buckley[374] reported *S. leiperi* from the leche antelope and as a pseudoinfection in man from Lake Bangweulu, northern Zambia. Eggs of *S. leiperi* and *S. margrebowiei* were readily detected in leche, puku, waterbuck, and reedbuck droppings from Chobe Game Reserve and eastern Caprivi, without *S. mattheei*, and what were probably *S. leiperi* eggs as a single infection in waterbuck from Kazangula ranch in west Rhodesia.[417] The known distribution of *S. leiperi* thus includes that of *S. margrebowiei* (except Katanga), but probably including the extreme western corner of Rhodesia through to northern Zambia and central southern Tanzania. The snail hosts of *S. leiperi* are *Bulinus (Physopsis) globosus* and *B. (Ph.) africanus*. Figure 2-42 shows the geographical distribution of *S. bovis, S. mattheei*, and *S. leiperi*.[383]

5. *Schistosoma rodhaini Brumpt, 1931*

This is a member of the *Schistosoma mansoni* complex, with lateral spined eggs, and is a schistosome of rodents and the dog in Zaire. It has been reported from man in that country.[391,428] *S. rodhaini* causes an intestinal type of schistosomiasis. Its eggs are polymorphic, with either terminal or subterminal spine. The opposite end is rounded off or knob-like and may be twisted to one side of the longitudinal axis (Figure 2-17). *Biomphalaria* spp. are the intermediate snail hosts, the same as in the case of *S. mansoni*.

S. rodhaini was also reported from five localities in the Kampala area of Uganda, and was also reported near Kisumu in Kenya.[368]

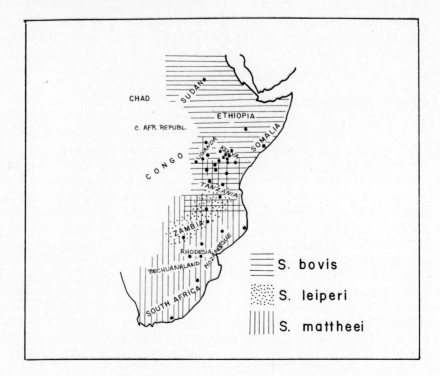

FIGURE 2-42.   Distribution of the schistosome species affecting domestic ruminants in the eastern half of Africa. (Redrawn after Dinnik, J. A. and Dinnik, N. N., *Bull. Epizoot. Dis. Afr.*, 13, 341, 1965.)

D. Potential Sources of Zoonotic Schistosomiasis

Other mammalian schistosomes whose zoonotic relationships are not known or have not been sufficiently studied may be mentioned. Very likely they produce dermatitis in man, and possibly a nonpatent infection: *Bivitellobilharzia loxodontae* (Figure 2-43) is a parasite of the elephant in the former Belgian Congo. Although the natural snail host of *Bivitellobilharzia loxodontae* has not been identified, *Lymnaea palustris* from Germany was experimentally infected by Vogel and Minning[433] who also described the schistosome as a new genus and species.

Among the schistosomes of India the following are listed in this group: *Schistosoma indicum* Montgomery, 1906 from the snail *Indoplanorbis exustus* is a parasite of the horse, water buffalo, sheep, and camel;[427] *Schistosoma incognitum* Chandler, 1926 (syn. *S. suis*) from *Lymnaea luteola* is a parasite of the pig and dog; *S. nasalis* from *Lymnaea luteola* and *L. acuminata* is a bovine schistosome; *Orientobilharzia nairi* (Mudaliar and Ramanuja-Chari, 1945) Bhlalerao, 1947 is a parasite of the elephant;[411] and *Orientobilharzia datti* Dutt and Srivastava 1952[384] occurs in water buffalo and cattle.

Schistosoma incognitum was described by Chandler[378] from one human case in India. It was later known from the pig and dog in that country. In addition to India, *S. incognitum* was later reported from small rodents in Thailand[400] and in small rodents in central Sulawesi, and Java, Indonesia.[375,376]

Orientobilharzia turkestanicum Skrjabin, 1913, Price, 1929 is a bovine schistosome that occurs in Russian Turkistan, China, Iraq, and Iran. It has also been reported in the cat in Russian Kasakhstan, and in equines and the camel in Iraq. Watson and Najim[435] incriminate the cercariae of this schistosome in some of the dermatitis cases reported in Iraq.

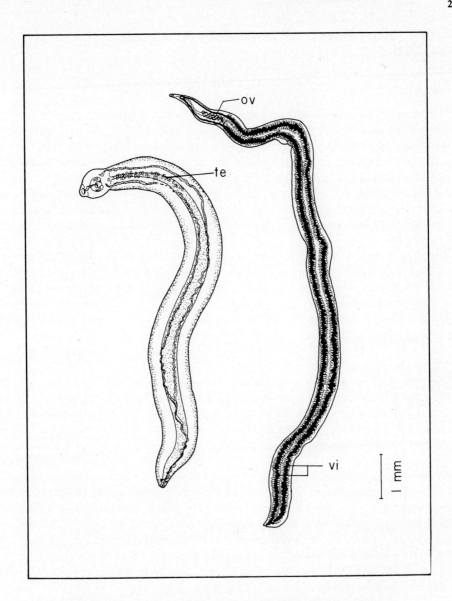

FIGURE 2-43. Adult male (left) and female (right) *Bivitellobilharzia loxodontae* from African elephants, ov, ovary; te, testes; and vi, double vitellaria. (Adapted from Vogel, H. and Minning, W., *Arch. Schiffs Trop. Hyg.*, 44, 562, 1940.)

Orientobilharzia harinasuti Kruatrachue, Bhaibulaya and Harinasuta, 1965 in a parasite of water buffaloes in southern Thailand. Its egg has a lateral spine similar to *S. mansoni*, but the snail host is a lymnaeid snail, *Lymnaea rubiginosa*,[398] rather than a planorbid snail. The adult worms live in the portal and mesenteric veins of the mammalian hosts.

The genus *Orientobilharzia* was erected in 1955 by Dutt and Srivastava (in Srivastava),[426] for the mammalian species which had previously been included in the genus *Ornithobilharzia* Odhner, 1912. In addition to *O. datti* Dutt and Srivastava, 1952; *O. turkestanicum* (Skrjabin, 1913) Price, 1929; *O. harinasutai*, and *O. nairi* (Mudaliar and Ramajuju-Chari, 1945) Bhalerae, 1947; it includes the species *O. bomfordi* (Montgomery, 1926) Price, 1929. The latter species could also be of zoonotic importance. Large outbreaks of dermatitis among rice farmers were encountered in northern Iran, and it was determined that they are caused by the cercariae of *O. turkestanicum* [422] (Figure 2-31).

XIII. CONTROL

The necessity to control schistosomiasis has become more eminent after its public health significance has been documented. Much information has been gathered on control measures. Among the possible control measures are use of chemotherapeutic agents to treat patients, prevention of contact between humans and water bodies likely to contain infected snails, or interrupting the life cycle of the parasites (schistosomes) at some point or the other, through sanitary measures and provision of piped water. Another measure which has been experimented with in many endemic areas is control or destruction of the snail hosts by various methods, namely environmental, chemical, and biological.

Although it is recognized that there are still gaps in our knowledge of the measures applicable for control, we now have available the tools for effectively controlling the disease. Thorough precontrol ecological and epidemiological studies of the snails, humans, and other mammalian hosts, as well as studies on the dynamics of transmission of the disease, are essential for determining the strategy of control that will suit the specific epidemiological conditions prevailing in a certain area.

A. Chemotherapeutic Treatment

The treatment includes that of the individual as well as mass chemotherapy. The latter is defined as the drug treatment of all those members of a defined community or a defined area who are shown to be infected with schistosomiasis on preliminary parasitological screening. Davis[437] discussed the problems involved in chemotherapeutic control, these problems being: epidemiological problems, problems associated with the parasite, with the human host, and with specific drugs, their actions and toxicity, and operational problems of personnel, time and money, of parasitological techniques, and of the attitudes of local health authorities and administrators toward the place of schistosomiasis in the rank order of diseases. The use of chemotherapy, either alone or in combination with other control measures, is currently an important component of programs for the control of the disease.

No one method of control or even one combination of methods will be applicable to all areas. The choice of the control measure or measures is decided upon by precontrol epidemiological studies.

B. Environmental Control of Schistosomiasis

Environmental control measures comprise the elimination or reduction of human-water contacts. The cost in St. Lucia of provision of a piped water supply in addition to a few swimming pools and laundry facilities, is $5 to $6 U.S. per capita at design capacity.[438] This figure was estimated at $7 to $11 U.S. at a later date.[439] Environmental control measures also include the control of snail infection through effective sewage disposal systems; they limit the introduction of schistosome eggs into fresh-water and moist habitats harboring the snail hosts. Activated sludge and oxidation pond treatment followed by chlorination are effective measures through their impact on the schistosome eggs, cercariae, and the snail populations which the oxidation ponds harbor.

There are a number of environmental control measures of the snail hosts, the objectives of which are to make the snail habitats untenable or less suitable for the snail hosts, and in this way eliminate or reduce the snail populations. Such measures have a good antecedent in the excellent results obtained in malaria control by eliminating, modifying, or managing the habitats of the larvae of various mosquito vectors. Some

A

B

FIGURE 2-44.   Some control measures against schistosomiasis. Chemical control (A, B, C, and D). (A) The use of very low concentrations of Frescon in an irrigation scheme near Arusha, Tanzania. The molluscicide is placed in a barrel inside the enclosure. Note a tubing carrying the molluscicide solution to the headwaters of the main canal. (B) Application of low concentration of copper sulphate (placed in jute bags) at the canal sluice in the Gezira Irrigation scheme, Sudan. (C) The molluscicide dinitrocyclo-hexyl-phenol is mixed with sawdust in a barrel, before its application in canals and drains in Egypt. (D) Use of a screen ("mechanical barrier"), and a molluscicide in experimental areas in the Sudan. Environmental control in Venezuela (E) and (F). (E) Use of latrines (F) Use of laundry facilities and showers in the center of the village.

FIGURE 2-44C

FIGURE 2-44D

of the control measures are designed for snail control in natural streams, in reservoirs and lakes, and in irrigation schemes. Engineering measures for irrigation schemes include the channel design, its operation, proper drainage, and weed control. Canal linings are recommended for several reasons, among which are seepage control, erosion control, reduction of maintenance, weed control, and reduction or elimination of snail

FIGURE 2-44E

FIGURE 2-44F

habitats. Covering irrigation canals and underground drains have also been recommended and the latter was introduced in some parts of Egypt; it is estimated that the system adds 1 acre in 8 to agricultural land. This means that in addition to being a snail and transmission control measure, it will probably meet the costs of the program.

C. Chemical and Biological Control

Chemical control of the snail hosts is by the use of molluscicides, which were discussed in Chapter 1. Acceptable data are now available which show the significant role which molluscicides play, alone, or in combination with other measures, in the control of schistosomiasis. WHO[439] reviewed the data from some control programs, and the review showed that in projects in Tanzania, Egypt, Japan, Rhodesia, and Ghana, where molluscicides alone were used, there was a marked impact on incidence of the disease. In Brazil near Belo Horizonte, in Rhodesia (Kyle Catchment), and in Egypt at Warraq El-Arab, there has also been evidence of decrease in prevalence. Molluscicides were also effective in combination with chemotherapy in Malagasy, Cameroon,

Khuzestan in Iran, Rhodesia, Fayoum Governorate in Egypt, and near Mochi at Arusha (Figure 2-44), and at Misungui in Tanzania.

For biological control of the snails the use of several competitors, predators, and pathogens (see Section VI) has been recommended. The snails *Marisa Cornvarietis* and *Helisoma* spp. have been used in laboratory and field experiments. Competitive displacement of *Biomphalaria glabrata* by *B. straminea* has been observed in several natural habitats in Brazil.

REFERENCES

HISTORICAL BACKGROUND

1. **Baelz, E.**, Ueber einige neue Parasiten des Menschen, *Berl. Klin. Wochenschr.*, 20, 234, 1883.
2. **Bilharz, Th.**, Fernere Beobachtungen über das die Pfortader des Menschen bewohnende *Distomum haematobium* und sein Verhaltniss zu gewissen Pathologischen Bildungen, *Z. Wiss. Zool. Abt. A*, 4, 72, 1852.
3. **Cobbold, T. S.**, On some new forms of Entozoa, *Trans. Linn. Soc. London*, 22, 363, 1859.
4. **Cort, W. W.**, The cercaria of the Japanese blood fluke, *Schistosoma japonicum* Katsurada, *Univ. Calif. Berkeley Publ. Zool.*, 18, 485, 1919.
5. **Cort, W. W.**, Notes on the eggs and miracidia of the human schistosomes, *Univ. Calif. Berkeley Publ. Zool.*, 18, 509, 1919.
6. **Farooq, M.**, Historical development, in *Epidemiology and Control of Schistosomiasis (Bilharziasis)*, Ansari, N., Ed., S. Karger, Basel, 1973, 1.
7. **Faust, E. C. and Meleney, H. E.**, Studies on schistosomiasis japonica, *Am. J. Hyg., Monogr. Ser.* 3, 1924.
8. **Fujii, Y.**, Katayama disease (with note), *Chugai Iji Shimpo*, 691, 1847.
9. **Fujinami, A.**, Further discussion on the pathology of Katayama disease and preliminary report concerning its parasite, *Geibi-Iji*, 100, 1904.
10. **Fujinami, A. and Nakamura, H.**, A few specimens demonstrated to show the life history of *Schistosoma japonicum*, *Nippon Byori Gakai Kai Shi*, 1, 1911.
11. **Garrison, P. E.**, The prevalence and distribution of the animal parasites of man in the Philippine Islands, with a consideration of their possible influence upon the public health, *Philipp. J. Sci.*, 3(B), 191, 1908.
12. **Girgis, R.**, *Schistosomiasis*, Bale-Sons-Danielsson, London, 1934.
13. **Katsurada, F.**, The schistosomiasis japonica in Saga Prefecture, *Okayama Igakai Zasshi*, No. 175, 1904.
14. **Kurimoto, T.**, Description of the eggs of a new parasite, *Tokyo Igakai Zasshi*, 7 (22—23), 1893.
15. **Leiper, R. T.**, Report on the results of the Bilharzia Mission in Egypt, 1915. 1. Transmission, *J. R. Army Med. Corps*, 25, 1, 1915.
16. **Leiper, R. T. and Atkinson, E. L.**, Observations on the spread of Asiatic schistosomiasis, *Br. Med. J.*, 1, 201, 1915.
17. **Malek, E. A.**, Ecology of schistosomiasis, in *Studies in Disease Ecology*, Vol. 2, May, J. M., Ed., Hafner, New York, 1961, 261.
18. **Manson, P.**, Tropical Diseases. A Manual of the Diseases of Warm Climates, Cassell, London, 1898, 605.
19. **Miyagawa, Y.**, Route of penetration of *Schistosoma japonicum* from skin to the portal vein, and morphology of the young worm at the time of penetrating the skin, *Tokyo Iji Shinshi*, No. 1736, 1911.
20. **Miyairi, K. and Suzuki, M.**, Contribution to the development of *Schistosoma japonicum*, *Tokyo Iji Shinshi*, No. 1836, 1913.
21. **Miyairi, K. and Suzuki, M.**, Der Zwischenwirt des *Schistosomum japonicum* Katsurada, *Mitt. Med. Fak. Kaiserl.*, 1, 187, 1914.
22. **Ruffer, M. A.**, Note on the presence of *Bilharzia hematobia* in Egyptian mummies of the twentieth Dynasty, 1220—1000 B.C., *Br. Med. J.*, 1, 16, 1910.
23. **Sambon, I. W.**, Remarks on *Schistosomum mansoni*, *J. Trop. Med. Hyg.*, 10, 303, 1907.

24. **Silva, M.,** Piraja da, Contribution to the study of schistosomiasis in Bahia, Brazil, *J. Trop. Med. Hyg.*, 12, 159, 1909.

25. **Soto, V. R.,** Naturaleza de la Disenteria en Caracas, Tesis de doctorado No. 63, Universidad Central de Venezuela, Caracas, 1906.

26. **Weinland, D. F.,** An essay on the tapeworms of man, giving a full account of their nature, organization, and embryonic development, the pathological symptoms they produce, and the remedies which have proved successful in modern practice, to which is added an appendix, containing a catalogue of all species of helminthes hitherto found in man, Cambridge, Mass., pp. 1—93, 1858.

27. **Woolley, P. G.,** The occurrence of *Schistosoma japonicum* vel cattoi in the Philippine Islands, *Philipp. J. Sci.*, 1, 83, 1906.

28. **Yamagiwa, K.,** Contribution to the etiology of Jacksonian epilepsy. (The pathological changes of the cerebral cortex caused by distoma eggs.), *Tokyo Igakai Zasshi,* 3 (18), 1889.

29. **Abdallah, A.,** Assignment report, Schistosomiasis Control in the Syrian Arab Republic, WHO unpublished document, EM/SCHISTO/42, World Health Organization, Alexandria, Egypt, 1968.

30. **Abdel-Azim, M. and Gismann, A.,** Bilharziasis survey in southwestern Asia, covering Iraq, Israel, Jordan, Lebanon, Saudi-Arabia, and Syria; 1950—1951, *Bull. W. H. O.,* 14, 403, 1956.

31. **Alcay, L., Marill, F., Musso, J., and Castryck, R.,** Découvert d'un foyer de bilharziose vesicale autochtone en Algerie, *Bull. Soc. Pathol. Exot.,* 32, 608, 1939.

32. **Alves, W.,** Report on Bilharziasis Surveys in Liberia, April, May, and June, 1955, World Health Organization, Brazzaville, Congo, 1955.

33. **Ayad, N.,** Bilharziasis survey in British Somaliland, Eritrea, Ethiopia, Somalia, the Sudan and Yemen,, *Bull. W. H. O.,* 14, 1, 1956.

34. **Azar, J. E.,** La bilharziose en Tunisie, Rapport de mission, 1967, WHO unpublished document, EM/BIL/41, World Health Organization, Geneva, 41, 1968.

35. **Azar, J. E., Luttermoser, G. W., and Schacher, J. F.,** First report of a focus of schistosomiasis in Lebanon, *Am. J. Trop. Med. Hyg.,* 10, 709, 1961.

36. **Azevedo, J. F. de,** Rapport sur les bilharzioses humaines au Cameroun Francais, Working document, World Health Organization/Bilharziasis African Conference, No. 49, Brazzaville, Congo, 1956.

37. **Azevedo, J. F. de,** Rapport sur les Bilharzioses Humaines au Maroc Espagnol, Working document, World Health Organization/Bilharziasis African Conference No. 47, Brazzaville, Congo, 1956.

38. **Azevedo, J. F. de,** O homen nos tropicos. A spectos bioecologicos. Junta de investigácoés do Ultra—mar, *Estados Ensaios e Documentos,* Lisboa, No. 114, 1964.

39. **Azevedo, J. F. de, Colaco, A. T. F. and Costa Faro, M. M. da,** As bilharzioses humanas no Sul de Save (Mocambique), *An. Inst. Med. Trop. Lisbon,* 11, 5, 1954.

40. **Azevedo, J. F. de, Costa Faro, M. M. da, Morais, T. de, and Dias, J. A. P. de,** As bilharzioses humanas em Manica e Sofala (Mocambique), *An. Inst. Med. Trop. Lisbon,* 14, 5, 1957.

41. **Azevedo, J. F. de, Da Silva, J. B., Cotto, A. de M., Coelho, M. F., and Colaco, A. T. F.,** O foco Portugues de schistosomiase, *An. Inst. Med. Trop., Lisbon,* 5, 175, 1948.

42. **Bailey, D. R. and Davis, A.,** The prevalence and intensity of infection with *Schistosoma haematobium* in primary school-children of Tanga, Tanzania, *East Afr. Med. J.,* 47, 106, 1970.

43. **Baruffa, G.,** Ambulant treatment of urinary schistosomiasis with Astiban, TWSB-6, *Trans. R. Soc. Trop. Med. Hyg.,* 45, 143, 1962.

44. **Bergerot, J.,** Le foyer de bilharziose de Djonet, pays Ajjer (Sahara algerien), *Arch. Inst. Pasteur Alger.,* 13, 47, 1935.

45. **Berry, E. G.,** Investigations of the Snail Intermediate Hosts of Bilharziasis in Libya, WHO unpublished document, PA/230.64, World Health Organization, Geneva, 1964.

46. **Blair, D. M.,** Bilharziasis survey in British West and East Africa, Nyasaland, and the Rhodesias, *Bull. W. H. O.,* 15, 203, 1956.

47. **Brygoo, E. R.,** Les bilharzioses humaines à Madagascar, *Tananarive,* 2nd ed., 1969.

48. **Buck, A. A., Spruyt, D. J., Wade, M. K., Deressa, A., and Feyssa, E.,** Schistosomiasis in Adwa. A report on an epidemiological pilot study, *Ethiopian Med. J.,* 3, 93, 1965.

49. **Burch, T. A.,** Quarterly Report, Laboratory of Tropical Diseases, National Institutes of Health, Bethesda, Md., June 30, 1953.

50. **Buttner, A.,** Le complexe "Mollusque-schistosome" au Brasil, *Bull. W. H. O.,* 18, 909, 1958.

51. **Coumbaras, A.,** Bilharziasis in South Tunisia, particular study at Nefzaoua, Kebili:Douz region, *Arch. Inst. Pasteur Tunis,* 37, 313, 1960.

52. **Coumbaras, A.,** Bilharziasis in South Tunisia, particular study at Nefzaoua, Kebili:Douz region, *Arch. Inst. Pasteur Tunis,* 38, 255, 1961.

53. **Cowper, S. G.,** Schistosomiasis in Mauritius, *Trans. R. Soc. Trop. Med. Hyg.,* 47, 564, 1953.

54. **Davies, A. M.,** Preliminary account of an outbreak of bilharzia in the Beth-Shean Valley, *Harefuah,* 49, 10, 1955.

55. **Delas, A., Deschiens, R., Ngalle, E. S., and Poirier, A.,** La bilharziose à *Schistosoma intercalatum* au Cameroun. Etude epidemiologique preliminaire, *Bull. Soc. Path. Exot.,* 61, 625, 1968.

56. **Deschiens, R.,** Problème sanitaire des bilharzioses dans les Territoires de l'Union Francaise. Frequence, mollusques vecteurs, conditions étiologiques, *Bull. Soc. Pathol. Exot.,* 44, 631, 1951.

57. **Deschiens, R.,** *Le Problème Sanitaire des Bilharzioses dans les Territoires de l'Union Francaise, Monogr. V, Soc. Path. Exot.,* Masson, Paris, 1952.

58. **Deschiens, R., Delas, A., Ngalle, E., and Poirier, A.,** La repartition geographique des bilharzioses humaines au Cameroun, *Bull. Soc. Pathol. Exot.,* 61, 772, 1968.

59. **Deschiens, R., Lamy, L., and Mauze, J.,** Repártition géographique et fréquence de la bilharziose intestinale en Guadeloupe, *Bull. Soc. Pathol. Exot.,* 46, 810, 1953.

60. **Duke, B. O. L. and McCullough, F. S.,** Schistosomiasis in the Gambia, II. The epidemiology and distribution of urinary schistosomiasis, *Ann. Trop. Med. Parasitol.,* 48, 287, 1954.

61. **Ferguson, F. F.,** Occurrence of *Schistosoma mansoni* in Puerto Ricans, *Public Health Rep.* 80, 339, 1965.

62. **Gadgil, R. K. and Shah, S. H.,** Human schistosomiasis in India. Discovery of an endemic focus in Bombay State, *Indian J. Med. Sci.,* 6, 760, 1952.

63. **Gaud, J.,** Les bilharzioses en Afrique Occidentale et en Afrique Centrale, *Bull. W. H. O.,* 13, 209, 1955.

64. **Gaud, J., Arfaa, F., and Zeini, A.,** Observation on the biology of *Bulinus truncatus* in Khouzistan, Iran, *Ann. Parasitol. Hum. Comp.,* 37, 232, 1962.

65. **Gaud, J. and Fauré, S.,** Variations dans le temps des index d'infestation humaine dans la bilharziose vésicale marocaine, *Bull. Instit. Hyg. Maroc.,* 6, 55, 1945.

66. **Gilles, H. M., et al.,** *Schistosoma haematobium* infection in Nigeria. II. Infection at a primary school in Ibadan, *Ann. Trop. Med. Parasitol.,* 59, 441, 1965.

67. **Gilles, H. M., et al.,** *Schistosoma haematobium* infection in Nigeria, III. Infection in boatyard workers at Epe, *Ann. Trop. Med. Parasitol.,* 59, 451, 1965.

68. **Gillet, J.,** Les Bilharzioses en Guinée Portugaise. Working document, World Health Organization/ Bilharziasis African Conference, No. 44, Brazzaville, Congo, 1956.

69. **Gillet, J. and Wolfs, J.,** Les bilharzioses humaines au Congo Belge et au Ruanda Urundi, *Bull. W. H. O.,* 10, 315, 1954.

70. **Gobert, E.,** Note sur le bilharziose en Tunisie, *Arch. Inst. Pasteur Tunis,* 23, 348, 1934.

71. **Gordon, R. M., Davey, T. II., and Peaston, H. II.,** The transmission of human bilharziasis in Sierra Leone, with an account of the life-cycle of the schistosomes concerned, *S. mansoni* and *S. haematobium, Ann. Trop. Med. Parasitol.,* 28, 323, 1934.

72. **Gürsel, A.,** Turkiye de Bilharzioz, *Turk Iji. Tecr. Biyol. Derg.,* 16, 195, 1956.

73. **Guyon, C.,** Les Bilharzioses, *Rapp. Final Première Conf. Technique OCCGEAC,* 1, 133, 1965.

74. **Hadidjaja, P., et al.,** *Schistosoma japonicum* and intenstinal parasites of the inhabitants of Lake Lindu, Sulawesi (Celebes); a preliminary report, *Southeast Asian J. Trop. Med. Public Health,* 3, 594, 1972.

75. **Harinasuta, C. and Krauatrachue, M.,** Schistosomiasis in Thailand, correspondence, *Trans. R. Soc. Trop. Med. Hyg.,* 54, 280, 1960.

76. **Henderson, A. C.,** Schistosomiasis mansoni. A survey of its incidence at Luampa Hospital, *Med. J. Zambia,* 2, 167, 1969.

77. **Hertwig, F. and Oberdoerster, F.,** Studies on some problems of parasitic diseases in China. 1. Schistosomiasis, *Z. Tropenmed. Parasitol.,* 11, 324, 1960.

78. **Hira, P. R.,** Schistosomiasis at Lake Kariba, Zambia. 1. Prevalence and potential intermediate snail hosts at Siavonga, *Trop. Geogr. Med.,* 22, 323, 1970.

79. **Hunter, G. W., Dillahunt, J. A., and Dalton, H. C.,** The epidemiology of schistosomiasis japonica in the Philippine Islands and Japan. I. Surveys for schistosomiasis japonica on Mindoro, P. I., *Am. J. Trop. Med.,* 30, 411, 1950.

80. **Ito, J., et al.,** Studies on schistosomiasis japonica in Shizuoka Prefecture, III. An epidemiological survey on the new endemic area, Fujikawa-Cho, *Jpn. J. Parasitol.,* 11, 393, 1962.

81. **Janz, G. J. and Carvalho, A. M. de,** Bilharziose en Angola, *An. Inst. Med. Trop. Lisbon,* 13, 597, 1956.

82. **Janz, G. J. and Carvalho, A. M. de,** Missao de prospeccao de endemias em Angola. Nota preliminar sobre a existencia de bilharziose mansoni no Alto Zambeze (Angola), *An. Inst. Med. Trop. Lisbon,* 14, 377, 1957.

83. **Kagan, I. G., Negron, H., Arnold, J. C., and Ferguson, F. F.,** A skin test survey for the prevalence of schistosomiasis in Puerto Rico, Washington Public Health Service Publication No. 1525, 1966.

84. **Kubasta, M.,** Schistosomiasis mansoni in the Harar Province, *Ethiopian Med. J.,* 2, 260, 1964.

85. **Kuntz, R. E., Malakatis, G. M., Lawless, D. K., and Strome, C. P. A.,** Medical mission to the Yemen, Southwest Arabia, 2. A cursory survey of the intestinal, protozoa and helminth parasites in the people of the Yemen, *Am. J. Trop. Med. Hyg.,* 2, 13, 1953.

86. **Lariviere, M., Aretas, R., Raba, A., and Charnier, M.,** Index d'infestation bilharzienne au Senegal (cercles de Thies et de Kaolack), *Bull. Med. Afr. Occid. Franc.,* 3, 239, 1958.

86a. **Lariviere, M. and Diallo, S.,** *Afr. Med.,* 6, 475, 1967.

87. **Lemma, A.,** Bilharziasis in the Awash Valley. 1. An epidemiological study with special emphasis on its possible future and public health importance, *Ethiopian Med. J.,* 7, 147, 1969.

88. **MacLean, G., Webbe, G., and Msangi, A. S.,** A report on a Bilharzia and molluscan survey in the Tanga District of Tanganyika, *East Afr. Med. J.,* 35, 7, 1958.

89. **McMullen, D. B. I. and Buzo, Z. J.,** Report of the Preliminary Survey by the Bilharziasis Advisory Team, 1959, V. Togo, WHO unpublished document, MHO/PA/55.60, World Health Organization, Geneva, 1960.

90. **Maldonado, J. F. and Oliver-Gonzalez, J.,** The prevalence of *Schistosoma mansoni* in certain localities of Puerto Rico. A three-year study, *Am. J. Trop. Med. Hyg.,* 7, 386, 1958.

91. **Maldonado, J. F. and Oliver-Gonzalez, J.,** The prevalence of intestinal parasitism in six selected areas of Puerto Rico- 5 years afterwards, *Bol. Asoc. Med. P.R.,* 54, 133, 1962.

92. **Malek, E. A.,** Bilharziasis control in pump schemes near Khartoum, Sudan and an evaluation of the efficacy of chemical and mechanical barriers, *Bull. W. H. O.,* 27, 41, 1962.

93. **Malek, E. A.,** Report on precontrol studies of bilharziasis in St. Lucia. PAHO/WHO mimeographed document, World Health Organization, Geneva, 1963.

94. **Martinez Larre, M. and Fuente, J. de la,** Esquistosomiasis en la Republica Dominicana, *Rev. Med. Dominicana,* 8, 44, 1953.

95. **Meira, J. A.,** Schistosomiasis mansoni. A survey of its distribution in Brazil, *Bull. W. H. O.,* 2, 31, 1949.

96. **Miller, M. J.,** A survey of *Schistosoma haematobium* infections in man in Liberia, *Am. J. Trop. Med. Hyg.,* 6, 712, 1957.

97. **Morais, T. de,** Human bilharziasis in the Zambezia, Mozambique, *An. Inst. Med. Trop. Lisbon,* 13, 69, 1956.

98. **Morais, T. de,** As bilharzioses humanas no Distrito do Niassi (Africa Oriental Portuguesa), *An. Inst. Med. Trop. Lisbon,* 14, 145, 1957.

99. **Morais, T. de,** As bilharzioses humanas no Distrito de Cabo Delgado (A.O.P.), *An. Inst. Med. Trop. Lisbon,* 14, 455, 1957.

100. **Morais, T. de,** As bilharzioses humanas no Distrito de Mocambique (A.O.P.), *An. Inst. Med. Trop. Lisbon,* 14, 461, 1957.

101. **Morais, T. de,** Personal communication, 1958.

102. **Nagaty, H. F.,** A Survey of Bilharziasis and Other Parasitic Infections in Somalia, WHO unpublished document, EM/BIL/27, World Health Organization, Alexandria, Egypt, 1963.

103. **Nelson, G. S.,** *Schistosoma mansoni* infection in the West Nile District of Uganda. I. The incidence of *S. mansoni* infection, *East Afr. Med. J.,* 35, 311, 1958.

104. **Olivier, L. J.,** Assignment Report on Bilharzia Control, India, WHO unpublished document, SEA/BILHARZ/4, World Health Organization, Geneva, 1961.

105. **Olivier, L., Vaughn, C. M., and Hendricks, J. R.,** Schistosomiasis in an endemic area in the Dominican Republic, *Am. J. Trop. Med. Hyg.,* 1, 680, 1952.

106. **Onori, E., McCullough, F. S., and Rosei, L.,** Schistosomiasis in the Volta region of Ghana, *Ann. Trop. Med. Parasitol.,* 57, 59, 1963.

107. **Pellon, B. and Teixeira, I.,** Distribuicao Geografica da Esquistosomose Mansonica no Brasil, Division of Sanitation, Ministry of Education and Health, Rio de Janeiro, 1950.

108. **Pellon, B. and Teixeira, I.,** Inquerito Helmintologico Escolar em 5 Estados das Regioes Leste, Sul e Centro-oeste, Division of Sanitation, Ministry of Education and Health, Curitiba, Brazil, 1953.

109. **Pesigan, T. P.,** The schistosomiasis problem in the Philippines, its public health and other aspects, *St. Tomas J. Med.,* 8, 1, 1953.

110. **Pesigan, T. P.,** Studies on *Schistosoma japonicum* infection in the Philippines. I. General considerations and epidemiology, *Bull. W. H. O.,* 18, 345, 1958.

111. **Pitchford, R. J.,** Influence of living conditions on bilharziasis infection in Africans in the Transvaal, *Bull. W. H. O.,* 18, 1088, 1958.

112. **Purnell, R. E.,** A survey of the intestinal helminths of primary schoolchildren in Mwanza, Tanzania, *East Afr. Med. J.,* 44, 31, 1967.

113. **Ranque, J. and Rioux, J. A.,** La schistosome urinaire dans la palmeraie de Faya-Largeau (Nord-Tchad). Considérations épidémiologiques, *Med. Afr. Noire,* 10, 287, 1963.

114. **Ransford, O. N.,** Schistosomiasis in the Kota Kota District of Nyasaland, *Trans. R. Soc. Trop. Med. Hyg.,* 41, 617, 1948.

115. Republic of the Sudan, Reports of the Medical Services, Ministry of Health, 1949, 1951-1952; 1952-1953 and 1954-1955, Annual Reports, Sudan Government, Khartoum.
116. **Ripert, C., Carteret, P., and Gayte, M. J.,** Étude épidémiologique des bilharzioses intestinale et urinaire dans la region du lac de retenue de la Lufira (Katanga). Prevalence de l'infestation d'après l' étude de l'elimination des oeufs dans les excreta, *Bull. Soc. Pathol. Exot.,* 62, 571, 1969.
117. **Roberts, J. I.,** A protozoological and helminthological survey of three races in Nairobi, Kenya, *J. Trop. Med. Hyg.,* 52, 49, 1949.
118. **Rosanelli, J. D.,** Some observations on vesical schistosomiasis in Acholi District, Uganda. Work carried out at the Uganda Medical Departments Hospital at Gulu and in Acholi District, *East Afr. Med. J.,* 37, 113, 1960.
119. **Russell, H. B. L.,** The Pilot Mobile Health Team, Ethiopia, Final report, WHO unpublished document, EM/PHA/62, World Health Organization, Alexandria, Egypt, 1958.
120. **Saugrain, J.,** La bilharziose en Republique Centralafricaine, *Med. Trop.,* 27, 156, 1967.
121. **Scott, J. A.,** The incidence and distribution of the human schistosomes in Egypt, *Am. J. Hyg.,* 25, 566, 1937.
122. **Shah, S. N. and Gadgil, R. K.,** Human schistosomiasis in India. III. Note on the clinical survey of endemic focus, *Indian J. Med. Res.,* 43, 689, 1955.
123. **Smithers, S. R.,** The occurrence of schistosomiasis mansoni in the Gambia, *Ann. Trop. Med. Parasitol.,* 51, 359, 1957.
124. **Sornmani, S.,** Schistosomiasis in Thailand, *Proc. 4th Southeast Asian Seminar Parasitol. Trop. Med.,* Harinasuta, C., Ed., Southeast Asian Minister of Education Council, Bangkok, 1969, 71.
125. **U.S. Army Japan,** Parasitiological studies in the Far East. XIV. Summary of the Common Intestinal and Blood Parasites of the Japanese, Japan Logistical Command Bull. No. 4, 406th Medical General Laboratory, Tokyo, 1951.
126. **Vermeil, C.,** Present state of the schistosomal-malcological research in Tunisia. The focus of vesical bilharziasis of the boundary of Gafsa, *Arch. Inst. Pasteur Tunis,* 34, 167, 1957.
127. **Vermeil, C., Tournoux, P., Tocheport, G., Noger, C., and Schmitt, P.,** Premieres donnes sur l'etat actuel des bilharzioses au Fezzan (Lybie), *Ann. Parasitol. Hum. Comp.,* 27, 499, 1952.
128. **Webbe, G.,** A bilharzia and molluscan survey in the Handeni and Korogwe Districts of Tanganyika, *J. Trop. Med. Hyg.,* 62, 37, 1959.
129. **White, P. C., Pimental, O., and Garcia, F. C.,** Distribution and prevalence of human schistosomiasis in Puerto Rico in 1953, *Am. J. Trop. Med. Hyg.,* 6, 715, 1957.
130. **Wright, W. H.,** Bilharziasis as a public-health problem in the Pacific, *Bull. W. H. O.,* 2, 581, 1950.
131. **Wright, W. H.,** Geographical distribution of schistosomes and their intermediate hosts, in *Epidemiology and Control of Schistosomiasis,* Ansari, N., Ed., S. Karger, Basel, 1973, 32.
132. **Yokogawa, M.,** Control of schistosomiasis in Japan, in *Proceedings of a Symposium on the Future of Schistosomiasis Control,* Miller, M. J., Ed., Tulane University, New Orleans, 1972, 129.

THE PARASITES

133. **Alves, W.,** The eggs of *Schistosoma bovis, S. mattheei,* and *S. haematobium, J. Helminthol.,* 23, 127, 1949.
134. **Armstrong, J. C.,** Mating behavior and development of schistosomes in the mouse, *J. Parasitol.,* 51, 605, 1965.
135. **Austin, F. G., Stirewalt, M. A., and Danziger, R. E.,** *Schistosoma mansoni:* stimulatory effect of rat skin lipid fractions on cercarial penetration behavior, *Exp. Parasitol.,* 31, 217, 1972.
136. **Brooks, C. P.,** A comparative study of *Schistosoma mansoni* in *Tropicorbis havanensis* and *Australorbis glabratus, J. Parasitol.,* 39, 159, 1953.
137. **Bueding, E. and Fisher, J.,** Biochemical effects of schistosomicides, *Ann. N.Y. Acad. Sci.,* 160, 536, 1969.
138. **Bueding, E. and Fisher, J.,** Biochemical effects of niridazole on *Schistosoma mansoni, Mol. Pharmacol.,* 6, 523, 1970.
139. **Bueding, E. and MacKinnon, J. A.,** Hexokinases of *Schistosoma mansoni, J. Biol. Chem.,* 215, 495, 1955.
140. **Chernin, E.,** Behavioral responses of miracidia of *Schistosoma mansoni* and other trematodes to substances emitted by snails, *J. Parasitol.,* 56, 287, 1970.
141. **Coles, G. C.,** Variations in malate dehydrogenase isoenzymes of *Schistosoma mansoni, Comp. Biochem Physiol.,* 38B, 35, 1971.
142. **Coles, G. C.,** Oxidative phosphorylation in adult *Schistosoma mansoni, Nature (London),* 240, 488, 1972.

143. **Coles, G. C.**, The metabolism of schistosomes: a review, *Int. J. Biochem.*, 4, 319, 1973.

144. **Dawood, M. M. and Gismann, A.**, Schistosomiasis (Bilharziasis) and vectors in Africa and adjacent regions, in *World Atlas of Epidemiology*, Falk-Verlag, Hamburg, Germany, 1956.

145. **Deramee, O. D., Thienpont, A., Fain, A. and Jadin, J.**, Sur un foyer de bilharziose canine a Schistosoma rodhaini Brumpt en Ruanda-Urundi, *Ann. Soc. Belge Med. Trop.*, 33, 207, 1953.

146. **Fischer, A. C.**, A study of the schistosomiasis of the Stanleyville district of the Belgian Congo, *Trans. R. Soc. Trop. Med. Hyg.*, 28, 277, 1934.

147. **Glaudel, R. J. and Etges, F. J.**, Toxic effects of freshwater turbellarians on schistosome miracidia, *J. Parasitol.*, 59, 74, 1973.

148. **Haenens, G. D. and Santele, A.**, Sur un cas humain de *Schistosoma rodhaini* trouve' aux environs d'Elizabethville, *Ann. Soc. Belge Med. Trop.*, 35, 497, 1955.

149. **Hairston, N. G.**, The dynamics of transmission, in *Epidemiology and control of schistosomiasis (Bilharziasis)*, Ansari, N., Ed., S. Karger, Basel, 1973, 250.

150. **Khalil, L. F.**, On the capture and destruction of miracidia by *Chaetogaster limnaei* (Oligochaeta), *J. Helminthol.*, 35, 269, 1961.

151. **Knight, W. B., Ritchie, L. S., Liard, F., and Chiriboga, J.**, Cereariophagic activity of guppy fish (Lebistes reticulatus) detected by cercariae labeled with radioselenium (75 Se), *Am. J. Trop. Med. Hyg.*, 19, 620, 1970.

152. **Kuntz, R. E.**, Biology of the schistosome complexes, *Am. J. Trop. Med. Hyg.*, 4, 383, 1955.

153. **MacInnis, A. J.**, Responses of *Schistosoma mansoni* miracidia to chemical attractants, *J. Parasitol.*, 51, 731, 1965.

154. **MacInnis, A. J.**, Identification of chemicals triggering cercarial penetration responses of *Schistosoma mansoni*, *Nature (London)*, 224, 1221, 1969.

155. **Malek, E. A.**, Susceptibility of the snail *Biomphalaria biossyi* to infection with certain strains of *Schistosoma mansoni*, *Am. J. Trop. Med.*, 30, 887, 1950.

156. **Malek, E. A.**, Natural and experimental infection of some bulinid snails in the Sudan with *Schistosoma haematobium*, *Proc. 6th Intl. Congr. Trop. Med. Malaria*, 2, 5, 1959.

157. **Malek, E. A.**, Susceptibility of tropicorbid snails from Louisiana to infection with *Schistosoma mansoni*, *Am. J. Trop. Med. Hyg.*, 16, 715, 1967.

158. **Malek, E. A.**, Studies on bovine schistosomiasis in the Sudan, *Ann. Trop. Med. Parasitol.*, 63, 501, 1969.

159. **Meyer, F., Meyer, H., and Bueding, E.**, Lipid metabolism in the parasitic and free living flatworms, *Schistosoma mansoni* and *Dugesia dorotocephala*, *Biochim. Biophys. Acta*, 210, 257, 1970.

160. **Michelson, E. H.**, The protective action of *Chaetogaster limnaei* on snails exposed to *Schistosoma mansoni*, *J. Parasitol.*, 50, 441, 1964.

161. **Oliver-Gonzalez, J.**, The possible role of the guppy, *Lebistes reticulatus*, on the biological control of schistosomiasis mansoni, *Sciences*, 104, 605, 1940.

162. **Pelligrino, J., DeMaria, M., and De Moura, M. F.**, Observations on the predatory activity of *Lebistes reticulatus* (Peters, 1859) on cercariae of *Schistosoma mansoni*, *Am. J. Trop. Med. Hyg.*, 15, 337, 1966.

163. **Pitchford, R. J.**, Differences in the egg morphology and certain biological characteristics of some African and Middle Eastern schistosomes, genus *Schistoma* with terminal spined eggs, *Bull. W. H. O.*, 32, 105, 1965.

164. **Rowan, W. B.**, Daily periodicity of *Schistosoma mansoni* cercariae in Puerto Rican waters, *Am. J. Trop. Med. Hyg.*, 7, 374, 1958.

165. **Sahba, G. H. and Malek, E. A.**, Hermaphroditic female *Heterobilharzia americana*, *J. Parasitol.*, 63, 947, 1977.

166. **Sahba, G. H. and Malek, E. A.**, Unisexual infections with *Schistosoma haematobium* in the white mouse, *Am. J. Trop. Med. Hyg.*, 26, 331, 1977.

167. **Schwetz, J. and Stijns, J.**, Sur la redécouverte de Schistosoma rodhaini Brumpt, et la decouverte de son hôte définitif, *C. R. Soc. Biol.*, 145, 1255, 1951.

168. **Shiff, C. J.**, Influence of light and depth on location of *Bulinus (Physopsis) globosus* by miracidia of *Schistosoma haematobium*, *J. Parasitol.*, 55, 108 1969.

169. **Shiff, C. J., Cmelik, S. H. W., Ley, E. H., and Kriel, R. L.**, The influence of human skin lipid on the cercarial penetration responses of *Schistosoma haematobium* and *Schistosoma mansoni*, *J. Parasitol.*, 58, 476, 1972.

170. **Smith, T. M. and Brooks, T. J.**, Lipid fractions in adult *Schistosoma mansoni*, *Parasitology*, 59, 293, 1969.

171. **Stirewalt, M. A.**, Important features of schistosomes, in *Epidemiology and Control of Schistosomiasis*, Ansari, N., Ed., S. Karger, Basel, 1973.

172. **Sudds, R. H.**, Observations of schistosome miracidial behavior in the presence of normal and abnormal hosts and subsequent tissue studies of these hosts, *J. Elisha Mitchell Sci Soc.*, 76, 121, 1960.

173. **Webbe, G. and James, C.**, A comparison of two geographical strains of *Schistosoma haematobium*, *J. Helminthol.*, 45, 271, 1971.

174. **Wright, C. A.**, The schistosome life-cycle, in *Bilharziasis*, Mostofi, F. K., Ed., Springer-Verlag, Berlin, 1967, 3.

175. **Wright, C. A., Southgate, V. R., and Knowles, R. J.**, What is *Schistosoma intercalatum* Fisher, 1934?, *Trans. R. Soc. Trop. Med. Hyg.*, 66, 28, 1972.

THE SNAIL HOSTS

176. **Abbott, R. T.**, Handbook of medically important mollusks of the Orient and the Western Pacific, *Bull. Mus. Comp. Zool. Harv. U.*, 100, 245, 1948.

177. **Barbosa, F. S.**, Aspects of the ecology of the intermediate hosts of *Schistosoma mansoni* interfering with the transmission of bilharziasis in northeastern Brazil, in *Bilharziasis, Ciba Foundation Symposium*, Wolstenholme, G. E. W. and O'Connor, M., Eds., Churchill, London, 1962, 23.

178. **Barbosa, F. S. and Coelho, M. V.**, Qualidades de vetor dos hospedeiros de *S. mansoni* no Nordeste do Brazil. I. Suscetibilidade de *A. glabratus* e *T. centimetralis* a infestacao por *S. mansoni*, *Publ. Avulsas Inst. Aggeu Magalhaes Recife, Braz.*, 3, 55, 1954.

179. **Barbosa, F. S. and Coelho, M. de V.**, Alguns aspectos epidemiologicos relacionados com a transmissao de esquistossomose em Pernambuco, Brasil, *Publ. Avulsas Inst. Aggeu Magalhaes Recife, Braz.*, 5, 31, 1956.

180. **Barbosa, F. S. and Olivier, L.**, Studies on the snail vectors of bilharziasis mansoni in northeastern Brazil, *Bull. W. H. O.*, 18, 895, 1958.

181. **Berrie, A. D.**, Fish ponds in relation to the transmission of bilharziasis in East Africa, *East Afr. Agric. For. J.*, 31, 276, 1966.

182. **Berrie, A. D.**, Snail problems in African schistosomiasis, in *Advances in Parasitology*, Vol. 8, Dawes, B., Ed., Academic Press, New York, 1970, 43.

183. **Brown, D. S.**, A review of the freshwater mollusca of Natal and their distribution, *Ann. Natal Mus.*, 18, 477, 1967.

184. **Brygoo, E. R.**, La température et la repartition des bilharzioses humaines à Madagascar, *Bull. Soc. Pathol. Exot.*, 60, 433, 1967.

185. **Claugher, D.**, The transport and laboratory culture of snail intermediate hosts of *Schistosoma haematobium*, *Ann. Trop. Med. Parasitol.*, 54, 333, 1960.

186. **Davis, G. M.**, The systematic relationship of *Pomatiopsis lapidaria* and *Oncomelania hupensis formosana* (Prosobranchia: Hydrobiidae), *Malacol. Int. J. Malacol.*, 6, 1, 1967.

187. **Deane, L. M., Martins, R. S., and Lobo, M. B.**, Um foco ativo de esquistossomose mansonica em Jacarepagua, Distrito Federal, *Rev. Bras. Malariol., Doencas Trop.*, 5, 249, 1953.

188. **DeLourdes Sampaio Xavier, M., de Azevedo, J. Fraga, and Avelino, I.**, Importance d'Oscillatoria formosa Bory dans la culture au laboratoire des mollusques vecteurs du *Schistosoma haematobium*, *Bull. Soc. Pathol. Exot.*, 61, 52, 1968.

189. **Dias, E.**, Estudos preliminares sobre a esquistossomose mansoni no Municipio de Bambui, Estado de Minas Gerais, *Rev. Bras. Malariol., Doencas Trop.*, 5, 211, 1953.

190. **Dias, E.**, Isolamento e selecao de microorganismos de planorbideos utilizaveis em ensaios de luta biologica contra estes invertebrados, *Hospital (Rio de Janeiro)*, 47, 111, 1955.

191. **Duke, B. O. L. and McCullough, F. S.**, Schistosomiasis in the Gambia. 2. The epidemiology and distribution of urinary schistosomiasis, *Ann. Trop. Med. Parasitol.*, 48, 287, 1954.

192. **Faust, E. C. and Hoffman, W. A.**, Studies on schistosomiasis mansoni in Puerto Rico. Biological studies. 1. The extramammalian phases of the life-cycle, *P. R. J. Public Health Trop. Med.*, 10, 1, 1934.

193. **Gillet, J. and Wolfs, J.**, Les bilharzioses humaines au Congo Belge et au Ruanda-Urundi, *Bull. W. H. O.*, 10, 315, 1954.

194. **Gordon, R. M., Davey, T. H., and Peaston, H.**, The transmission of human bilharziasis in Sierra Leone, with an account of the life cycle of the schistosomes concerned, *S. mansoni* and *S. haematobium*, *Ann. Trop. Med. Parasitol.*, 28, 323, 1934.

195. **Greany, W. H.**, Schistosomiasis in the Gerzira irrigated area of the Anglo-Egyptian Sudan. 1. Public health and field aspects, *Ann. Trop. Med. Parasitol.*, 46, 250, 1952.

196. **Gumble, A., Otori, Y., Ritchie, L. S., and Hunter, G. W.**, The effect of light, temperature and pH on the emergence of *Schistosoma japonicum* cercariae from *Oncomelania nosophora*, *Trans. Am. Microsc. Soc.*, 76, 87, 1957.

197. **Harrison, A. D. and Farina, T. D. W.**, A naturally turbid water with deleterious effects on the egg capsules of planorbid snails, *Ann. Trop. Med. Parasitol.*, 59, 327, 1965.

198. **Harrison, A. D., Nduku, W., and Hooper, A. S. C.,** The effects of a high magnesium-to-calcium ratio on the egg-laying rate of an aquatic planorbid snail, *Biomphalaria pfeifferi, Ann. Trop. Med. Parasitol.,* 60, 212, 1966.

199. **Harrison, A. D. and Shiff, C. J.,** Factors influencing the distribution of some species of aquatic snails, *S. Afr. J. Sci.,* 62, 25, 1966.

200. **Kuntz, R. E.,** *Schistosoma mansoni* and *S. haematobium* in the Yemen, South-west Arabia; with a report of an usual factor in the epidemiology of schistosomiasis mansoni, *J. Parasitol.,* 38, 24, 1952.

201. **Lietar, J.,** Biologie et ecologie des mollusques vecteurs de bilharziose a Jadotville, *Ann. Soc. Belge Med. Trop.,* 36, 919, 1956.

202. **McClelland, W. F. J.,** A method of breeding *Bulinus (Physopsis) nasutus* in the laboratory, *Ann. Trop. Med. Parasitol.,* 58, 265, 1964.

203. **McMullen, D. B., Hubendick, B., Pesigan, T. P., and Bierstein, P.,** Observations made by the World Health Organization schistosomiasis team in the Philippines, *J. Philipp. Med. Assoc.,* 30, 615, 1954.

204. **McQuay, R. M.,** Susceptibility of a Louisiana species of *Tropicorbis* to infection with *Schistosoma mansoni, Exp. Parasitol.,* 1, 184, 1952.

205. **Malek, E. A.,** Susceptibility of the snail *Biomphalaria boissyi* to infection with certain strains of *Schistosoma mansoni, Am. J. Trop. Med.,* 30, 887, 1950.

206. **Malek, E. A.,** Anatomy of *Biomphalaria boissyi* as related to its infection with *Schistosoma mansoni, Am. Midl. Nat.,* 54, 394, 1955.

207. **Malek, E. A.,** Factors conditioning the habitat of bilharziasis intermediate hosts of the family Planorbidae, *Bull. W. H. O.,* 18, 785, 1958.

208. **Malek, E. A.,** The ecology of schistosomiasis, in *Studies in Disease Ecology,* Vol. 2, May, J. M., Ed., Hafner, New York, 1961, 261.

209. **Malek, E. A.,** The biology of mammalian and bird schistosomes, *Bull. Tulane Univ. Med. Fac.,* 20, 181, 1961.

210. **Malek, E. A.,** Bilharziasis control in pump schemes near Khartoum, Sudan and an evaluation of the efficacy of chemical and mechanical barriers, *Bull. W. H. O.,* 27, 41, 1962.

210a. **Malek, E. A.,** Report on Precontrol Studies of Bilharziasis in St. Lucia, PAHO/WHO Document, World Health Organization, Washington, D. C., 1963.

211. **Malek, E. A.,** Susceptibility of tropicorbid snails from Louisiana to infection with *Schistosoma mansoni, Am. J. Trop. Med. Hyg.,* 16, 715, 1967.

212. **Malek, E. A.,** Studies on "tropicorbid" snails (*Biomphalaria*: Planorbidae) from the Caribbean and Gulf of Mexico areas, including the Southern United States, *Malacol. Int. J. Malacol.,* 7, 183, 1969.

212a. **Malek, E. A. and Cheng, T. C.,** *Medical and Economic Malacology,* Academic Press, New York, 1974.

213. **Mandahl-Barth, G.,** Intermediate hosts of *Schistosoma* African *Biomphalaria* and *Bulinus, W. H. O. Monogr. Ser.* 37, 1958.

214. **Mandahl-Barth, G.,** Intermediate hosts of *Schistosoma* in Africa. Some recent information, *Bull. W. H. O.,* 22, 565, 1960.

215. **Mandahl-Barth, G.,** The species of the genus *Bulinus,* intermediate hosts of *Schistosoma, Bull. W. H. O.,* 33, 33, 1965.

216. **Martinez Larre, M. and Ravelo de la Fuente, J. de J.,** Esquistosomiasis en la Republica Dominican , *Rev. Med. Dominica,* 8, 44, 1953.

217. **Michelson, E. H.,** An acid-fast pathogen of freshwater snails, *Am. J. Trop. Med. Hyg.,* 10, 423, 1961.

218. **Michelson, E. H.,** The effects of temperature on growth and reproduction of *Australorbis glabratus* in the laboratory, *Am. J. Hyg.,* 73, 66, 1961.

219. **Mills, E. A., MacHattie, C., and Chadwick, C. R.,** *Schistosoma haematobium* and its life-cycle in Iraq, *Trans. R. Soc. Trop. Med. Hyg.,* 30, 317, 1936.

220. **Newton, W. L.,** The comparative tissue reaction of two strains of *Australorbis glabratus* to infection with *Schistosoma mansoni, J. Parasitol.,* 38, 362, 1952.

221. **Olivier, L.,** A note on schistosomiasis in Eastern Japan, *Am. J. Trop. Med.,* 28, 867, 1948.

222. **Olivier, L., Vaughn, C. M., and Hendricks, J. R.,** Schistosomiasis in an endemic area in the Dominican Republic, *Am. J. Trop. Med. Hyg.,* 1, 680, 1952.

223. **Pan, C. T.,** Studies on the host-parasite relationship between *Schistosoma mansoni* and the snail *Australorbis glabrata, Am. J. Trop. Med. Hyg.,* 14, 931, 1965.

224. **Paraense, W. L. and Santos, J. M.,** Um ano de observacoes sobre esquistossomose em planorbideos da Lagoa Santa, *Rev. Bras. Malariol., Doencas Trop.,* 5, 253, 1953.

225. **Pesigan, T. P., et al.,** Studies on *Schistosoma japonicum* infection in the Philippines. 2. The molluscan host, *Bull. W. H. O.,* 18, 481, 1958.

226. **Rasmussen, O.,** Biological control of *Biomphalaria pfeifferi* by *Helisoma duryi, Proc. 3rd. Intl. Congr. Parasitol. Munich,* Sec. GI, 1598, 1974.
227. **Richards, C. S.,** Two new species of *Hartmannella* amoebae infecting freshwater mollusks, *J. Protozool.,* 15, 651, 1968.
228. **Richards, C. S.,** Genetics of a molluscan vector of schistosomiasis, *Nature, (London),* 227, 806, 1970.
229. **Richards, C. S. and Merritt, J. W.,** Genetic factors in the susceptibility of juvenile *Biomphalaria glabrata* to *Schistosoma mansoni* infection, *Am. J. Trop. Med. Hyg.,* 21, 425, 1972.
230. **Ritchie, L. S.,** The biology and control of the amphibious snails that serve as intermediate hosts for *Schistosoma japonicum, Am. J. Trop. Med. Hyg.,* 4, 426, 1955.
231. **Ruiz-Tiben, E., Palmer, J. R., and Ferguson, F. F.,** Biological control of *Biomphalaria glabrata* by *Marisa cornuarietis* in irrigation ponds of Puerto Rico, *Bull. W. H. O.,* 41, 329, 1969.
232. **Schutte, C. H. J. and Frank, G. H.,** Observations on the distribution of fresh-water mollusca and chemistry of the natural waters in the south-eastern Transvaal and adjacent northern Zwaziland, *Bull. W. H. O.,* 30, 389, 1964.
233. **Smithers, S. R.,** On the ecology of schistosome vectors in the Gambia, with evidence of their role in transmission, *Trans. R. Soc. Trop. Med. Hyg.,* 50, 354, 1956.
234. **Standen, O. D.,** Experimental infection of *Australorbis glabratus* with *Schistosoma mansoni.* 1. Individual and mass infection of snails, and the relationship of infection to temperature and season, *Ann. Trop. Med. Parasitol.,* 46, 48, 1952.
235. **Stirewalt, M. A.,** Effect of snail maintenance temperatures on development of *Schistosoma mansoni, Exp. Parasitol.,* 3, 504, 1954.
236. **Sturrock, R. F.,** The influence of temperature on the biology of *Biomphalaria pfeifferi* (Krauss), an intermediate host of *Schistosoma mansoni, Ann. Trop. Med. Parasitol.,* 60, 100, 1966.
237. **Sudds, R. H.,** Observations of schistosome miracidial behavior in the presence of normal and abnormal snail hosts and subsequent tissue studies of these hosts, *J. Elisha Mitchell, Sci. Soc.,* 76, 121, 1960.
238. **Van der Schalie, H.,** Vector snail control in Qalyub, Egypt, *Bull. W. H. O.,* 19, 263, 1958.
239. **Van Eeden, J. A.,** Tendense in die verspreiding van die slaklussengashere van Bilharzia in die Transvaal, *Tydskr. Natuurwet.,* 5, 152, 1965.
240. **von Brand, T., Baernstein, H. D., and Mehlman, B.,** Studies on the anoerobic metabolism and the aerobic carbohydrate consumption of some freshwater snails, *Biol. Bull. (Wood's Hole, Mass.),* 98, 266, 1950.
241. **Webbe, G.,** The transmission of *Schistosoma haematobium* in an area of Lake Province, Tanganyika, *Bull. W. H. O.,* 27, 59, 1962.
242. **White, P. C., Pimental, D., and Garcia, F. C.,** Distribution and prevalence of human schistosomiasis in Puerto Rico in 1953, *Am. J. Trop. Med. Hyg.,* 6, 715, 1957.
243. **Wright, C. A.,** The crowding phenomenon in laboratory colonies of freshwater snails, *Ann. Trop. Med. Parasitol.,* 54, 224, 1960.
244. **Wright, W. H.,** Bilharziasis as a public health problem in the Pacific, *Bull. W. H. O.,* 2, 581, 1950.
245. **Wright, W. H., McMullen, D. B., Bennett, H. J., Bauman, P. M., and Ingalls, J. W.,** The epidemiology of schistosomiasis japonica in the Philippine Islands and Japan. III. Surveys of endemic areas of schistosomiasis japonica in Japan, *Am. J. Trop. Med.,* 27, 417, 1947.

EPIDEMIOLOGY

246. **Barbosa, F. S.,** Epidemiologia, in *Esquistossomose Mansoni,* da Cunha, A. S., Ed., Sarvier, Univ. Sao Paulo, Brazil, 1970, 31.
247. **Cheever, A. W.,** A quantitative post-mortem of schistosomiasis mansoni in man, *Am. J. Trop. Med. Hyg.,* 17, 38, 1968.
248. **Cook, J. A., Baker, S. T., Warren, K. S., and Jordan, P.,** A controlled study of morbidity of schistosomiasis mansoni in St. Lucian children, based on quantitative egg excretion, *Am. J. Trop. Med. Hyg.,* 23, 625, 1974.
249. **Farooq, M., Hairston, N. G., and Samaan, S. A.,** The effect of area-wide snail control on the endemicity of bilharziasis in Egypt, *Bull. W. H. O.,* 35, 369, 1966.
250. **Hairston, N. G.,** On the mathematical analysis of schistosome populations, *Bull. W. H. O.,* 33, 45, 1965.
251. **Jordan, P.,** Schistosomiasis and disease, in *Proc. Symp. Future of Schistosomiasis Control,* Miller, M. J., Ed., Tulane University, New Orleans, 1972, 17.
252. **Kagan, I. G. and Pellegrino, J.,** A critical review of immunological methods for the diagnosis of bilharziasis, *Bull. W. H. O.,* 25, 611, 1961.

253. **Kloetzel, K.,** Splenomegaly, in schistosomiasis mansoni, *Am. J. Trop. Med. Hyg.,* 11, 472, 1962.
254. **Macdonald, G.,** The dynamics of helminthic infections with special reference to schistosomes, *Trans. R. Soc. Trop. Med. Hyg.,* 59, 489, 1965.
255. **Malek, E. A.,** Bilharziasis control in pump schemes near Khartoum, Sudan and an evaluation of the efficacy of chemical and mechanical barriers, *Bull. W. H. O.,* 27, 41, 1962.
256. **Malek, E. A.,** Correlation analyses of cercarial exposure, worm load and egg content in stools of dogs infected with *Heterobilharzia americana, Z. Tropenmed. Parasitol.,* 20, 333, 1969.
256a. **Malek, E. A.,** Report on Precontrol Studies of Bilharziasis in St. Lucia, PAHO/WHO Document, World Health Organization, Washington D.C., 1963.
257. **Martin, L. K. and Beaver, P. C.,** Evaluation of Kato thick-smear technique for quantitative diagnosis of helminth infections, *Am. J. Trop. Med. Hyg.,* 17, 282, 1968.
258. **Pesigan, T. P., Farooq, M., Hairston, N. G., Jauregui, E. G., Garcia, B. C., Santos, B. C., and Besa, A. A.,** Studies on *Schistosoma japonicum* infection in the Philippines. I. General considerations and epidemiology, *Bull. W. H. O.,* 18, 345, 1958.
259. **Ritchie, L. S.,** An ether sedimentation technique for routine stool examination, *Bull. U.S. Army Med. Dept.,* 8, 326, 1948.
260. **Rowan, W. B.,** The ecology of schistosome transmission foci, *Bull. W. H. O.,* 33, 63, 1965.
261. **Sandt, D. G.,** Direct filtration for recovery of *Schistosoma mansoni* cercariae in the field, *Bull. W. H. O.,* 48, 27, 1973.
262. **Scott, J. A.,** The regularity of egg output of helminth infestations, with special reference to *Schistosoma mansoni, Am. J. Hyg.,* 27, 155, 1938.
263. **Scott, J. A.,** Egg counts as estimates of intensity of infection with *Schistosoma haematobium, Texas Rep. Biol. Med.,* 15, 425, 1957.
264. **Stimmel, C. M. and Scott, J. A.,** The regularity of egg output of *Schistosoma haematobium, Texas Rep. Biol. Med.,* 14, 440, 1956.

IMMUNITY

265. **Amin, M. A., Nelson, G. S., and Saoud, M. F. A.,** Studies on heterologous immunity in schistosomiasis. 2. Heterologous schistosome immunity in rhesus monkeys, *Bull. W. H. O.,* 38, 19, 1968.
266. **Amin, M. A. and Nelson, G. S.,** Studies on heterologous immunity in schistosomiasis. 3. Further observations on heterologous immunity in mice, *Bull. W. H. O.,* 41, 225, 1969.
267. **Chandler, A. C.,** Immunity in parasitic diseases, *J. Egypt, Med. Assoc.,* 36, 811, 1953.
268. **Clarke, V. de V.,** Evidence of the development in man of acquired resistance to infection of *Schistosoma* spp., *Cent. Afr. J. Med.,* 12, 1, 1966.
269. **Deelder, A. M., Klappe, H. T. M., van den Aardweg, G. J. M. J., and van Meerbeke, E. H. E. M.,** *Schistosoma mansoni:* demonstration of two circulating antigens in infected hamsters, *Exp Parasi tol.,* 40, 189, 1976.
270. **Fischer, A. C.,** A study of schistosomiasis of the Stanleyville district of the Belgian Congo, *Trans. R. Soc. Trop. Med. Hyg.,* 28, 277, 1934.
271. **Gothe, K. M.,** Beobachtungen uber die Immunitat bei *Schistösoma haematobium* Infektionen, *Z. Tropenmed. Parasitol.,* 14, 512, 1963.
272. **Houba, V., Koech, D. K., Sturrock, R. F., Butterworth, A. E., Kusel, J. R., and Mahmoud, A. A. F.,** Soluble antigens and antibodies in sera from baboons infected with *Schistosoma mansoni, J. Immunol.,* 117, 705, 1976.
273. **Hussein, M. F., Saeed, A. A., and Nelson, G. S.,** Studies on heterologous immunity in schistosomiasis. 4. Heterologous schistosome immunity in cattle, *Bull. W. H. O.,* 42, 745, 1970.
274. **Jackson, T. F. H. G.,** Intermediate host antigens associated with the cercariae of *Schistosoma haematobium, J. Helminthol.,* 50, 45, 1976.
275. **Jackson, T. F. H. G. and De Moor, P. P.,** A demonstration of the presence of anti-snail antibodies in individuals infected with *Schistosoma haematobium, J. Helminthol.,* 50, 59, 1976.
276. **Kagan, I. G.,** Mechanisms of immunity in trematode infection, in *Biology of Parasites,* Soulsby, E. J. L., Ed., Academic Press, New York, 1966, 277.
277. **Lin, S. S., Ritchie, L. S., and Hunter, G. W.,** Acquired immunologic resistance against *Schistosoma japonicum, J. Parasitol.,* 40(Suppl). 42, 1954.
278. **Madwar, M. A. and Voller, A.,** Circulating soluble antigens and antibody in schistosomiasis, *Br. Med. J.,* 435, 1975.
279. **Malek, E. A.,** Some factors influencing the worm recovery rates from white mice infected with *Schistosoma mansoni, Proc. 3rd Intl. Congr. Parasitol.,* 2, 804, 1974.
280. **Ozawa, M.,** Experimental studies on acquired immunity to schistosomiasis japonica, *Jpn. J. Exp. Med.,* 8, 79, 1930.

281. **Perez, H., Clegg, J. A., and Smithers, S. R.,** Acquired immunity to *Schistosoma mansoni* in the rat: measurement of immunity by the lung recovery technique, *Parasitology,* 69, 349, 1974.
282. **Sadun, E. H. and Lin, S. S.,** Studies on the host parasite relationship to *Schistosoma japonicum.* IV. Resistance acquired by infection, by vaccination, and by the injection of immune serum in monkeys, rabbits, and mice, *J. Parasitol.,* 45, 543, 1959.
283. **Sher, A., Mackenzie, P., and Smithers, S. R.,** Decreased recovery of invading parasites from the lungs as a parameter of acquired immunity of schistosomiasis in the mouse, *J. Infect. Dis.,* 130, 626, 1974.
284. **Smith, M. A., Clegg, J. A., and Webbe, G.,** Cross-immunity to *Schistosoma mansoni* and *S. haematobium* in the hamster, *Parasitology,* 73, 53, 1976.
285. **Smithers, S. R.,** Recent advances in the immunology of schistosomiasis, *Br. Med. Bull.,* 28, 49, 1972.
286. **Smithers, S. R. and Terry, R. J.,** Acquired resistance to experimental infections of *Schistosoma mansoni* in the albino rat, *Parasitology,* 55, 711, 1965.
287. **Smithers, S. R. and Terry, R. J.,** Resistance to experimental infection with *Schistosoma mansoni* in Rhesus monkeys induced by the transfer of adult worms, *Trans. R. Soc. Trop. Med. Hyg.,* 61, 517, 1967.
288. **Smithers, S. R. and Terry, R. J.,** Immunity in schistosomiasis, *Ann. N.Y. Acad. Sci.,* 160, 826, 1969.
289. **Taylor, M. G., James, E. R., Nelson, G. S., Bickle, Q., Dunne, D. W., and Webbe, G.,** Immunization of sheep against *Schistosoma mattheei* using either irradiated cercariae or irradiated schistosomula, *J. Helminthol.,* 50, 1, 1976.
290. World Health Organization, Immunology of schistosomiasis, Memoranda, *Bull. W. H. O.,* 51, 553, 1974.

PATHOLOGY AND CLINICAL MANIFESTATIONS

291. **Alves, W.,** The distribution of Schistosoma eggs in human tissues, *Bull. W. H. O.,* 18, 1092, 1958.
292. **Andrade, Z. A.,** Aspectos experimentais da esplenomegalia da esquistossomose. *Rev. Inst. Med. Trop. Sao Paulo,* 4, 249, 1962.
293. **Andrade, Z. A. and Andrade, S. G.,** Patologia do baco na esquistossomose hepatosplenica, *Rev. Inst. Med. Trop. Sao Paulo,* 7, 218, 1965.
294. **Andrade, Z. A. and Cheever, A. W.,** Clinical and pathological aspects of schistosomiasis in Brazil, in *Bilharziasis,* Mostofi, F. K., Ed., Springer-Verlag, Berlin, 1967, 157.
295. **Barlow, C. H.,** Is there dermatitis in Egyptian schistosomiasis?, *Am. J. Hyg.,* 24, 587, 1936.
296. **Bhagwandeen, S. B.,** Schistosomiasis and carcinoma of the bladder in Zambia, *S. Afr. Med. J.,* 50, 1616, 1976.
297. **Bogliolo, L.,** Subsidios para o conhecimento da forma hepato-esplenica e da forma toxemica de esquistossomose mansonica, Ministry of Education and Health, Rio de Janeiro, Brasil, 1958.
298. **Bruce, J. I., Pezzlo, F., McCarty, J. E., and Yajima, Y.,** Migration of *Schistosoma mansoni* through mouse tissue. Ultrastructure of host tissue and integument of migrating larvae following cercarial penetration, *Am. J. Trop. Med. Hyg.,* 19, 959, 1970.
299. **Carter, R. A. and Shaldon, S.,** The liver in schistosomiasis, *Lancet,* 2, 1003, 1959.
300. **Cheever, A. W.,** Quantitative comparison of the intensity of *Schistosoma mansoni* infections in man and experimental animals, *Trans. R. Soc. Trop. Med. Hyg.,* 63, 781, 1969.
301. **Cheever, A. W., DeWitt, W. B., and Warren, K. S.,** Repeated infection and treatment of mice with *Schistosoma mansoni:* functional, anatomic and immunologic observations, *Am. J. Trop. Med. Hyg.,* 14, 239, 1965.
302. **Clarke, V. de V., Warburton, B., and Blair, D. M.,** The Katayama syndrome: report on an outbreak in Rhodesia, *Cent. Afr. J. Med.,* 16, 123, 1970.
303. **Clegg, J. A. and Smithers, S. R.,** Death of schistosome cercariae during penetration of the skin. II. Penetration of mammalian skin by *Schistosoma mansoni, Parasitology,* 58, 111, 1968.
304. **DeWitt, W. B. and Warren, K. S.,** Hepato-splenic schistosomiasis in mice, *Am. J. Trop. Med. Hyg.,* 8, 440, 1959.
305. **Dimmette, R. M. and Sproat, H. F.,** Rectosigmoid polyps in schistosomiasis. I. General clinical and pathological considerations, *Am. J. Trop. Med. Hyg.,* 4, 1057, 1955.
306. **Edington, G. W. and Gilles, H. M.,** *Pathology in the Tropics,* Williams & Wilkins, Baltimore, 1969.
307. **Elwi, A. M.,** Pathological aspects of bilharziasis in Egypt, in *Bilharziasis,* Mostofi, F. K., Ed., Springer-Verlag, Berlin, 1967, 39.
308. **Ghandour, A. M. and Webbe, G.,** A comparative study of the death of schistosomula of *Schistosoma haematobium* and *Schistosoma mansoni* in the skin of mice and hamsters, *J. Helminthol.,* 50, 39, 1976.
309. **Gutekunst, R. R., Browne, H. G., and Meyers, D. M.,** Influence of *Schistosoma mansoni* eggs on growth of monkey heart cells, *Exp. Parasitol.,* 17, 194, 1965.

310. **Hoeppli, R.,** Histological observation in experimental schistosomiasis japonica, *Chin. Med. J.,* 46, 1179, 1932.

311. **Lehman, J. S., Farid, Z., Bassily, S., and Kent, D. C.,** Hydronephrosis, bacteriuria, and maximal urine concentration in urinary bilharziasis, *Ann. Intern. Med.,* 75, 49, 1971.

312. **Lewis, F. A. and Colley, D. G.,** Modification of the lung recovery assay for schistosomula and correlations with worm burdens in mice infected with *Schistosoma mansoni, J. Parasitol.,* 63, 413, 1977.

313. **Lichtenberg, F. von, Sadun, D. H., Cheever, A. W., Erickson, D. G., Johnson, A. J., and Boyce, H. W.,** Experimental infection with *Schistosoma japonicum* in chimpanzees. Parasitologic, clinical, serologic and pathological observations, *Am. J. Trop. Med. Hyg.,* 20, 850, 1971.

314. **Lichtenberg, F. von, Smith, J. H., and Cheever, A. W.,** The Hoeppli phenomenon in schistosomiasis. Comparative pathology and immunopathology, *Am. J. Trop. Med. Hyg.,* 15, 886, 1966.

315. **Magalhaes Filho, A., Krupp, I. M., and Malek, E. A.,** Localization of antigen and presence of antibody in tissues of mice infected with *Schistosoma mansoni* as indicated by fluorescent antibody techniques, *Am. J. Trop. Med. Hyg.,* 14, 84, 1965.

316. **Maged, A.,** The L-shaped ureter: a rare complication of bilharziasis, *Ain Shams Med. J.,* 23, 11, 1972.

317. **Makar, N.,** Some clinico-pathological aspects of urinary bilharziasis, in *Bilharziasis,* Mostofi, F. K., Ed., Springer-Verlag, Berlin, 1967, 45.

318. **Most, H. and Levine, D. I.,** Schistosomiasis in American tourists, *JAMA,* 186, 453, 1963.

319. **Perquis, P., Fillaudeau, G., Montbarbon, J. P., Deleymarie, J., and Grange, G.,** Lithiase urinaire et bilharziose vesicale (a propos de 20 observations), *Med. Afr. Noire,* 19, 913, 1972.

320. **Rabinowitz, D.,** The Katayama syndrome, the early allergic stage of bilharziasis, *S. Afr. Med. J.,* 32, 658, 1958.

321. **Sadun, E. H., Lichtenberg, F. von, Cheever, A. W., and Erickson, D. G.,** Schistosomiasis mansoni in the chimpanzee. The natural history of chronic infections after single and multiple exposures, *Am. J. Trop. Med. Hyg.,* 19, 258, 1970.

322. **Sadun, E. H., Lichtenberg, F. von, Cheever, A. W., Erickson, D. G., and Hickman, R. L.,** Experimental infection with *Schistosoma haematobium* in chimpanzees. Parasitologic, clinical, serologic, and pathological observations, *Am. J. Trop. Med. Hyg.,* 19, 427, 1970.

323. **Standen, O. D.,** The penetration of the cercariae of *Schistosoma mansoni* into the skin and lymphatics of the mouse, *Trans. R. Soc. Trop. Med. Hyg.,* 47, 292, 1953.

324. **Stirewalt, M. A.,** Penetration of host skin by cercariae of *Schistosoma mansoni.* I. Observed entry into skin of mouse, hamster, rat, monkey, and man, *J. Parasitol.,* 42, 565, 1956.

325. **Symmers, W. St. C.,** Note on a new form of liver cirrhosis due to the presence of ova of *Bilharzia haematobia, J. Path. Bacteriol.,* 9, 237, 1903.

326. **Warren, K. S.,** The pathology of schistosome infections, *Helminthol. Abstr. Ser. A,* 42, 592, 1973.

327. **Warren, K. S. and DeWitt, W. B.,** Production of portal hypertension and esophageal varies in the mouse, *Proc. Soc. Exp. Biol. Med.,* 98, 99, 1958.

DIAGNOSIS

328. **Bell, D. R.,** A new method for counting *Schistosoma mansoni* eggs in faeces with special reference to therapeutic trials, *Bull. W. H. O.,* 29, 525, 1963.

329. **Coudert, J., Garin, J. P., Ambroise-Thomas, P., and Pothier, M. A.,** Premiers resultáts a propòs du diagnostic serologique de la bilharziose par immunofluorescence sur coupes a la congelatión de *Schistosoma mansoni, Ann. Parasitol. Hum. Comp.,* 42, 483, 1967.

330. **Kagan, I. G.,** Serologic diagnosis of schistosomiasis, *Bull. N.Y. Acad. Med.,* 44, 262, 1968.

331. **Kagan, I. G.,** Advances in the immunodiagnosis of parasitic infections, *Z. Parasitenkd.,* 45, 163, 1974.

332. **Kato, K. and Miura, M.,** Comparative examinations, *Jpn. J. Parasitol.,* 3, 35, 1954.

333. **Katz, N., Chaves, A., and Pellegrino, J.,** A simple device for quantitative stool thick-smear technique in schistosomiasis mansoni, *Rev. Inst. Med. Trop. Sao Paulo,* 14, 397, 1972.

334. **Knight, W. B., Hiatt, R. A., Cline, B. L., and Ritchie, L. S.,** A modification of the formol-ether concentration technique for increased sensitivity in detecting *Schistosoma mansoni* eggs, *Am. J. Trop. Med. Hyg.,* 25, 818, 1976.

335. **Layrisse, M., Martinez, C. T., and Ferrer, H. F.,** A simple volumetric device for preparing stool samples in the cellophane thick-smear technique, *Am. J. Trop. Med. Hyg.,* 18, 553, 1969.

336. **Martin, L. K. and Beaver, P. C.,** Evaluation of Kato thick-smear technique for quantitative diagnosis of helminth infections, *Am. J. Trop. Med. Hyg.,* 17, 382, 1968.

337. **Reis, A. P., Katz, N., and Pellegrino, J.,** Immuno-diffusion tests in patients with *Schistosoma mansoni* infection, *Rev. Inst. Med. Trop. Sao Paulo,* 12, 245, 1970.

338. **Ritchie, L. S.,** An ether sedimentation technique for routine stool examinations, *Bull. U.S. Army Med. Dept.,* 8, 326, 1948.

339. **Warren, K. S., Kellermeyer, R. W., Jordan, P., Littel, A. S., Cook, J. A., and Kagan, I. G.,** Immunologic diagnosis of schistosomiasis: Parts I, II, III, *Am. J. Trop. Med. Hyg.,* 22, 189, 1973.

340. **Wilson, M., Sulzer, A. J., and Walls, K. W.,** Modified antigens in the indirect immuno-fluorescence test for schistosomiasis, *Am. J. Trop. Med. Hyg.,* 23, 1072, 1974.

341. World Health Organization, Memoranda, Immunology of Schistosomiasis, Bull. *W. H. O.,* 51, 553, 1974.

TREATMENT

342. **Axton, J. H. M. and Garnett, P. A.,** A trial of oral oxamniquine in the treatment of *Schistosoma* infection in children, *S. Afr. Med. J.,* 50, 1051, 1976.

343. **Berberian, D. A., Freele, H., Rosi, D., Dennis, E. W., and Archer, S.,** A comparison of oral and parenteral activity of hycanthone and lucanthone in experimental infections with *Schistosoma mansoni, Am. J. Trop. Med. Hyg.,* 16, 487, 1967.

344. **Blair, D. M.,** Lucanthone hydrochloride. A review, *Bull. W. H. O.,* 18, 989, 1958.

345. **Christopherson, J. B.,** Intravenous injections of antimonium tartaratum in bilharziasis, *Br. Med. J.,* 2, 652, 1918.

346. **Cook, J. A., Jordan, P., and Armitage, P.,** Hycanthone dose-response in treatment of schistosomasis mansoni in St. Lucia, *Am. J. Trop. Med. Hyg.,* 25, 602, 1976.

347. **Davis, A.,** Personal communication, 1978.

348. **Davis, A. and Bailey, D. R.,** Metrifonate in urinary schistosomiasis, *Bull. W. H. O.,* 41, 209, 1969.

349. **Ferraz, M. P. T., Borges, D. R., Vilela, M. P., and Braziliano, C. J. C.,** Psicose sintomatica apos o uso de hicantone, *Rev. Paul. Med.,* 81, 275, 1973.

350. **Foster, R. and Cheetham, B. L.,** Studies with the schistosomicide oxamniquine (UK-4271)I. Activity in rodents and in vitro, *Trans. R. Soc. Trop. Med. Hyg.,* 67, 674, 1973.

351. **Foster, R., Cheetham, B. L., and King, D. F.,** Studies with the schistosomicide oxamniquine (UK-4271). II. Activity in primates, *Trans. R. Soc. Trop. Med. Hyg.,* 67, 685, 1973.

352. **Foster, R., Mesmer, E. T., Cheetham, B. L., and King, D. F.,** The control of immature *Schistosoma mansoni* in mice by UK 3883, a novel 2-aminomethyl-tetrahydroquinoline derivative, *Ann. Trop. Med. Parasitol.,* 65, 221, 1971.

353. **Jewsbury, J. M., Cooke, M. J., and Weber, M. C.,** Field trial of metrifonate in the treatment and prevention of schistosomiasis infection in man, *Ann. Trop. Med. Parasitol.,* 71, 67, 1977.

354. **Lees, R. E. M.,** Suppressive treatment of schistosomiasis mansoni with spaced doses of Lucanthone Hydrochloride, *Trans. R. Soc. Trop. Med. Hyg.,* 62, 782, 1968.

355. **McMahon, J. E.,** Non-intensive chemotherapy in bilharziasis with Lucanthone Hydrochloride (Preliminary Report), *Trans. R. Soc. Trop. Med. Hyg.,* 64, 433, 1970.

356. **McMahon, J. E.,** Oxamniquine (UK-4271) in *Schistosoma haematobium* infections, *Ann. Trop. Med. Parasitol.,* 70, 121, 1976.

357. **Moore, J. A.,** Teratogenicity of hycanthone in mice, *Nature (London),* 239, 107, 1972.

358. **Pellegrino, J., Katz, N., and Scherrer, J. F.,** Oogram studies with hycanthone, a new antischistosomal agent, *J. Parasitol.,* 53, 55, 1967.

359. **Plestina, R., Davis, A., and Bailey, D. R.,** Effect of metrifonate on blood cholinesterases in children during the treatment of schistosomiasis, *Bull. W.H.O.,* 46, 747, 1972.

360. **Rosi, D., Peruzzotti, G., Dennis, E. W., Berberian, D. A., Freele, H., Tullar, B. F., and Archer, S.,** Hycanthone, a new active metabolite of lucanthone, *J. Med. Chem.,* 10, 867, 1967.

361. **Shoeb, S. M., et al.,** Treatment of bilharziasis with hycanthone, *Ain Shams Med. J.,* 22, 759, 1971.

362. **Silva, L. C., Sette, H., Chamone, D. A. F., Saez-Alquezar, A., Punskas, J. A., and Raia, S.,** Further clinical trials with oxamniquine (UK 4271). A new anti-schistosomal agent, *Rev. Inst. Med. Trop. Sao Paulo,* 17, 307, 1975.

363. **Ständen, O. D.,** Schistosomicidal drugs: an evaluation, in *Proceedings of a Symposium on the Future of Schistosomiasis Control,* Miller, M. J., Ed., Tulane University, New Orleans, 1972, 31.

364. **Yarinsky, A., Hernandez, P., Ferrari, R. A., and Freele, H. W.,** Effects of hycanthone against two strains of *Schistosoma japonicum* in mice, *Jpn. J. Parasitol.,* 21, 101, 1972.

ZOONOTIC SCHISTOSOMIASIS

365. **Alves, W.,** The distribution of schistosome eggs in human tissues, *Bull. W. H. O.,* 18, 1092, 1958.

366. **Bearup, A. J.,** Life cycle of *Austrobilharzia terrigalensis* Johnston, 1917, *Parasitology,* 46, 470, 1956.

367. **Bearup, A. J. and Langsford, W. A.,** Schistosome dermatitis in association with rice growing in the Northern Territory of Australia, *Med. J. Aust.,* 1, 521, 1966.

368. **Berrie, A. D. and Goodman, J. D.**, The occurrence of *Schistosoma rodhaini* Brumpt in Uganda, *Ann. Trop. Med. Parasitol.*, 56, 297, 1962.

369. **Biocca, E.**, Observations on the morphology and biology of the Sardinian strain of *Schistosoma bovis* and the human dermatitis which it provokes, *Parassitologia (Rome)*, 2, 47, 1960.

370. **Blair, D. M.**, The occurrence of terminal-spined eggs, other than those of *Schistosoma haematobium*, in human beings in Rhodesia, *Cent. Afr. J. Med.*, 12, 103, 1966.

371. **Bonsel, J., Stam, A. B., and van Thiel, P. H.**, Schistosome dermatitis in the Hague, *Ned. Tijdschr. Geneeskd.*, 102, 938, 1958.

372. **Brumpt, E.**, Action des hotes definitifs sur l'evolution et sur le selection des sexes de certains helminthes heberges par eux. Experiences sur des schistosomes, *Ann. Parasitol. Hum. Comp.*, 14, 541, 1936.

373. **Buckley, J. J. C.**, On a dermatitis in Malaya caused by the cercariae of *Schistosoma spindale* Montgomery, 1906, *J. Helminthol.*, 16, 117, 1938.

374. **Buckley, J. J. C.**, A helminthological survey in Northern Rhodesia, *J. Helminthol.*, 21, 111, 1946.

375. **Carney, W. P., Brown, R. J., Van Peenen, P. F. D., Ibrahim, P. B., and Koesharjono, C. R.**, *Schistosoma incognitum* from Cikurai, West Java, Indonesia, *Int. J. Parasitol.*, 7, 361, 1977.

376. **Carney, W. P., Van Peenen, P. F. D., Brown, R. J., and Sudomo, M.**, *Schistosoma incognitum* from mammals of Central Sulawesi, Indonesia, *Proc. Helminth. Soc. Wash.*, 44, 150, 1977.

377. **Cawston, F. G.**, Schistosome resembling *S. japonicum* in South Africa, *J. Trop. Med. Hyg.*, 33, 292, 1930.

378. **Chandler, A. C.**, A new schistosome infection of man, with notes on other human fluke infections in India, *Indian J. Med. Res.*, 14, 179, 1926.

379. **Chu, G. W. T. C.**, First report of the presence of a dermatitis-producing marine larval schistosome in Hawaii, *Science*, 115, 151, 1952.

380. **Cort, W. W.**, Schistosome dermatitis in the United States (Michigan), *JAMA*, 90, 1027, 1928.

381. **Cort, W. W.**, Studies on schistosome dermatitis. XI. Status of knowledge after more than twenty years, *Am. J. Hyg.*, 52, 251, 1950.

382. **De Meillon, B. and Stoffberg, N.**, "Swimmer's itch" in South Africa, *S. Afr. Med. J.*, 28, 1062, 1954.

383. **Dinnik, J. A. and Dinnik, N. N.**, The schistosomes of domestic ruminants in Eastern Africa, *Bull. Epizoot. Dis. Afr.*, 13, 341, 1965.

384. **Dutt, S. C. and Srivastava, H. D.**, On the morphology and life history of a new mammalian bloodfluke - *Ornithobilharzia datti*, n. sp. (preliminary report), *Parasitology*, 42, 144, 1952.

385. **Edwards, D. K. and Jansch, M. E.**, Two new species of dermatitis producing schistosome cercariae from Cultus Lake, British Columbia, *Can. J. Zool.*, 33, 182, 1955.

386. **Fain, A.**, Etudes sur les schistosomes d'oiseaux au Ruanda - Urundi (Congo Belge). Un nouveau schistosome du Tantale ibis (*Ibis ibis* Linn) *Gigantobilharzia tantali* n. sp., *Ann. Parasitol. Hum. Comp.*, 30, 321, 1955.

387. **Fain, A.**, Une nouvelle bilharziose des oiseaux. La trichobilharziose nasale. Remarque sur l'importance des schistosomes d'oiseaux en pathologie humaine. Note preliminaire, *Ann. Soc. Belge Med. Trop.*, 35, 323, 1955.

388. **Fain, A.**, Sur une furcocercaire du groupe Ocellata produisant expérimentalement la dermatite des nageurs a Astrida (Ruanda-Urundi), *Ann. Soc. Belge Med. Trop.*, 35, 701, 1955.

389. **Fairley, N. H.**, The early spontaneous cure of bilharziasis (*S. spindale*) in monkey (*Macacus sinicus*), and its bearing on species immunity, *Indian J. Med. Res.*, 14, 685, 1927.

390. **Graefe, G., Aspock, H., and Picher, O.**, Auftreten von Bade-Dermatitus in Osterreich und Moglichkeiten ihrer Bekampfung, *Zentraibl. Bakteriol. Hyg., I Ab. Orig. A*, 225, 398, 1973.

391. **Haenens, G. d' and Santele, A.**, Sur un cas humain de *Schistosoma rodhaini* trouvé aux environs d'Elisabethville, *Ann. Soc. Belge Med. Trop.*, 35, 497, 1955.

392. **Harinasuta, C., Kruatrachue, M., and Sornmani, S.**, A study of *Schistosoma spindale* in Thailand, *J. Trop. Med. Hyg.*, 68, 125, 1965.

393. **Hunter, G. W., III**, Studies on schistosomiasis. XIII. Schistosome dermatitis in Colorado, *J. Parasitol.*, 46, 231, 1960.

394. **Hunter, G. W., III.**, Schistosome cercarial dermatitis and other rare schistosomes that may infect man, in *Pathology of Protozoal and Helminthic Diseases*, Marcial-Rojas, R. A., Ed., Krieger, Huntington, N.Y., 1971, 450.

395. **Jellison, W. L., et al.**, Schistosome dermatitis in Montana, *Northwest Sci.*, 26, 10, 1952.

396. **Kagan, I. G.**, Experimental infections of rhesus monkeys with *Schistosomatium douthitti* (Cort, 1914), *J. Infect. Dis.*, 93, 200, 1953.

397. **Kisner, C. D., Stoffberg, N., and DeMeillon, B.**, Human infection with *Bilharzia bovis*, *S. Afr. Med. J.*, 27, 357, 1953.

398. **Kruatrachue, M., Bhaibulaya, M., and Harinasuta, C.,** *Orientobilharzia harinasutai* sp. nov., a mammalian blood-fluke, its morphology and life cycle, *Ann. Trop. Med. Parasitol.,* 59, 181, 1965.

399. **Lapierre, J. and Hein, T. V.,** A case of triple schistosome infection with *Schistosoma mansoni, S. haematobium* and *Rhodobilharzia margrebowiei, Ann. Parasitol. Hum. Comp.,* 46, 301, 1973.

400. **Lee, H. F. and Wykoff, D. E.,** Schistosomes from wild rats in Thailand, *J. Parasitol.,* 52, 323, 1966.

401. **Leigh, W. H.,** The morphology of *Gigantobilharzia huttoni* (Leigh, 1953) an avian schistosome with marine dermatitis-producing larvae, *J. Parasitol.,* 41, 262, 1955.

402. **Leigh, W. H.,** Brown and white pelicans as hosts for schistosomes of the genus *Gigantobilharzia, J. Parasitol.,* 43 (Suppl.), 35, 1957.

403. **Le Roux, P. L.,** Remarks on the habitats and pathogenesis of *Schistosoma mattheei;* together with notes on the pathological lesions observed in infected sheep, *15th Annual Report of the Director of Veterinary Services,* Union of South Africa, Johannesburg, 1929, 347-406.

404. **Le Roux, P. L.,** A preliminary note on *B. margrebowei,* a new parasite of ruminants and possibly man in Northern Rhodesia, *J. Helminthol.,* 11, 57, 1933.

405. **Le Roux, P. L.,** A new mammalian schistosome (*Schistosoma leiperi* sp. nov.) from herbivora in Northern Rhodesia, *Trans. R. Soc. Trop. Med. Hyg.,* 49, 293, 1955.

406. **Malek, E. A.,** The biology of mammalian and bird schistosomes, *Bull. Tulane Univ. Med. Fac.,* 20, 181, 1961.

407. **Malek, E. A.,** Studies on bovine schistosomiasis in the Sudan, *Ann. Trop. Med. Parasitol.,* 63, 501, 1969.

408. **Malek, E. A.,** Further studies on mammalian susceptibility to experimental infection with *Heterobilharzia americana, J. Parasitol.,* 56, 64, 1970.

409. **Malek, E. A. and Armstrong, J. C.,** Infections with *Heterobilharzia americana* in primates, *Am. J. Trop. Med. Hyg.,* 16, 708, 1967.

410. **Meyer, P. O. and Dubois, G.,** Dermatite humaine causee par des furcocercaires ocellees dans les bains publics de Zurich, *Bull. Soc. Neuchatel. Sci. Nat.* 77, 81, 1954.

411. **Mudaliar, S. V. and Ramonujachari, G.,** *Schistosoma nairi* n. sp. from an elephant, *Indian Vet. J.,* 22, 1, 1945.

412. **Nelson, G. S.,** Schistosome infections as zoonoses in Africa, *Trans. R. Soc. Trop. Med. Hyg.* 54, 301, 1960.

413. **Olivier, L.,** Observations on the migration of avian schistosomes in mammals previously unexposed to cercariae, *J. Parasitol.,* 39, 237, 1953.

414. **Penner, L. R.,** The possibilities of systemic infection with dermatitis producing schistosomes, *Science,* 93, 327, 1941.

415. **Pitchford, R. J.,** Cattle schistosomiasis in man in the Eastern Transvaal, *Trans. R. Soc. Trop. Med. Hyg.,* 53, 285, 1959.

416. **Pitchford, R. J.,** Observations on a possible hybrid between the two schistosomes *Schistosoma haematobium* and *S. mattheei, Trans. R. Soc. Trop. Med. Hyg.,* 55, 44, 1961.

417. **Pitchford, R. J.,** Some preliminary observations on schistosomes occurring in antelope in central southern Africa, *Rhodesian Vet. J.,* 4, 57, 1974.

418. **Pitchford, R. J.,** Preliminary observations on the distribution, definitive hosts and possible relation with other schistosomes, of *Schistosoma margrebowiei,* Le Roux, 1933 and *Schistosoma leiperi,* Le Roux, 1955, *J. Helminthol.,* 50, 111, 1976.

419. **Porter, A.,** Notes on the structure and life history of *Schistosoma spindalis* (Montgomery), observed in South Africa, *S. Afr. J. Sci.,* 23, 661, 1926.

420. **Raper, A. B.,** *Schistosoma bovis* infection in man, *East Afr. Med. J.,* 28, 50, 1951.

421. **Russel, W. W.,** Probable marine schistosome dermatitis in Northern Queensland, *Med. J. Austr.,* 1, 327, 1972.

422. **Sahba, G. H. and Malek, E. A.,** Dermatitis caused by cercariae of *Orientobilharzia turkestanicum* in the Caspian Sea area of Iran, *Am. J. Trop. Med. Hyg.,* 28, 912, 1979.

423. **Salome, B. Z.,** Schistosome dermatitis in the Netherlands, *Doc. Med. Geogr. Trop.,* Amsterdam, 6, 30, 1954.

424. **Soliman, K. N.,** *Schistosoma bovis*-shaped eggs associated with *S. haematobium* infestations in Egypt, *J. Egypt. Med. Assoc.,* 39, 630, 1956.

425. **Sprent, J. F. A.,** Helminth "Zoonoses": An analysis, *Helminthol. Abstr.,* 38, 333, 1969.

426. **Srivastava, H. D.,** Blood-flukes, *Proc. 47th Indian Sci. Congr.,* 2, 1, 1960.

427. **Srivastava, H. D. and Dutt, S. C.,** Life history of *Schistosoma indicum* Montgomery, 1906, a common blood fluke of Indian ungulates, *Curr. Sci.,* 20, 273, 1951.

428. **Stijns, J.,** Sur les rongeurs hotes naturels de *Schistosoma rodhaini* Brumpt, *Ann. Parasitol. Hum. Comp.,* 27, 385, 1952.

429. **Stunkard, N. W. and Hinchliffe, M. C.**, The morphology and life history of Microbilharzia variglandis (Miller & Northup, 1926) Stunkard & Hinchliffe, 1951, avian blood fluke whose larvae cause "swimmer's itch" of ocean beaches, *J. Parasitol.*, 38, 248, 1952.

430. **Szidat, L. and de Szidat, U. C.**, Eine neue Dermatitis erzeugende cercarie der Trematoden - Familie Schistosomida aus *Tropicorbis peregrinus* (D'Orbigny) des Rio Quequen, *Z. Parasitenkd.*, 20, 359, 1960.

431. **Ulmer, M. J.**, Schistosome dermatitis at Lake Okoboji, Iowa, *J. Parasitol.*, 44, (Suppl.), 13, 1958.

432. **van den Berghe, L.**, Les schistosomes humaines et animals au Katanga, *Ann. Soc. Belge Med. Trop.*, 14, 313, 1934.

433. **Vogel, H. and Minning, W.**, Bilharziose bei Elefanten, *Arch. Schiffs Trop. Hyg.*, 44, 562, 1940.

434. **Walkiers, J.**, Five cases of schistosomiasis with spineless eggs from upper Uele, *Ann. Soc. Belge Med. Trop.*, 8, 21, 1928.

435. **Watson, J. M. and Najim, A. T.**, Studies on bilharziasis in Iraq. Part II. Observations on schistosome dermatitis, *J. Iraqi Med. Prof.*, 4, 4, 1956.

436. World Health Organization, Joint WHO/FAO Expert Committee on Zoonosis. Second Report, *W. H. O., Tech. Rep. Ser.*, 169, 1959.

CONTROL

437. **Davis, A.**, Mass chemotherapy: problems and prospects in *Proceedings of a Symposium on the Future of Schistosomiasis Control,* Miller, M. J., Ed., Tulane University, New Orleans, 1972, 40.

438. **Unrau, G. O.**, Control of schistosomiasis in St. Lucia, in *Proceedings of a Symposium on the Future of Schistosomiasis Control,* Miller, M. J., Ed., Tulane University, New Orleans, 1972, 93.

439. World Health Organization, "Schistosomiasis Control", *W.H. O. Tech. Rep. Ser.*, 515, 1973.

INDEX

A

C

E